**RELIGION, HISTORY, AND PLACE
IN THE ORIGIN OF SETTLED LIFE**

RELIGION, HISTORY, AND PLACE IN THE ORIGIN OF SETTLED LIFE

edited by **IAN HODDER**

THE RESEARCH ON WHICH THIS VOLUME
IS BASED WAS FUNDED IN PART BY THE
JOHN TEMPLETON FOUNDATION

UNIVERSITY PRESS OF COLORADO
Boulder

© 2018 by University Press of Colorado

Published by University Press of Colorado
245 Century Circle, Suite 202
Louisville, Colorado 80027

All rights reserved
First paperback edition 2019
Printed in the United States of America

 The University Press of Colorado is a proud member of the Association of University Presses.

The University Press of Colorado is a cooperative publishing enterprise supported, in part, by Adams State University, Colorado State University, Fort Lewis College, Metropolitan State University of Denver, Regis University, University of Colorado, University of Northern Colorado, Utah State University, and Western State Colorado University.

∞ This paper meets the requirements of the ANSI/NISO Z39.48-1992 (Permanence of Paper).

ISBN: 978-1-60732-736-3 (cloth)
ISBN: 978-1-60732-940-4 (pbk)
ISBN: 978-1-60732-737-0 (ebook)
https://doi.org/10.5876/9781607327370

Library of Congress Cataloging-in-Publication Data

Names: Hodder, Ian, editor.
Title: Religion, history and place in the origin of settled life / edited by Ian Hodder.
Description: Boulder, Colorado : University Press of Colorado, 2018. | Includes bibliographical references and index.
Identifiers: LCCN 2017045683| ISBN 9781607327363 (cloth) | ISBN 9781607329404 (pbk.) | ISBN 9781607327370 (ebook)
Subjects: LCSH: Social archaeology—Middle East. | Dwellings—Middle East—Religious aspects. | Economic anthropology—Middle East. | Spatial analysis (Statistics) in archaeology. | Excavations (Archaeology)—Middle East. | Neolithic period—Middle East. | ?Catal Mound (Turkey) | Antiquities, Prehistoric—Turkey—?Catal Mound.
Classification: LCC DS56 .R45 2018 | DDC 939.4—dc23
LC record available at https://lccn.loc.gov/2017045683

Cover illustration, a reconstruction of neolithic mural from Çatalhöyük, from Wikimedia Commons. Photograph by Omar Hoftun.

This volume is dedicated to the memory of Klaus Schmidt.

Contents

List of Figures   *ix*
List of Tables   *xiii*

Introduction: Two Forms of History Making in the Neolithic of the Middle East
  Ian Hodder   3

1. Simulating Religious Entanglement and Social Investment in the Neolithic
   F. LeRon Shults
     and Wesley J. Wildman   33

2. Creating Settled Life: Micro-Histories of Community, Ritual, and Place—the Central Zagros and Çatalhöyük
   Wendy Matthews   64

3. Long-Term Memory and the Community in the Later Prehistory of the Levant
   Nigel Goring-Morris
     and Anna Belfer-Cohen   99

4. Establishing Identities in the Proto-Neolithic: "History Making" at Göbekli Tepe from the Late Tenth Millennium cal BCE
   Lee Clare, Oliver Dietrich, Jens Notroff,
     and Devrim Sönmez   115

5. Re-presenting the Past: Evidence from Daily Practices and Rituals at Körtik Tepe
   Marion Benz, Kurt W. Alt, Yilmaz S. Erdal, Feridun S. Şahin, and Vecihi Özkaya   *137*

6. Sedentism and Solitude: Exploring the Impact of Private Space on Social Cohesion in the Neolithic
   Güneş Duru   *162*

7. "Every Man's House Was His Temple": Mimetic Dynamics in the Transition from Aşıklı Höyük to Çatalhöyük
   Mark R. Anspach   *186*

8. Interrogating "Property" at Neolithic Çatalhöyük
   Rosemary A. Joyce   *212*

9. The Ritualization of Daily Practice: Exploring the Staging of Ritual Acts at Neolithic Çatalhöyük, Turkey
   Christina Tsoraki   *238*

10. Virtually Rebuilding Çatalhöyük History Houses
    Nicola Lercari   *263*

    List of Contributors   *283*
    Index   *285*

Figures

0.1.	Chronological relationships between sites in the Middle East and Turkey	4
0.2.	Distribution of main late Epipaleolithic and Neolithic sites in the Near East	5
1.1.	Stock-and-flow diagram for conversion model between low-investment and high-investment lifestyles	40
1.2.	Core of the Black Box: causal architecture of the Neolithic transition	41
1.3.	Six outputs from the Black Box	49
1.4.	Closest solution to the Target transition pathway for the growing population of Çatalhöyük	53
1.5.	Behavior of Social Intensity	54
1.6.	Impact on HI_prop of varying the Stress parameter	55
1.7.	Impact on HI_prop of varying the Technology parameter	56
1.8.	Probability Distribution Function and Cumulative Distribution Function for HI_prop and SocialInt	57
2.1.	Location of Neolithic sites in the Zagros	66
2.2.	Origins of early settled life	71
2.3.	Çatalhöyük, Turkey	74
2.4.	Sheikh-e Abad, Iran	77
2.5.	Jani, Iran	79

2.6.	Repeated building at Bestansur	84
2.7.	Placing the dead	87
3.1.	Section of Locus 15 at early Epipaleolithic Ohalo II	101
3.2.	Plan of the Loci 51, 62, and 131 architectural complex at Eynan	103
4.1.	Aerial view of Göbekli Tepe excavations	116
4.2.	Göbekli Tepe: schematic plan of excavations in the southeast hollow and on the southwest mound	117
4.3.	Two examples of T-pillars with indications of reuse and alteration	124
4.4.	Distribution of calibrated radiocarbon ages from Enclosure H at Göbekli Tepe	125
4.5.	Enclosure H in the northwest hollow	126
4.6.	Pillar 66 and Pillar 54 in Enclosure H	127
4.7.	Excavations in Enclosure D	129
4.8.	Reconstruction of the southeast hollow (main excavation area)	130
5.1.	Trenches of Körtik Tepe, southeastern Anatolia	139
5.2.	Built hearth of the early Holocene settlement	141
5.3.	Orientation of the individuals according to age-class	142
5.4.	Results of the carbon and nitrogen isotope analyses show no clusters of segregating groups	143
5.5.	Repaired chlorite vessel	146
5.6.	Burial of female juvenile individual	147
5.7.	Vessels with motifs	148
5.8.	Decorated chlorite platelets from Körtik Tepe	149
6.1.	Continuity of buildings at Aşıklı Höyük	173

6.2.	Settlement consists of two distinct areas	174
6.3.	Five phases of renewal of Building T	175
8.1.	One of the more typical oval marks on the exterior of a sherd	223
8.2.	Showing the internal facet that is the trace of the anvil	223
9.1.	Stone artifact clusters in the main room of Building 77	244
9.2.	Frequency of worked stone in Building 77, by phase	245
9.3.	Quern found embedded in a platform in Building 77	246
9.4.	Floor plan of Building 77 showing the location of stone clusters and other finds	247
9.5.	Cluster 17509	248
9.6.	Degree of weathering on fractured edges of grinding tool fragments found in Cluster 17509	249
9.7.	Large quern found in burned fill of Building 77	251
10.1.	Overlaying view of 3D reconstructions	265
10.2.	Highly evocative 3D reconstruction of Çatalhöyük history house F.V.I	266
10.3.	View of 3D models and immersive interaction	267
10.4.	"Shrine" 10 sequence	269
10.5.	Overlaying view of CAD drawings and house-based history making in the "Shrine" 10 sequence	271
10.6.	Comparative view of Mellaart's visual restoration of "Shrine" 10.VIB and a 3D reconstruction	272

10.7. Overlaying and comparative views of Mellaart's visual restoration of "Shrine" 10.VIA and "Shrine" 10.VIB and a 3D reconstruction   274

10.8. Comparative view of Mellaart's visual restoration of "Shrine" 10.VII and isometric drawing   276

Tables

1.1.	Black Box parameters and variables constituting the causal architecture of the model	43
1.2.	Additional parameters impacting output variables but not Black Box variables	50
1.3.	Parameter settings yielding the closest approximation to the Target transition pathway through the model's parameter space from a low to a high proportion of HI people	53
4.1.	Socio-ritual contexts at Göbekli Tepe and Çatalhöyük East	119
4.2.	AMS-radiocarbon ages from Enclosure H in the northwest hollow	128
8.1.	Stratigraphic relationships of buildings and spaces from which sherds were recorded	225
8.2.	Results of analysis of technological style represented in middens associated with Buildings 75 to 10	226
8.3.	Results of analysis of technological style represented in middens associated with Building 53 and Building 42	227
8.4.	Results of analysis of technological style represented in midden in Spaces 279 and 226	228
8.5.	Pots with recorded paddle mark	232
9.1.	Frequency of stone materials in Building 77	243

**RELIGION, HISTORY, AND PLACE
IN THE ORIGIN OF SETTLED LIFE**

Introduction

Recent data excavated from the Middle East challenge many of the narratives to which we have become accustomed regarding the origins of agriculture and settled life. The notion of a Neolithic Revolution has been replaced by a very long-term gradual process (Maher, Richter, and Stock 2012). The old Levant-centered sequence has been replaced by polycentric models that see early complexity and domestication of plants and animals in diverse locations. Large ritual centers and elaborate sites have been discovered in northern Mesopotamia in the pre-Pottery Neolithic A (PPNA) well before the appearance of fully domesticated resources in the pre-Pottery Neolithic B (PPNB). The assumed primacy of "it's the economy, stupid" has been replaced by a singular focus of "it all began with ritual."

This volume responds to these exciting new challenges by exploring one aspect of the new narrative that needs to be built, based on emerging evidence for the importance of history making from the later Epipaleolithic through the Neolithic in the Middle East (figures 0.1, 0.2). This focus also allows some integration and bridging between subsistence-based and symbol-based approaches. There has been much discussion (summarized recently by Arbuckle 2015) of whether there is evidence of resource depletion in the later Epipaleolithic as humans shifted to the exploitation of a wider range of resources that required increased time and effort to extract and manage. From 25,000 BCE onward there is evidence of investment in

Two Forms of History Making in the Neolithic of the Middle East

Ian Hodder

DOI: 10.5876/9781607327370.c000

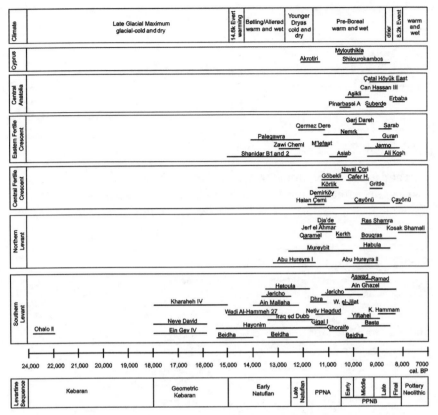

FIGURE 0.1. *Chronological relationships between sites in the Middle East and Turkey. Source: Zeder 2011.*

tools such as grinding stones and sickles, and storage and more stable settlement gradually appear. Woodburn (1980) made a distinction between immediate and delayed returns for labor. As humans increasingly intensified resource extraction and invested in tools, equipment, and land, the return for labor became delayed. The complex hunter-gatherers of the late Pleistocene and early Holocene in the Middle East increasingly encountered delayed returns for their labor. The group had to be held together over the period between investment and return. History making was thus key. Through the period from the late Epipaleolithic to the PPNB, subsistence intensification and history making had to go hand in hand.

Equally, however, as humans invested in subsistence practices that demanded more labor, they increasingly depended on wider networks to obtain resources

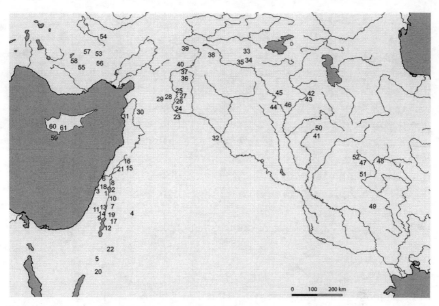

FIGURE 0.2. *Distribution of main late Epipaleolithic and Neolithic sites in the Near East.*
Source: Zeder 2011.

such as materials for tools, to obtain collaborative labor, and to build social ties that could buffer downturns in local production. Gamble (1998) and Coward (2010) have documented the increased emphasis on networks and cultural interchange in the later Epipaleolithic and early Neolithic at the regional scale. At the local scale, one effective way of building strong networks is to build (actual or fictive) relations through ancestors. The emergence from the Natufian onward of a concern with the deposition, circulation, and re-deposition of skulls and other human remains allows the building of community (Kuijt 2000). The greater the temporal depth achieved in the building of ties to ancestral remains, the wider the network of affiliated individuals. Both within and between houses, the burial of human remains allows history making and thus the establishment of various scales of community building and subsistence co-reliance.

The initiative for this volume was a project funded by the John Templeton Foundation titled "The Primary Role of Religion in the Origin of Settled Life: The Evidence from Çatalhöyük and the Middle East" (ID: 22893). The project culminated in an international conference held at Çatalhöyük on August 2–3, 2014, at which several of the papers published in this volume were presented.

There was a follow-up session at the Society for American Archaeology meetings in San Francisco in 2015. Both the conference and the session were called "Religion, History, and Place in the Origin of Settled Life." The main question contributors were asked to assess was whether there was widespread evidence that delayed-return agricultural systems emerged in tandem with an increased focus on history making. In the Çatalhöyük project there had long been recognition of repetitive practices within houses at the site (Hodder and Cessford 2004), and more recently the term *history house* had been coined (Hodder and Pels 2010). But could such emphases be identified elsewhere? What is the timing of the emergence of a concern with history making in place? At what point in regional sequences do such features emerge, and with what does their appearance correlate? And in what context does history making most clearly emerge, public ritual buildings or domestic houses?

WHAT IS MEANT BY HISTORY MAKING?

Throughout this volume we will come across many examples of continuity. For example, a building is continually rebuilt in the same place. What does it take to move from such evidence to the claim for history making?

It is first necessary to consider whether the continuities were produced by material constraints. Settlement may at different times be attracted to a particular water source or fertile patch of land, resulting in a palimpsest of occupation in the same place but in which there is no historical or cultural connection. Similarly, a new house may be built exactly onto the firm foundations of the walls of earlier buildings to provide stability. Or a new house may be sunk into the pit created by an earlier semi-subterranean house to save energy in excavating a new pit. Or houses may be built on the imprint of earlier houses because the settlement is so packed that there is no room to change house location. In these cases we cannot assume that historical ties were being created through time.

Thus we need to start with exploring whether the functional requirements of, for example, building technologies produced the continuities observed. Was a tell matrix so soft and mixed that stable buildings could only be constructed on wall stubs? In assessing whether there was any social meaning in building continuity, it is also important to explore variation through time and place. For example, Düring (2006) has shown that in the upper levels at Çatalhöyük there is a weakening of the earlier focus on a strict and exact repetition of houses on the same footprint. This change may have occurred in part because of changes in building technique in the upper levels, but there were also important social changes that produced greater house independence (Hodder 2014). Similarly, Kotsakis (1999) has argued that different parts of the site of Sesklo saw different relations between buildings through time, some areas showing repeated building on the same footprint and others showing horizontal shifting. For the Balkans, Tringham (2000) has discussed the different ways houses replaced each other during the Neolithic in terms of meaningful social and cultural practices.

Often, the repetition of the layout of activities in buildings is too great to be determined by wall settings. There are then two broad possibilities in terms of memory construction or history making. The first is that the repetition of practices within buildings is the result of habituated behavior. Many archaeologists, influenced ultimately by the work of Bourdieu (1977), have documented the ways in which practices become routinized and habituated at the non-discursive level. In other words, we know it is right to put the hearth in this location rather than that one because it has always been done that way. Our daily bodily movements get accustomed to certain routines, and we cannot discursively explain why. This is a type of history making in that the body is "remembering" earlier practices and there is continuity in the overall system of meanings and practices. Thus at Çatalhöyük there is a long-term practice of keeping northern parts of main rooms clean while allowing refuse to build up in southern "dirty" areas (Hodder and Cessford 2004). This habituated practice at Çatalhöyük may not have been consciously interpreted and explained, but it was part of a larger set of oppositions between adult burial and rich symbolism in the north and child burial and food preparation in the south. People knew that "it had always been done this way" even if they could not explain why. This type of history making is very embodied and may not be conscious.

A second possible interpretation of culturally meaningful continuities is that they are the result of commemorative behavior in which people consciously build social memories and historical links into the past (Connerton 1989). In the case of habituated behavior, ritual and other acts may become routinized and codified but there is no specific memory of events and histories, while in the commemorative case a link is remembered to a specific event or person. Here the onus is on the archaeologist to demonstrate specificity of memory construction (Van Dyke and Alcock 2008). This can often be achieved by studying the curation, circulation, and deposition of objects. There are many examples from Çatalhöyük. For example, in the sequence of two buildings constructed on the same footprint, Buildings 59 and 60, there is an example of an obsidian projectile point kept/owned in a house for the duration of fourteen wall re-plasterings. Elsewhere we have evidence of much longer-term curation of objects. Space 279 midden in Level 4040 I had an inscribed Canhasan III point otherwise typical of the aceramic levels of the site, perhaps indicating an heirloom (we do not usually find evidence of postdepositional processes that could have relocated such an object across so many levels of occupation). In Building 1, a pit was dug down to retrieve an installation or relief from the west wall of the main room (Hodder and Cessford 2004). In the sequence of buildings numbered from earliest to latest—65, 56, 44, 10—Boz and Hager (2013) found that teeth that fit into the jaw of an individual originally buried in Building 65 were found with another burial in Building 56 directly above it. This suggests that those living in Building 56 were constructing relationships or memories with the individuals buried within the earlier building below.

If commemorative history making is an important social practice, breaks and discontinuities will likely be marked and embedded in ritual. Throughout this volume we shall see examples of the careful cleaning and dismantling of buildings, burning, holding of special feasts during abandonment. Buildings are often intentionally and carefully filled. The foundation of new buildings may again be marked by burials, feasts, special events. These types of symbolic emphases certainly suggest a focus on history making. They may more immediately suggest forgetting, de-commissioning, but even so, these practices refer to a larger process in which links to the past have become important, if also dangerous and contested.

In this chapter, history making thus refers to continuities produced both by habituated practices and by commemorative links to the past. We shall see that there is often a related link to ownership. The building of historical links to the past in specific places may be associated with the assertion of rights to land or to animals, to buildings or ancestors. Again, this link needs to be scrutinized carefully. The continued use of a distinctive type of mudbrick in making a column of houses may suggest ownership of a clay source, or it may simply reflect habituated behavior. The repeated use of a particular part of the landscape for sheep grazing may suggest ownership, but the degree of exclusivity needs to be evaluated.

Another important consideration is the degree of temporal depth over which practices endure. Ancestral bones and relics may quickly become generic, the specific names and individuals long forgotten. It remains unclear over how many generations histories were constructed in the Neolithic of the Middle East. The depth may well have varied. Certainly at Aşıklı Höyük the longevity of habituated practices in individual houses is many hundreds of years (see chapters 6 and 7). At Çatalhöyük a good case can be made for commemorative memories and histories that go back up to 100 years. Refinements in dating techniques are allowing new insight in these areas (Bayliss et al. 2015).

Commemorative history making involves constructing a link between the present and a specific event in the past. At times there is remarkable precision and specificity at Çatalhöyük in the way people dug down in exactly the right place to retrieve a skull; clearly in these cases the locations of specific earlier burials had been remembered. Also at Çatalhöyük, a plastered female skull was found held in the arms of an adult woman (Hodder 2006). The head had been plastered and painted at least four times, and the specificity of the arrangements in the graves suggests that specific links were being built between the two individuals. This is in contrast to the frequent amassing of plastered skulls and their more communal or generic nature in the Levant (Bonogofsky 2004). At some point skulls may have become generic ancestors or merged into myth.

Myth can be distinguished from history in that although myths may be rife with origin stories, the time of their occurrence is in a relatively undifferentiated distant past. There is the time of myth and the time of the present, but they are not specifically connected. This may be true of the treatment of skulls in the Levant. It is also tempting to follow Clare and colleagues (chapter 4) and argue that the T-shaped pillars at Göbekli represent mythical ancestors. However, Clare and colleagues also argue that the stone enclosures at the site were involved in history making. The onus is on the archaeologist to differentiate myth making from history making. It is only when specific links between past and present can be seen to have been constructed that we are warranted to talk of history making.

In general terms, it is possible to argue for a close link between religion and history making. Definitions of religion in relatively non-complex societies are fraught with difficulties. These have been discussed at length in previous volumes (Hodder 2010, 2014). Religion often seems to have to do with "the beyond" or the "transcendent." It is this latter definition that links religion and history making. For Bloch (2008), as discussed by Benz and colleagues (chapter 5), religion is about creating links through time that extend beyond the everyday. Religion creates a "transcendental social," an imagined communal identity of a social entity. It might be expected, then, that as the demands on social community and continuity increase, so history making would be elaborated in ritual and religious contexts.

In this volume the larger question of the role of religion in the shift to more intensive agricultural systems is explored by Shults and Wildman (chapter 1). They develop a systems-dynamics model that is novel in a number of ways, particularly its inclusion of what they term religious variables dealing with cognitive, moral, ritual, and social dimensions. They model the shift from low-investment to high-investment systems, comparable to the change from societies more engaged with immediate returns to those more involved in delayed returns for their labor. Their model integrates in an exciting way a wide range of religious and social variables, including contestation, into economic and subsistence variables. The result is a fascinating interdisciplinary exercise that fits much of the data we have from Çatalhöyük. In particular, the rapid changes seen in their figure 5 (figure 1.5) correspond with the marked cultural, social, and ritual changes we see halfway through the sequence at the site. In more general terms, their model demonstrates the centrality of religious transformation to the overall growth of mega-sites such as Çatalhöyük. The results support the claim made by Whitehouse and Hodder (2010) that during the occupation of the site there was a shift from more "imagistic" to "doctrinal" modes of religiosity.

THE EVIDENCE FOR HISTORY MAKING IN THE NEOLITHIC OF THE MIDDLE EAST

So what, then, is the evidence for these various forms of history making during the late Epipaleolithic and Neolithic of the Middle East?

Of course, there were repetitive practices earlier in the Paleolithic. These involved repeated seasonal uses of the landscape in such a way that certain sites that provided shelter, such as caves, were returned to over long periods of time. For example, Ksar Akil in Lebanon has 23 m of deposit covering the period from the Middle Paleolithic through the Early and Upper Middle Paleolithic to the Kebaran Epipaleolithic. In the upper levels there was a "fine and complex stratigraphy" (Bergman 1987, 3). Kebara Cave also has deposits that span from the Middle Paleolithic through to Natufian, or from roughly 60,000 to 10,000 BCE. The Middle Paleolithic deposits show repeated use of part of the cave for hearths, while an inner part of the cave was used as a dump area (Goldberg 2001). The hearth area has deep deposits of overlapping hearths, each of which results from several episodes of combustion (Meignen et al. 2000, 14). These multiphase hearths indicate long periods of repetitive use in the same depression (Meignen et al. 2000, 15), and similar processes are found in other sites in the Middle East—there is an abundance of fire installations vertically superimposed (Meignen et al. 2000, 16). But the placing of these hearths was not exact. Rather, there was a zone in the cave where, over a long period of time, people made hearths. Each hearth involved a few re-firings, but the hearths themselves created a vertical palimpsest of overlaps. There was general use of a part of the cave for hearths but no specific backward reference. For Upper Paleolithic examples from the Levant, see Goring-Morris and Belfer-Cohen (chapter 3).

The Kebaran in the Levant has lowland aggregation sites of twenty-five to fifty people and upland camps of fourteen to seventeen people, and there may have been seasonal cycles of aggregation and dispersal. Little architecture has been excavated, but there was possibly twice a year occupation in the early Kebaran at Ohalo II about 19,000 years ago (Nadel 1990). At Ohalo II the huts have multiple floors with trash between them. Burial beneath floors probably occurred in the Kebaran at Kharaneh IV and Ein Gev (Valla 1991). At Ein Gev I in the Jordan Valley in Israel, there is a fourteenth millennium BCE Kebaran site on the east side of the Sea of Galilee (Arensburg and Bar-Yosef 1973). A hut was found dug into the slope of a hill. "The hut was periodically occupied as indicated by six successive layers which accumulated within it" (Arensburg and Bar-Yosef 1973, 201). Each layer had a floor 5–7 m in diameter littered with artifacts and bones, covered by a sandy layer that included artifacts. In section the floors clearly repeat each other, and from one of the middle floors a grave was cut. There is no evidence of specific repetitions of feature or artifact placements, but this example clearly indicates some specific backward reference in the location of a house structure, even in the absence of permanent occupation. In chapter 2, Matthews discusses the evidence of repeated building at Zawi Chemi Shanidar, ca. 11,150–10,400 BCE, where a round house, Structure 1, was repeatedly constructed three times in the same place.

In the Natufian there is some degree of sedentism (Bar-Yosef and Valla 2013). 'Ain Mallaha has animals and birds from all seasons (Valla 1991), and there are commensals (such as the house mouse), indicating sedentism. Settlements occur in the hill zones of Israel, Jordan, Lebanon, and Syria; and related sites are found to the north in Mureybet and Abu Hureyra. The later Natufian starts at the same time as the Younger Dryas climatic deterioration. In the Levant in the later Natufian, many but not all hamlets dispersed and became more mobile (Bar-Yosef 2001). But in the Taurus in southeastern Turkey and adjacent areas, the response to the Younger Dryas may have been greater sedentism at sites such as Hallan Çemi (Bar-Yosef 2004).

There were both base camps and shorter-term intermittent sites in the Natufian. In the shorter-term sites there is little evidence of repetitive practices. In the Natufian site of Hatula there is a Natufian layer and then (PPNA) Khiamian and Sultanian occupation. The Natufian layer is about 0.8 m deep, and there are no houses or burials. This site is interpreted as an accumulation of short halts related to a specialized task, probably hunting gazelle (Ronen and Lechevallier 1991). This shows that palimpsest sites that do not involve placed continuity and memory construction did occur. In the short-term or seasonal encampment at Beidha, the Natufian had two to five distinct layers within 0.6 m of deposit (Byrd 1989). There were hearths and roasting areas, but no visible architecture and no burials were found.

In the Natufian at Hayonim Cave, some structures had paved floors. In one of the structures, Locus 4, there were two stages of paving and built-up hearths (Bar-Yosef 1991), although this structure then became a kiln for burning lime and then a bone tool workshop—so this is not a long sequence of repetitive use. In Stratum B there were five stages of Natufian activity within only 1 m of deposit (Bar-Yosef and Goren 1973), and there is little evidence of repetitive use of the same place or layout.

Even in substantial Natufian sites, there may be little evidence of structured repetition. Valla (1991) notes that it is often difficult to follow coherent levels of habitation in Natufian sites and difficult to show the absolute contemporaneity of buildings. In Square M1 of the tell site at Jericho in the "proto-Neolithic," there were 4 m of occupation, including a large number of beaten floors, but no evidence of repeated behavior. At Abu Hureyra 1, Moore and colleagues (2000, 105) describe "numerous, superimposed, thin floor surfaces," but there was little sense of repetition or continuity. Large numbers of small fires and artifacts are described, and the deposits sound more like midden than house floors.

However, in the early Natufian site of Wadi Hammeh 27 in the central Jordan Valley, a sounding "has revealed three successive constructional phases, overlying a human burial and associated burials." These are phases of circular stone built houses. The evidence "shows a continuity in spatial arrangement of constructed features through successive phases" (Edwards 1991, 125). The earliest evidence of Natufian occupation at Hayonim Cave is Grave XIII, "which was covered by the floor of Locus 3"—that is, by one of the structures with undressed stone walls (Bar-Yosef 1991, 86).

At 'Ain Mallaha there is definitely super-positioning of houses. In the "ancient level," Houses 131, 51, and 62–73 succeed others on the same spot (Perrot 1966). In the "recent level," there is another sequence of houses dug into each other (26, 45, 22). In the final Natufian at Mallaha, each major building had a succession of floors, one on top of another, with no sterile layers between (i.e., no abandonment fill) (Samuelian, Khalaily, and Valla 2006).

There is a rough repeated pattern in the layout of hearths and other structures in some of the Mallaha buildings. Sometimes this is very specific. According to Perrot (1966), in dwelling No. 1 at Mallaha there was a rectangular hearth on the north side with a human skull just by it to the south. The dwelling was then filled in, and from the surface a pit was dug to make a grave on the same alignment to the south. Piles of stones also occurred on the same alignment. The southern grave was then covered with stones to make a "tomb" that was visible. All this suggests structured use of space through time and memory construction in the reference to earlier practices.

By the end of the Natufian there is evidence of the removal of the human skull after death, although in the absence of evidence for circulation and reuse, this does not by itself indicate the construction of historical links to ancestors. Skull removal may have had other roles such as healing, divination, and so on. There were quite a lot of skeletons within the houses at Mallaha, but the stratigraphical positioning is often unclear in Valla (1991). According to the reanalysis by Boyd (1995), the 131-51-62-73 sequence of buildings started with twelve skeletons beneath the floor of 131. He draws attention to the continuity of activity in the same place starting with a set of burials (see, however, the critical discussion by Goring-Morris and Belfer-Cohen in chapter 3). Burials also occurred in pits outside houses at Mallaha. In Building 203 in the final Natufian at Mallaha, a building without housing facilities turned into a house with two distinct phases of floors, separated by a grave. In the final stage of occupation of this building, a corpse was deposited on the floor. Then the bones were rearranged during later habitation (Valla et al. 2002). At Natufian Hayonim Cave, graves were dug into several structures that then went out of use (Bar-Yosef 1991), but in some of the occupation phases at this site there were also graves outside the structures.

As described above, for societies in which temporal depth and memory construction are important, ending and starting buildings are likely to be significant events surrounded in ritual. Did such practices already occur in the Natufian? At Mallaha in both "ancient" and "recent levels," the fills of buildings are full of artifacts. Boyd (1995) argues that the material on the floors of Building 131 at Mallaha may have been there as part of abandonment or founding rituals. The floors at Wadi Hammeh 27 are also filled over with artifact-rich deposits (Edwards 1991). This could be a specific abandonment and dumping process, but Hardy-Smith and Edwards (2004) suggest that at Wadi Hammeh 27 and other sites, the pattern results from a lack of focus on differentiated activities in space.

In the ruins of one house at Mallaha there were several boar heads (Valla 1991), which could indicate ritualized abandonment processes. In what he called Abri 26 at Mallaha, Perrot (1966) found a child's skeleton and necklace on the abandoned floor. Complete basalt artifacts were found "discarded" or cached on interior floors at Wadi Hammeh 27 (Edwards 1991), but it is not clear whether they were just abandoned in a context of use or whether this was ritualized in some way. Goring-Morris and Belfer-Cohen (2001, 260–62) describe a number of possible cases of Natufian "ritual caches" of stone tools that could represent special deposition of some form.

Goring-Morris and Belfer-Cohen (chapter 3) discuss numerous other cases of history making in the Natufian. Of particular interest is the evidence that during the shift to more dispersed occupation in the later Natufian, people returned to earlier settlement sites to inter the dead there (see also Bar-Yosef and Valla 2013). They also discuss the history making evident in the circulation of human skulls, and they raise the broader point that distinct local cultural traits passed down from generation to generation indicate identity marking while at the same implying locally specific memories and constructed histories.

In the PPNA in the Levant, settlements were 0.2 to 2.5 hectares in size and are thus three to eight times larger than the largest Natufian sites (Bar-Yosef 2001). The houses were often oval and semi-subterranean, with internal hearths and plaster floors. In northern Syria, too, mounds were often long-lived. Jerf el Ahmar had at least ten building levels comprising about 800 years of settlement (Akkermans 2004, 287). PPNA and related sites were also often much more structured than most Natufian sites. Nadel (1998, 9) has noted that "in Natufian and other Epipaleolithic sites, it is common to find the entire range of typological variability in each site, and even in each locus . . . However, in PPNA cases, it is common to find typological differences between assemblages from contemporaneous loci at a site." Goodale and Kuijt (2006) have noted a similar shift in the way sites are formed as a result of their work at 'Iraq ed-Dubb in Jordan. Here a late Natufian occupation "had fairly non-delineated use of space compared to a more delineated use of space during the PPNA."

There is much more evidence of repeated use of the same space or house in the PPNA throughout the region. In the small site of Hallan Çemi by the upper reaches of the Tigris there are no human burials and no storage pits, and there appears to be little evidence of overall continuity from layer to layer in the location of buildings (Rosenberg and Redding 2000). However, within the latest that was examined in greater detail, the floors of buildings had surfaces of thin sand and plaster mixture "and were . . . resurfaced multiple times" (Rosenberg and Redding 2000, 45). But Qermez Dere in northern Iraq has good evidence of rebuilding in the same place (Watkins 2004, 2006).

In Phase II at Mureybet on the Middle Euphrates, there were round houses superimposed on an "Epi-Natufian" house xxxvii: "Trois niveaux d'habitation en maisons rondes se superposent directement à la maison xxxvii de la phase IB. Il s'agit manifestement de la reutilization du meme espace d'habitat en continuité directe avec la période épinatoufienne" (Cauvin 1979, 26). In part of the site, there were five levels of occupation in this phase.

At Jericho (Kenyon 1981) in Trench D II there is a huge amount of very repetitive surfaces adjacent to the tower in PPNA—between the tower and adjacent circular enclosures. But it is inside the walls that one sees most of the residential continuity in PPNA and PPNB deposits. For example, in Trench E there was continuity in E 5 in PPNA and E 135 to E 146 to E 161 in PPNB. On the whole, walls were cut down further than at Çatalhöyük. In PPNA in Squares E I, E II, and E V, there were twenty-four main building phases. In most cases there were only two to four floors for each building phase: "Some of the houses lasted through several phases, but usually with rebuildings almost from the base of the walls. Associated with most of the phases was usually a long succession of surfaces, particularly in the courtyard areas linking the various buildings" (Kenyon 1981, 269). In Square MI in Phase xxxvii a house MH was built "which has a very long life, lasting from phase xxxvii until xlv or xlvi" (Kenyon 1981, 220). "The interior of house ME, unlike MC, shows a number of renewals of the floors, two of them with a considerable make-up of cobbles; associated with the floors are occupation levels" (Kenyon 1981, 228). The ME to MH sequence covers fourteen phases, from xxxiii to xlvi.

The greater delineation of space in PPNA sites has been noted, and this is relevant to abandonment and foundation processes. There is more evidence of refuse management practices, with separate middens and more cleaning out of houses on abandonment (Hardy-Smith and Edwards 2004; Goodale and Kuijt 2006). At Hallan Çemi, plant and animal remains and groundstone were rare inside buildings, although there were fragments of copper ore and evidence of obsidian knapping inside (Rosenberg and Redding 2000). But in PPNA Göbekli, the fill in Enclosure D was full of artifacts, animal bones, and other typical settlement debris (Schmidt 2002).

Evidence for abandonment and foundation deposits is also found at Jericho. In PPNA in Trenches E I, E II, and E V, there was one building with a central stone-lined post socket under which was an infant burial (Kenyon 1981), which may represent a foundation deposit. In Square MI in PPNA in Phase xlii in house MM, the clay floor had a foundation of cobblestones: "Set in the cobbles, but sealed by the clay floor, and therefore contemporary with the construction of the building, were two burials" (Kenyon 1981, 232). At Jericho in MI in PPNA, there was perhaps a repeated pattern of large numbers of burials occurring when a new house was founded, either in the foundation or in the first occupation. This is true of buildings MO, MM, and MH. MJ had a burial in its destruction level. Burials sometimes occurred beneath floors of houses at Jericho in PPNA (Kenyon 1981). Skull removal also occurred in the PPNA (Bar-Yosef 2001). At Jerf el Ahmar in northern Syria, in Village 1/east there was a sunken building with wooden posts to hold up the roof. At the bottom of one of these posts "two human skulls were found" (Stordeur 2000, 1). This begins to suggest the specific use of skulls to build histories in houses, although the use of skulls in this way may have been simply protective or magical. Yet the use of a human skull begins to suggest that links to the past and past individuals were of increasing salience.

In their account of Körtik Tepe in southeast Turkey, Benz and colleagues (chapter 5) focus on the building of community identities and histories. The strong overall commitment to the local place is suggested by isotopic studies of human skeletons that indicate largely local and collective food consumption. The designs on stone vessels are found across the site as a whole and indeed have similarities at other sites in the region. But there is also much evidence for smaller, perhaps house-based, history making. For example, there is evidence for renovation and reoccupation of individual houses. Outside fireplaces were placed repeatedly in the same areas. There are also burials placed beneath the floors of houses. The burials are often associated with the deliberate smashing of decorated stone vessels, with some fragments kept aside and not buried.

Clare and colleagues (chapter 4) also see the elaborate structures at Göbekli Tepe in terms of the building of public histories or cultic community. They note the frequent modifications of the PPNA circular enclosures with T-shaped pillars, the reuse and resurfacing of pillars, all of which suggests curation and history making. In addition, there is good evidence for intentional burial of the enclosures, with the tops of the pillars probably still visible in the hollows between later PPNB occupation. The closure of buildings was associated with feasting. The fact that these histories were contested is suggested by the digging of a pit in Enclosure C, linked to the destruction of its two central pillars.

Turning to the PPNB in the Levant, 'Ain Ghazal has frequent floor replasterings (Banning 2003), but perhaps the best evidence is from the extensive excavations and soundings at Jericho. As in the PPNA, walls are built on walls and there are repeated floors inside houses. So in E I, E II, and E V, in Phase xlvii "the levels in the northern room of the eastern range [of rooms] were gradually raised by a series of floors . . . The numerous floor levels suggest a prolonged period of use" (Kenyon 1981, 295). But the best evidence for repeated surfaces was in the outside, courtyard areas between buildings. The courtyards had alternating layers of clay or mud floors and spreads of charcoal (Kenyon 1981, 294). There were hearths in these areas but Kenyon did not plan them, and so it is not possible to say whether there was repetitive location of hearths in outside areas.

The main changes in house sequences at Jericho are that rooms were added, built out onto courtyard areas before retreating again. Despite the overall focus on continuity, there is more interruption than at Aşıklı Höyük and Çatalhöyük. At Jericho there were real breaks, with horizons of destruction, collapse, decay, with burials in fills and fireplaces dug into stumps of walls and into abandoned floors. But the basic pattern of repeated buildings reasserts itself after the break.

In Jordan at Beidha, "the inhabitants were extremely conservative in their siting of the different elements of the village" (Kirkbride 1966, 14). "The siting of the large houses was conservative," with sequences of houses in the same location (Kirkbride 1966, 17). There was also at the site "plenty of evidence for the long use of the Level II corridor buildings; most of them were rebuilt at least partially one or more times . . . The buildings [of Level III] underlie the Level II corridor units with great precision . . . Where investigations in depth have been made the walls and buttresses lie immediately below those of Level II in most cases" (Kirkbride 1966, 18). The plaster floor of one building had been re-laid five times. Each floor was composed of many thin layers of plaster, and there was continuity in bands of red painted plaster (Kirkbride 1966, 17). In one building at Beidha the total thickness of the multiple plaster layers was over 5.5 cm, and parallels were drawn with Çatalhöyük (Kirkbride 1966, 18).

At Abu Hureyra 2 "each house was usually constructed on the remains of an earlier one, and the form of that building largely determined the plan of its successor" (Moore, Hillman, and Legge 2000, 262). The rooms of the ruined house were filled in and the stubs of the walls cut down: "The houses in Trench E were rebuilt four, and the houses in Trench B no fewer than nine times" (Moore, Hillman, and Legge 2000, 266). Floors were renewed at least two to three times and sometimes up to ten times. Walls also had mud plaster or whitewash refreshed several times during a room's life. "The hearths were often set in the same place in successive houses" (Moore, Hillman, and Legge 2000, 265)—for example, the series of hearths in houses of Phases 2–7 in Trench B. "We conclude from this that the builders of a new house often remembered not only the plan but also the internal arrangements of its predecessor, and considered it appropriate to replicate both" (Moore, Hillman, and Legge 2000, 265). "We know, too, that in some instances they themselves were the descendants of the inhabitants of the earlier structures" (Moore, Hillman, and Legge 2000, 266), since some distinctive skeletal and dental traits that are probably genetically transmitted were identified in house burials.

On the Euphrates at Bouqras in the PPNB "the house plans recovered suggest a preference for a standard dwelling pattern with respect to the location of certain areas, perhaps connected with their different functions" (Akkermans et al. 1983, 340). Thus the house had a large room with a horseshoe-shaped oven placed diagonally in one of the corners, although not always in the same corner. There is clear evidence of continuity of building in one spot (De Contenson and Van Liere 1966; Akkermans et al. 1983). "The foundations of new walls were erected directly on the stumps of the old walls, or on the floors of the former rooms, parallel to and often against older walls" (Akkermans et al. 1983, 340), although previous walls were cut down to only a few mudbricks high. As a result, the site had ten architectural levels in only 4.5 m height of mound.

At Dja'de el Mughara on the Euphrates small, one-room houses were built of pisé on stone foundations and they had been repeatedly renewed, although there were insubstantial short-term structures as well (Akkermans 2004, 285). Repetition of houses continues into the Pottery Neolithic at sites such as Tell Sabi Abyad in northern Syria (Akkermans et al. 2006), as well as in later periods.

In the Zagros, Matthews (chapter 2) notes that locations may initially be defined as "places" during early phases of infrequent and low-intensity use. It is only through time that repetitive behaviors such as floor resurfacing start to occur and more established histories are built. Matthews notes that this increased association with place may have very mundane practical concerns, such as the dependence on herbivore dung for fuel, associated with the domestication of animals. In addition, there is evidence for increasing numbers of sites and intensification of land use, all contributing to, or motivated by, greater attachment to place. She also identifies a pattern found in many other sites and regions of the repetition of houses linked to burial practices; the history making in house building was closely tied to history making around human remains. The idea that the repetitions of house layout are more than just routinized practices is suggested by their embedding in specific practices, such as the choice of very white plaster or red paint for special locations, and by the placing of bull and other installations. History making is seen in both domestic and more public ritual buildings.

In southeastern Turkey at Çayönü, there seems at first sight to be much more evidence of conformity within phases than between phases, as houses change in form from Round to Grill to Channeled to Pebble paved to Cell to Large room. There is a striking homogeneity of building types in each building layer (Özdoğan and Özdoğan 1989, 72). Thus there seems to be more of a focus on horizontal similarity rather than vertical continuity. However, even here Özdoğan and Özdoğan (1989, 73) argue that "in every building layer, the foundations of the new building are always directly on top of the preceding one, without disturbing or re-using its stones." Some buildings are mentioned as having several rebuilds, and the Skull Building went through at least five major rebuilds.

At Aşıklı Höyük in central Turkey, dated to the late ninth and early eighth millennia BCE, "in one of the excavated rooms, 'room A' (trench 3K . . .) 13 floor levels have been recognized" (Düring 2006, 73). At this site there is also the possibility of variation between houses in memory construction. Only 35 percent of rooms have hearths at this site (Esin and Harmanakaya 1999), but there is clear continuity in those houses that do and do not have hearths (Duru, chapter 6). Given the relatively small percentage of buildings with hearths, this evidence suggests that some buildings passed down the practice of hearth use while others did not. Anspach (chapter 7) argues that the hearth itself may have been a major symbolic focus. He argues that buildings with hearths more commonly had burials beneath floors and had a special non-domestic status. There is also much continuity at the site in terms of the location of the major street and the "ritual complex" and the location of midden areas (in the deep sounding). History making thus seems to occur at a variety of scales at Aşıklı: at the house level, at the level of the hearth buildings, and at the level of the community as a whole. The emphasis on continuity of buildings seen at Aşıklı Höyük and Çatalhöyük is also found elsewhere in the Ceramic Neolithic in central Anatolia. Thus, Erbaba in the mid-seventh millennium is only 4 m high, but "in some cases the walls seem to have been constructed on top of earlier walls in the same alignment" (Düring 2006, 236), and up to ten successive floors occurred in a single room.

We have seen that there is much evidence for repetitive practices in houses and for memory construction in the PPNB and related groups in the Middle East and Turkey. There is also continued evidence for abandonment and foundation practices. At Beidha the large houses were kept scrupulously clean, but the corridor buildings, which may be basements (although in some cases the floors were plastered), have fills containing implements and waste (Kirkbride 1966). At Bouqras the fills between house levels were a mixture of midden and building material (Akkermans et al. 1983). At Jericho there was often "bricky debris," perhaps from the destruction of the previous phase, between phases, and rebuilds of houses. The walls were generally cut down much more that at Çatalhöyük.

Heads tend to be found in groups in the Levant, sometimes with features plastered on, but it is not clear how much they were circulated. There are male and female skulls as well as sub-adults, raising the question of whether the skulls represent ancestor veneration at all rather than apotropaic or other protective functions (Bonogofsky 2004). However, the depositional contexts of some skull deposition are suggestive of practices that may have involved backward or forward reference. The skull of a child was found between the stones of the foundations of Wall E 180 at Jericho (Kenyon 1981). In Phase lxi in a room in a house in E I, II, V, the cranium of an elderly man was set upright in the corner about 15 cm below floor level. In E III–IV a plastered skull was found in a building fill. Goring-Morris (2000, 119) argues that many PPNB burials definitely stratigraphically predated the construction of the overlying architectural features and floors. For example, "In at least three instances at Kfar HaHoresh burial pits clearly stratigraphically underlie and are sealed by plaster surfaces" (Goring-Morris 2000, 119). In some cases there is a lapse of time between burial or skull removal and the making of the floor. Thus buildings "remembered" the location of the burials or skulls. Sometimes there is evidence of markers above the burials or skulls. Goring-Morris suggests that constructing buildings in relation to earlier buildings may have started at Mallaha in the Levant (see above).

Stevanović (1997) has argued that buildings were intentionally burned as part of abandonment practices in the Neolithic of southeastern Europe. Verhoeven (2000) has made a similar case for the Middle East, although here he sees a link to death. Thus at the late seventh and early sixth millennium cal BCE site of Tell Sabi Abyad in northern Syria (Verhoeven 1999), in the "Burnt Village" there is evidence of intentional and ritual burning related to mortuary ritual (Verhoeven 2000). Two bodies were found with large clay objects. Verhoeven (2002) interprets other evidence of firing as intentional at Bouqras and Jerf al Ahmar. At Bouqras a localized fire was again associated with dead bodies, and at Jerf el Ahmar in the PPNA a burned body in a burned house had had its head removed.

The caching of lime plaster statuettes at ʿAin Ghazal is of interest as it seems that they were taken out of a context of use and deposited. Features on the feet of the statuettes suggest they were displayed upright, anchored to the floor, before being dismantled and placed in pits. The evidence suggests an ending or beginning act. There are also claims that the statuettes were broken and that heads had been removed (Rollefson 2000). Other examples of abandonment and foundation deposits and burial practices indicating a concern with temporal depth are found in southeastern Turkey and northern Syria.

Special abandonment practices are found at Çayönü—for example, in the Cell phase there is blocking of doorways, and intact artifacts are abandoned in cell rooms (Özdoğan and Özdoğan 1989). Charnel houses or buildings for the dead occur at Çayönü (the Skull Building) and at Abu Hureyra and Dja'de el Mughara (the Maison des Morts) in Syria (Akkermans 2004, 289).

The individual plastered skull found at Çatalhöyük had been circulated and reused, as evidenced by multiple layers of re-plastering and repainting of the skull (Hodder 2006). Talalay (2004) gives examples of "untethered heads" in Turkey, at Nevalı Çori, Köşk Höyük, and Cafer Höyük. She interprets a pillar at Göbekli as showing an animal holding a human head. At Köşk Höyük there is a plastered skull of a twenty-one- to twenty-four-year-old woman and another plastered female skull. There are detachable head figurines from Çatalhöyük but also from Hacılar and Höyücek (Talalay 2004). There is evidence of circulation and handing down of artifacts through time in much of the region in the PPNB. Recirculation and reuse of stones was found at Çayönü. Standing stones up to 2 m high were found in the plaza and in the Skull and Flagstone ceremonial buildings. "Some of the standing stones were intentionally broken and then buried under the subsequent reflooring of the plaza" (Özdoğan and Özdoğan 1989, 74).

At Jericho in the PPNB levels, a large bituminous block was found (Kenyon 1981, 306–7). It had been carefully flaked and was obtained from the Nebi Musa district 17 miles away. It was found in the foundation of Wall E223 of Phase lxv. But it exactly fit into a niche of the earlier Phase lxiv, where it probably stood on a stone set on a pillar of earth on which there were traces of plaster. So this stone had a role in Phase lxiv and was then reused in the foundation of lv. In Phase lxiii this same room had a distinctive green clay floor, all suggesting that this part of the building had a special character over three phases.

A clear example of abandonment practices that involved collecting artifacts from different contexts is provided by the deposition of groundstone artifacts in Building 77 at Çatalhöyük, discussed in this volume by Tsoraki (chapter 9). The deposition of large numbers of intentionally broken and not fully used grinding stones showing various stages of weathering and deriving from a variety of contexts (since the fragments rarely fit together) indicates a carefully staged and choreographed process. A large network of social relations was indexed and memorialized in this closing ceremony as the house was burned and buried.

In her chapter, Joyce considers the passing down of immaterial property from house to house at Çatalhöyük. Clearly, this type of history making is less visible archaeologically, but Joyce studies the passing down of practices in the manufacture of pottery. She finds that knowledge of how to make pots, or knowledge of and rights to acquire pots, distinguished some residents at the site from others. Because of the relative scarcity of ceramics in houses at the site, she focuses on sequences of middens, assuming that in some way these middens were used by a community of houses at the site. Such evidence concurs with other data from the site that suggest that in the upper levels, sequences of houses used similar brick recipes in house construction (Love 2013) or herded sheep in distinct locales (Pearson 2013). The passing down of ritual knowledge at Çatalhöyük is seen, for example, in the repetition of the same leopard reliefs in Mellaart's Shrine 44, Levels VII and VI.

The PPNB levels at Halula (Saña, Tornero, and Molist 2014) provide much evidence of columns of rectangular houses built in the same place during five to seven rebuilding events. There is continuity in the placement of internal features and a standardization of burial practices. The mitochondrial genetic relationships of those buried in houses suggest much mixing and homogeneity in the community as a whole; this evidence perhaps parallels that from Çatalhöyük (Pilloud and Larsen 2011), indicating that those buried in the same house were not more closely related than those across the settlement as a whole. Those living in Çatalhöyük houses were practical or fictive kin, suggesting that house histories were constructed rather than a direct reflection of biological descent. At Halula as at Çatalhöyük, the different houses were associated with specific herd management practices.

CONCLUSION

Overall, then, there is abundant evidence of an increasing concern with temporal depth in the pre-Neolithic and Neolithic societies of the Middle East and Turkey. Perhaps one of the reasons commentators have not foregrounded such evidence in discussions of early sedentism and agricultural intensification is that archaeologists tend to base their accounts on two-dimensional settlement plans. The emphasis has tended to be on the shapes of houses, the activities that take place in them, the spatial relationships between larger and smaller houses and between public and domestic buildings, and the spatial locations of burials. The discussion has been dominated by a two-dimensional perspective, despite the fact that many of the sites have complex and deep stratigraphies. The stratigraphies are seen as important in sorting out chronological sequences, but they are not themselves seen as social. It is for this reason that the development of 3D reconstructions, as discussed by Lercari in chapter 10, is of importance, especially when linked to interactive data exploration.

Lercari's work challenges any separation of analytical research from public engagement, arguing that collaborative research on the interpretation of 3D environments benefits from an open and inclusive approach. In his reconstruction of part of the "Shrine" 10 sequence at Çatalhöyük, he shows that a 3D visualization of a history house sequence assists considerably in interpreting the degree of continuity between houses over time.

The chapters in this volume indicate clearly that there is increasing 3D stratigraphic evidence for repetitive practices in houses and sometimes in outside areas (e.g., courtyard or midden areas at Jericho and Aşıklı Höyük), as well as in public spaces such as paved streets (at Aşıklı Höyük), through time in the late Pleistocene and early Holocene in the Middle East and Turkey. There is also increasing evidence of specific memory construction as houses are built over burials or skulls and other objects are circulated and passed down through time. The concern with time depth, history, and memory reaches its apogee in the PPNB at the same time domesticated plants appear, but it starts to emerge at least by later Kebaran and Natufian times, even in contexts in which sedentism is limited. It is difficult to explain the focus on temporal depth as the result of living in dense villages. For example, at the late ninth and early eighth millennium BCE site of Boncuklu on the Konya plain, houses are rebuilt in the same locations even though the settlement pattern is fairly open and dispersed (houses are not densely packed together) (Baird 2007). Rather, it seems that the emergence of greater temporal depth was a necessary condition for dense settled life, the delayed returns of intensive subsistence systems, and the shift to domesticated plants and animals, as well as for the staging of larger-scale feasts, exchanges, and marriages.

But it is clear that there are at least two types or scales of history making. Some authors in this volume focus on the repetition of large-scale public monuments in the same location, their use and reuse, their incorporation of earlier features, their careful abandonment and rebuilding. Other authors focus more on the history making evident in columns of individual houses. At times, as at Aşıklı Höyük, both types of history making are present. What is the relationship between these two forms?

Duru (chapter 6) provides an overall synthetic account in which theories about the Neolithic are linked to present-day political concerns in an interesting way. Duru argues that public ritual aggregation sites were important in establishing sedentism in the Middle East. Public history making played an important initial role. But he goes on to argue that these public histories came to be contested. The degree of history making in the construction of ritual buildings and streets at Aşıklı Höyük is remarkable but is matched by the almost compulsive attentiveness to the repetition of houses and hearths at the site. As is clear in his figure 1 (figure 6.1), houses are continually replaced in the same locations, and those with and without hearths are continually reproduced over centuries of occupation. Duru sees a long-term process whereby collective history making was overtaken and contested by sub-identities that he describes as individualized and private but that could also be seen as house-based. As more and more activities are brought into the house, the overall focus on shared rituals comes into conflict with house-based production and consumption so that in the end the PPNB pattern of collective villages dissipates in many, though not all, regions.

One could indeed see some sort of evolutionary process of this sort, but the evidence can also be interpreted as a long-term tension between community and house-based production, exchange, and consumption. After all, as we have seen, house-based history making is evident far back into the Epipaleolithic, whereas clear evidence of collective history making is not evident until the PPNA. It is widely and commonly asserted that these public buildings, such as at Göbekli, brought communities together and that they mitigated the tensions arising from increased population, more intensive and competitive production, and increasing social differentiation. There are numerous problems with this narrative (Banning 2011). For example, the "public" buildings are often insufficiently large to house an entire community, and at Göbekli it is possible that several were in use at the same time. The provision of restricted entries and a dromos suggests that these are in fact secluded places, restricted to groups smaller than the community as a whole. Throughout the southern and northern Levant, the special buildings are at the edges of settlements rather than centrally located. In central Anatolia, at Aşıklı Höyük, the public buildings are marginal to the "ancestral core" (see Duru, figure 6.2) and could only have housed a small fraction of the community as a whole. While it is certainly possible that the "public" buildings brought social cohesion to sub-groups within society, they also suggest in-groups versus out-groups. Their architectural structure, inward-looking and closed off, indicates more the participation of individuals in secret societies, men's houses (Flannery and Marcus 2012), and other exclusive sodalities rather than community-wide cohesion.

One type of history making during the late Epipaleolithic and Neolithic in the Middle East and Anatolia centers on houses, their repetition and renewal. It seems that this house-based history making involved both practical routines as well as the passing down of objects and immaterial knowledge and forms of ownership. The histories constructed were often fictive and imagined but stabilized and made concrete in the material practices of house building, burial, and the circulation of objects. This type of history making has the greater temporal depth, starting way back in the Epipaleolithic if not before. It can be linked to the gradual rise of economic and social systems in which there is a delayed return for labor input. This type of history making centers on the house because through this time period there is a largely domestic mode of production in which much of the production, processing, and consumption took place at the household level. Houses were linked together through (often fictive) descent that underpinned economic and social collaboration.

Gradually through time, however, a second form of history making emerged that allowed greater and wider collaboration within segments of the community as a whole. This second type allowed crosscutting sodalities to be constructed. These sodalities probably included a diversity of forms such as hunting societies, men's houses, secret societies, and medicine societies (Mills 2014). They functioned to link people together across house-based groups. They were often exclusive and highly ritualized. But they allowed any particular individual to call upon a wider array of support in times of hardship. They allowed cross-community sharing on a larger scale while at the same time promoting difference and contestation.

It is thus incorrect to see the increase of ritual and public monuments, of special buildings, as resulting from community cohesion. Rather, over millennia there was a tension between house-based and sodality-based forms of history making. These forms competed with each other, and in the end the "public" buildings that came to dominate in the PPNA decreased in importance and influence. Throughout, there is a competitive process of inventing and materializing histories, both between house-based and sodality-based groups and between different houses and sodalities. Both the house-based descent groups and the crosscutting sodalities that convened in "public" buildings invested in religious practices that involved history making. Both involved creating time depth and at the same time building networks. The greater the temporal depth, the more that people could be incorporated into networks of sharing and co-dependency. But also, the greater the temporal depth, the more could people invest in subsistence practices in which there were delayed returns for labor. The pattern of PPN villages, including the mega-sites of the PPNB, were knit together by religion and history tied to place. The chapters in this volume demonstrate that shifting from a 2D to a 3D perspective allows us to recognize the importance of historical depth in these early village societies.

Another reason for the lack of attention paid to history making in the period leading up to the Neolithic in the Middle East is that the focus on 2D house and settlement plans is matched by a focus on the spatial organization of regional exchange and settlement systems. Perhaps in part because our chronologies are so approximate, much research has concentrated on networks of interaction, interaction spheres, population aggregation, and settlement systems. Here, too, a 3D perspective is needed. Settlement systems develop over time, and sites such as Göbekli Tepe and perhaps also Çatalhöyük became attractors to population in their regions, acting as historical magnets. As noted earlier, Gamble (1998) and Coward (2010) have documented the increased emphasis on networks of interchange in the later Epipaleolithic and early Neolithic. Exchange relations are often historical in that the objects transferred carry histories with them, linking communities in cycles of interdependence, giving but also keeping (Weiner 1992). One way of creating networks is to build historical ties through actual or fictive ancestry. Another mechanism is to coalesce groups of people around the passing down of rights and duties in "house societies" (as discussed in chapter 8). The greater the temporal depth achieved in the building of ties, the wider the network of affiliated individuals. But there is a logical contradiction between networks that are flat and open and those that are apical and deep in time, in that the latter promote separation and distinction. Time depth creates attachment to place, sedentism, tendencies toward ownership, investment. In these ways and at the most general of levels, history making contributes to the production of sedentism and the origins of farming. It undermines the collective sharing and reciprocal exchange that appear to dominate the Epipaleolithic and early Neolithic.

REFERENCES

Akkermans, Peter. 2004. "Hunter-Gatherer Continuity: The Transition from the Epipaleolithic to the Neolithic in Syria." *BAR International Series* S1263: 281–94.

Akkermans, Peter, J.A.K. Boerma, Antje T. Clason, S. G. Hill, Erik Lohof, Christopher Meiklejohn, M. Le Miere et al. 1983. "Bouqras Revisited: Preliminary Report on a Project in Eastern Syria." *Proceedings of the Prehistoric Society* 49: 335–72. https://doi.org/10.1017/S0079497X00008045.

Akkermans, Peter, René Cappers, Chiara Cavallo, Olivier Nieuwenhuyse, Bonnie Nilhamn, and Iris N. Otte. 2006. "Investigating the Early Pottery Neolithic of Northern Syria: New Evidence from Tell Sabi Abyad." *American Journal of Archaeology* 110 (1): 123–56. https://doi.org/10.3764/aja.110.1.123.

Arbuckle, Benjamin S. 2015. "Large Game Depression and the Process of Animal Domestication in the Near East." In *Climate Change in Ancient Societies*, ed. Susanne Kerner, Rachael J. Dann, and Pernille Bangsgaard, 215–43. Chicago: University of Chicago Press.

Arensburg, Baruch, and Ofer Bar-Yosef. 1973. "Human Remains from Ein Gev I, Jordan Valley, Israel." *Paléorient* 1 (2): 201–6.

Baird, Douglas. 2007. "The Boncuklu Project: The Origins of Sedentism, Cultivation, and Herding in Central Anatolia." *Anatolian Archaeology* 13: 14–18.

Banning, Edward B. 2003. "Housing Neolithic Farmers." *Near Eastern Archaeology* 66 (1–2): 4–21. https://doi.org/10.2307/3210928.

Banning, Edward B. 2011. "So Fair a House." *Current Anthropology* 52 (5): 619–60. https://doi.org/10.1086/661207.

Bar-Yosef, Ofer. 1991. "The Archaeology of the Natufian Layer at Hayonim Cave." In *Natufian Culture in the Levant*, ed. Ofer Bar-Yosef and François R. Valla, 81–92. Ann Arbor, MI: International Monographs in Prehistory.

Bar-Yosef, Ofer. 2001. "From Sedentary Foragers to Village Hierarchies: The Emergence of Social Institutions." In *The Origin of Human Social Institutions*, ed. Walter Garrison Runciman, 1–38. Oxford: Oxford University Press.

Bar-Yosef, Ofer. 2004. "Guest Editorial: East to West—Agricultural Origins and Dispersal into Europe." *Current Anthropology* 45 (S4): 1–4. https://doi.org/10.1086/423970.

Bar-Yosef, Ofer, and Naama Goren. 1973. "Natufian Remains in Hayonim Cave." *Paléorient* 1 (1): 49–68.

Bar-Yosef, Ofer, and François R. Valla, eds. 2013. *Natufian Foragers in the Levant: Terminal Pleistocene Social Changes in Western Asia*. Ann Arbor, MI: International Monographs in Prehistory.

Bayliss, Alex, Fiona Brock, Shahina Farid, Ian Hodder, John Southon, and Royal E. Taylor. 2015. "Getting to the Bottom of It All: A Bayesian Approach to Dating the Start of Çatalhöyük." *Journal of World Prehistory* 28 (1): 1–26. https://doi.org/10.1007/s10963-015-9083-7.

Bergman, Christopher A. 1987. "Ksar Akil Lebanon: Vol. II." *BAR International Series* S329: 1–334.

Bloch, Maurice. 2008. "Why Religion Is Nothing Special but Is Central." *Philosophical Transactions of the Royal Society of London: Series B, Biological Sciences* 363 (1499): 2055–61. https://doi.org/10.1098/rstb.2008.0007.

Bonogofsky, Michelle. 2004. "Including Women and Children: Neolithic Modeled Skulls from Jordan, Israel, Syria, and Turkey." *Near Eastern Archaeology* 67 (2): 118–19. https://doi.org/10.2307/4132367.

Bourdieu, Pierre. 1977. *Outline of a Theory of Practice*. Cambridge: Cambridge University Press. https://doi.org/10.1017/CBO9780511812507.

Boyd, Brian. 1995. "Houses and Hearths, Pits and Burials: Natufian Mortuary Practices at Mallaha (Eynan), Upper Jordan Valley." In *The Archaeology of Death in the Ancient Near East*, ed. Stuart Campbell and Anthony Green, 17–23. Oxford: Oxbow Books.

Boz, Başak, and Lori Hager. 2013. "Intramural Burial Practices at Çatalhöyük." In *Humans and Landscapes of Çatalhöyük: Reports from the 2000–2008 Seasons*, ed. Ian Hodder, 413–40. Los Angeles: Cotsen Institute of Archaeology at UCLA.

Byrd, Brian F. 1989. *The Natufian Encampment at Beidha: Late Pleistocene Adaptation in the Southern Levant*. Aarhus, Denmark: Jutland Archaeological Society Publications.

Cauvin, Jacques. 1979. "Les Fouilles de Mureybet (1971–1974) et leur Signification pour les Origins de la Sedentarisation au Proche-Orient." *Annual of the American Schools of Oriental Research* 44: 19–48.

Connerton, Paul. 1989. *How Societies Remember*. Cambridge: Cambridge University Press. https://doi.org/10.1017/CBO9780511628061.

Coward, Fiona. 2010. "Small Worlds, Material Culture, and Near Eastern Social Networks." *Proceedings of the British Academy* 158: 449–79.

De Contenson, Henri, and Willem J. Van Liere. 1966. "Premier Sondage à Bouqras en 1965: Rapport Preliminaire." *Annales Archeologiques Arabes Syriennes* 16 (2): 181–92.

Düring, Bleda S. 2006. *Constructing Communities: Clustered Neighbourhood Settlements of the Central Anatolian Neolithic, ca. 8500–5500 Cal BC*. Leiden: Nederlands Instituut voor het Nabije Oosten.

Edwards, Phillip C. 1991. "Wadi Hammeh 27: An Early Natufian Site at Pella, Jordan." In *The Natufian Culture in the Levant*, ed. Ofer Bar-Yosef and François R. Valla, 123–48. Ann Arbor, MI: International Monographs in Prehistory.

Esin, Ufuk, and Savas Harmanakaya. 1999. "Aşıklı in the Frame of Central Anatolian Neolithic." In *The Neolithic in Turkey: The Cradle of Civilization: New Discoveries*, ed. Mehmet Özdoğan, Nezih Başgelen, and Peter Kuniholm, 115–32. Istanbul: Arkeoloji ve Sanat Yayınları.

Flannery, Kent, and Joyce Marcus. 2012. *The Creation of Inequality: How Our Prehistoric Ancestors Set the Stage for Monarchy, Slavery, and Empire*. Cambridge, MA: Harvard University Press. https://doi.org/10.4159/harvard.9780674064973.

Gamble, Clive. 1998. "Paleolithic Society and the Release from Proximity: A Network Approach to Intimate Relations." *World Archaeology* 29 (3): 426–49. https://doi.org/10.1080/00438243.1998.9980389.

Goldberg, Paul. 2001. "Some Micromorphological Aspects of Prehistoric Cave Deposits." *Cahiers d'Archéologie du CELAT* 10: 161–75.

Goodale, Nathan B., and Ian Kuijt. 2006. "Keeping Space with Agriculture: Differential Patterns of Human Behavior and Intra-Community Organization in the Near East." Working paper.

Goring-Morris, Adrian Nigel. 2000. "The Quick and the Dead." In *Life in Neolithic Farming Communities: Social Organization, Identity, and Differentiation*, ed. Ian Kuijt, 103–36. New York: Kluwer Academic/Plenum.

Goring-Morris, Adrian Nigel, and Anna Belfer-Cohen. 2001. "The Symbolic Realms of Utilitarian Material Culture: The Role of Lithics." In *Beyond Tools*, ed. Isabella Caneva, Christina Lemorini, Daniela Zampetti, and Paolo Biagi, 257–71. Berlin: Ex oriente.

Hardy-Smith, Tania, and Philip C. Edwards. 2004. "The Garbage Crisis in Prehistory: Artifact Discard Patterns at the Early Natufian Site of Wadi Hammeh 27 and the Origins of Household Refuse Disposal Strategies." *Journal of Anthropological Archaeology* 23 (3): 253–89. https://doi.org/10.1016/j.jaa.2004.05.001.

Hodder, Ian. 2006. *The Leopard's Tale*. London: Thames and Hudson.

Hodder, Ian, ed. 2010. *Religion in the Emergence of Civilization: Çatalhöyük as a Case Study*. Cambridge: Cambridge University Press. https://doi.org/10.1017/CBO9780511761416.

Hodder, Ian, ed. 2014. *Integrating Çatalhöyük: Themes from the 2000–2008 Seasons*. Los Angeles: Cotsen Institute of Archaeology at UCLA.

Hodder, Ian, and Craig Cessford. 2004. "Daily Practice and Social Memory at Çatalhöyük." *American Antiquity* 69 (1): 17–40. https://doi.org/10.2307/4128346.

Hodder, Ian, and Peter Pels. 2010. "History Houses: A New Interpretation of Architectural Elaboration at Çatalhöyük." In *Religion in the Emergence of Civilization: Çatalhöyük as a Case Study*, ed. Ian Hodder, 163–86. Cambridge: Cambridge University Press. https://doi.org/10.1017/CBO9780511761416.007.

Kenyon, Kathleen M. 1981. *The Architecture and Stratigraphy of the Tell*, vol. 3: *Excavations at Jericho*. London: British School of Archaeology in Jerusalem.

Kirkbride, Diana. 1966. "Five Seasons at the Pre-Pottery Neolithic Village of Beidha in Jordan." *Palestine Exploration Quarterly* 98 (1): 8–72. https://doi.org/10.1179/peq.1966.98.1.8.

Kotsakis, Kostas. 1999. "What Tells Can Tell: Social Space and Settlement in the Greek Neolithic." In *Neolithic Society in Greece*, ed. Paul Halstead, 66–76. Sheffield: Sheffield Academic Press.

Kuijt, Ian, ed. 2000. *Life in Neolithic Farming Communities: Social Organization, Identity, and Differentiation*. New York: Kluwer Academic/Plenum.

Love, Serena. 2013. "An Archaeology of Mudbrick Houses from Çatalhöyük." In *Substantive Technologies at Çatalhöyük: Reports from the 2000–2008 Seasons*, ed. Ian Hodder, 81–96. Los Angeles: Cotsen Institute of Archaeology at UCLA.

Maher, Lisa A., Tobias Richter, and Jay T. Stock. 2012. "The Pre-Natufian Epipaleolithic: Long-Term Behavioral Trends in the Levant." *Evolutionary Anthropology* 21 (2): 69–81. https://doi.org/10.1002/evan.21307.

Meignen, Liliane, Ofer Bar-Yosef, Paul Goldberg, and Steve Weiner. 2000. "Le Feu au paléolithique Moyen: Recherches sur les Structures de Combustion et le Statut des Foyers: L'exemple du Proche-Orient." *Paléorient* 26 (2): 9–22. https://doi.org/10.3406/paleo.2000.4706.

Mills, Barbara. 2014. "Relational Networks and Religious Sodalities at Çatalhöyük." In *Religion at Work in a Neolithic Society: Vital Matters*, ed. Ian Hodder, 159–86. Cambridge: Cambridge University Press.

Moore, Andrew Michael Tangye, Gordon C. Hillman, and Anthony J. Legge. 2000. *Village on the Euphrates: From Foraging to Farming at Abu Hureyra*. Oxford: Oxford University Press.

Nadel, Dani. 1990. "Ohalo II—a Preliminary Report." *Mitekufat Haeven* 23: 48–59.

Nadel, Dani. 1998. "A Note on PPNA Intra-Site Tool Variability." *Neo-Lithics* 1 (98): 8–10.

Özdoğan, Mehmet, and Asli Özdoğan. 1989. "Çayönü: A Conspectus of Recent Work." *Paléorient* 15 (1): 65–74. https://doi.org/10.3406/paleo.1989.4485.

Pearson, Jessica. 2013. "Human and Animal Diets as Evidenced by Stable Carbon and Nitrogen Isotope Analysis." In *Humans and Landscapes of Çatalhöyük: Reports from the 2000–2008 Seasons*, ed. Ian Hodder, 271–98. Los Angeles: Cotsen Institute of Archaeology at UCLA.

Perrot, Jean. 1966. "Le Gisement Natoufien de Mallaha (Eynan), Israel." *L'Anthropologie* 70: 437–84.

Pilloud, Marin A., and Clark Spencer Larsen. 2011. "'Official' and 'Practical' Kin: Inferring Social and Community Structure from Dental Phenotype at Neolithic Catalhoyuk, Turkey." *American Journal of Physical Anthropology* 145 (4): 519–30. https://doi.org/10.1002/ajpa.21520.

Rollefson, Gary. 2000. "Ritual and Social Structure at Neolithic 'Ain Ghazal." In *Life in Neolithic Farming Communities: Social Organization, Identity, and Differentiation*, ed. Ian Kuijt, 165–90. New York: Kluwer Academic/Plenum.

Ronen, Avraham, and Monique Lechevallier. 1991. "The Natufian of Hatula." In *The Natufian Culture in the Levant*, ed. Ofer Bar-Yosef and François R. Valla, 149–60. Ann Arbor, MI: International Monographs in Prehistory.

Rosenberg, Michael, and Richard W. Redding. 2000. "Hallan Çemi and Early Village Organization in Eastern Anatolia." In *Life in Neolithic Farming Communities: Social Organization, Identity, and Differentiation*, ed. Ian Kuijt, 39–61. New York: Kluwer Academic/Plenum.

Samuelian, Nicolas, Hamudi Khalaily, and François R. Valla. 2006. "Final Natufian Architecture at Eynan (Ain Mallaha): Approaching the Diversity behind Uniformity." In *Domesticating Space*, ed. Edward B. Banning and Michael Chazan, 35–42. Berlin: Ex Oriente.

Saña, Maria, Carlos Tornero, and Miquel Molist. 2014. "Property and Social Relationships at Tell Halula during PPNB." Paper presented at the conference Religion, History, and Place in the Origin of Settled Societies, Çatalhöyük, Turkey, August 2–3.

Schmidt, Klaus. 2002. "Göbekli Tepe—Southeastern Turkey: The Seventh Campaign, 2001." *Neo-Lithics* 1 (2): 23–25.

Stevanović, Mira. 1997. "The Age of Clay: The Social Dynamics of House Destruction." *Journal of Anthropological Archaeology* 16 (4): 334–95. https://doi.org/10.1006/jaar.1997.0310.

Stordeur, Danielle. 2000. "New Discoveries in Architecture and Symbolism at Jerf el Ahmar (Syria), 1997–1999." *Neo-Lithics* 1 (00): 1–4.

Talalay, Lauren E. 2004. "Heady Business: Skulls, Heads, and Decapitation in Neolithic Anatolia and Greece." *Journal of Mediterranean Archaeology* 17 (2): 139–63. https://doi.org/10.1558/jmea.17.2.139.65540.

Tringham, Ruth. 2000. "The Continuous House." In *Beyond Kinship: Social and Material Reproduction in House Societies*, ed. Rosemary A. Joyce and Susan D. Gillespie, 115–34. Philadelphia: University of Pennsylvania Press.

Valla, François R. 1991. "Les Natoufiens de Mallaha et L'espace." In *The Natufian Culture in the Levant*, ed. Ofer Bar-Yosef and François R. Valla, 111–22. Ann Arbor, MI: International Monographs in Prehistory.

Valla, François R., Hamoudi Khalaily, Nicolas Samuelian, and Fanny Bocquentin. 2002. "De la Predation à la Production: L'apport des Fouilles de Mallaha (1996–2001)." *Bulletin du Centre de Recherche Francais de Jérusalem* 10: 17–38.

Van Dyke, Ruth M., and Susan E. Alcock, eds. 2008. *Archaeologies of Memory*. New York: John Wiley and Sons.

Verhoeven, Marc. 1999. *An Archaeological Ethnography of a Neolithic Community*. Istanbul: Nederlands Instituut.

Verhoeven, Marc. 2000. "Death, Fire, and Abandonment." *Archaeological Dialogues* 7 (1): 46–83. https://doi.org/10.1017/S1380203800001598.

Verhoeven, Marc. 2002. "Ritual and Ideology in the Pre-Pottery Neolithic B of the Levant and Southeast Anatolia." *Cambridge Archaeological Journal* 12 (2): 233–58. https://doi.org/10.1017/S0959774302000124.

Watkins, Trevor. 2004. "Building Houses, Framing Concepts, Constructing Worlds." *Paléorient* 30 (1): 5–23. https://doi.org/10.3406/paleo.2004.4770.

Watkins, Trevor. 2006. "Architecture and the Symbolic Construction of New Worlds." In *Domesticating Space*, ed. Edward B. Banning and Michael Chazan, 15–24. Berlin: Ex Oriente, Berlin.

Weiner, Annette B. 1992. *Inalienable Possessions: The Paradox of Keeping-While Giving*. Berkeley: University of California Press. https://doi.org/10.1525/california/9780520076037.001.0001.

Whitehouse, Harvey, and Ian Hodder. 2010. "Modes of Religiosity at Çatalhöyük." In *Religion in the Emergence of Civilization: Çatalhöyük as a Case Study*, ed. Ian Hodder, 122–45. Cambridge, MA: Cambridge University Press. https://doi.org/10.1017/CBO9780511761416.005.

Woodburn, James. 1980. "Hunters and Gatherers Today and Reconstruction of the Past." In *Soviet and Western Anthropology*, ed. Ernest Gellner, 95–117. London: Duckworth.

Zeder, Melinda. 2011. "The Origins of Agriculture in the Near East." *Current Anthropology* 52 (S4): S221–S235. https://doi.org/10.1086/659307.

1

Simulating Religious Entanglement and Social Investment in the Neolithic

F. LeRon Shults and
Wesley J. Wildman

This chapter's first author was also a participant in the first two of the three-year projects at Çatalhöyük supported by the John Templeton Foundation, and he contributed essays to the two volumes that reported on those projects (Shults 2010, 2014a). Throughout those six years of multidisciplinary engagement, he was repeatedly invited to focus on the empirical material and asked to help illuminate the concrete findings unearthed at the site and in the surrounding area. This is not what philosophers of religion and theologians are trained to do. During long hours spent interacting with the scientists at Çatalhöyük and combing through concrete findings in the site research reports during Phase I and Phase II, he increasingly got in the habit of trying to tie his philosophical reflections to the hard data (and, apparently, of writing about himself in the third person).

When invited to return for Phase III, he reached out for help to the second author of this chapter, another philosopher of religion and theologian who is well-versed in simulation and modeling. We spent our time together at Çatalhöyük reviewing the data, engaging in even longer conversations with the scientists on-site, hammering out the causal architecture for the computer simulations we report on below, and exploring ways to calibrate these models by finding tie-downs to the archaeological data and expert opinion on the interpretation of the site.

The title of the project for Phase III was "The Primary Role of Religion in the Origin of Settled Life:

DOI: 10.5876/9781607327370.c001

The Evidence from Çatalhöyük and the Middle East." The general hypothesis guiding the project as a whole was that religion played a primary role in the transition to more sedentary forms of human civilization because it led to or intensified the production of "historical depth" and "attachment to place." We focused particularly on the third research question: How did the "history house" and the "history town" at Çatalhöyük emerge? As we explain in more detail below, our goal was to help answer this question by constructing computer simulations of the emergence of high levels of social investment and religious entanglement in the Neolithic.

We believe these simulations shed light on some of the relevant mechanisms at work in the creative tension between the two forms of "history making" (house-based and sodality-based) outlined by Ian Hodder in the introduction to this volume.

Another part of our assignment was to reflect on the way discussions about the role of religion in providing the conditions for historical depth and attachment to place in early sedentary communities might affect our self-understanding *today*—in late modern, pluralistic, globalizing contexts. Part of our motivation for pursuing this sort of research has to do with our interest in broader questions about the relation between religion and civilizational forms—questions that are of concern to a wider audience. The model described briefly here is part of the larger Modeling Religion Project at the Center for Mind and Culture (mindandculture.org/focus-areas/religion/), in which we propose to develop multi-modal simulations of the role of religion in social transitions from the Upper Paleolithic to modern times—and even to make experimental projections about the future of religion and spirituality.

In this context, however, we stay focused on this early shift from hunter-gatherer societies to relatively egalitarian sedentary societies, using the empirical findings from and theoretical arguments about Çatalhöyük to verify and validate our model. But why would an archaeologist (or any other scientist interested in Neolithic societies) care about this sort of computer simulation? In fact, the use of simulation techniques has been growing so rapidly that it has been referred to as the "third pillar" of science, alongside theory and experimentation (Yilmaz 2015). Across the social sciences, one increasingly finds scholars who are adapting and expanding these techniques (Squazzoni, Jager, and Edmonds 2014). To provide the background for understanding the uniqueness and potential value of our model, we begin with a brief overview of some other uses of modeling and simulation (M&S) in archaeology.

SIMULATION AND MODELING IN ARCHAEOLOGY

Nearly two decades before he became the leader of the Çatalhöyük excavations, Ian Hodder edited a volume called *Simulation Studies in Archaeology*, one of the earliest attempts to encourage the application of M&S techniques to archaeology (Hodder 1978). Although these techniques were still relatively new at the time, the contributors to that book clearly recognized their potential for facilitating conceptual clarity, hypothesis testing, and the development of new theories. In a summary of trends in computer modeling in archaeology written a decade later, James Bell observed that many of the early simulations were imitative, that is, adopted from other fields and not explicitly designed for the needs of archaeologists (Bell 1987).

The adaptation and use of simulations within archaeology increased significantly in the 1990s. Perhaps the best-known were the agent-based models developed as part of the project that focused on simulating the emergence of complex social forms in the Upper Paleolithic (Doran et al. 1994; Doran and Palmer 1995). In the last decade, archaeologists have taken advantage of even more advanced statistical techniques and increases in computer power (e.g., Costopoulos and Lake 2010; Gerbault et al. 2014). While M&S techniques cannot "prove" that a particular archaeological theory is true (what technique could?), they can help resolve long-standing debates by providing warrant for accepting one hypothesis over another. Despite the limitations and difficulties involved in constructing, calibrating, and validating computer simulations, the general consensus seems to be that this is an important and valuable trend in archaeology (Lake 2014; Crabtree and Kohler 2012).

M&S techniques have been applied across a variety of archaeological contexts and time frames, both before and after the Neolithic. For example, computer models of the environmental conditions and likely behaviors during the dispersal of hominids "out of Africa" have helped provide clarity for scholars involved in debates over the timing and extent of that exodus (Mithen and Reed 2002; Nikitas and Nikita 2005). Other models have dealt with sites settled and deserted long after Çatalhöyük, such as the Long House Valley region inhabited by the Kayenta Anasazi (Axtell et al. 2002). Another recent agent-based archaeological model has simulated the emergence, intensification, and eventual dispersion of prehispanic Pueblo societies (Kohler et al. 2012). For examples of systems-dynamics models, see Lowe's (1985) simulation of the collapse of the Mayan civilization and Jayyousi and Reynolds's (2014) simulation of the ancient urban center of Monte Albán, which integrates micro-, meso-, and macro-levels of systemic change.

The Neolithic transition itself has also been the subject of several computer models. Archaeobotanists have long debated whether the transition to agriculture (domestication of plants) was relatively rapid or occurred slowly and involved long periods of pre-domestic cultivation. While the latter hypothesis has been the more widely accepted, it was hard to account for the results of genome-wide multi-locus studies of crop origins that appear to suggest that domesticate crops are associated by monophyly with only a single geographic locality. The use of M&S techniques has resolved this problem by demonstrating that such results are virtually inevitable over time and are largely influenced by population size. Multiple-origin crops are actually more likely to produce monophyletic clades than are crops with a single origin—a counterintuitive, but now intelligible, conclusion made possible by computer simulations (Allaby, Fuller, and Brown 2008; Allaby, Brown, and Fuller 2010).

Another modeling tool that has been used by archaeologists interested in the Neolithic is the Global Land Use and technological Evolution Simulator (GLUES). For example, the results of one simulation study of the Neolithic transition in the Indus Valley corroborated the hypothesis of an independent South Asian Neolithic, that is, the emergence of agropastoralism on the Indian subcontinent independent of developments in the Levant (Lemmen and Khan 2012). Another study using GLUES attempted to resolve one of the long-standing arguments among archaeologists about the mechanisms of the transition during which farming and herding were introduced to Europe from the Near East and Anatolia. This simulation found that although both demic diffusion (by migration) and cultural diffusion (by trade) can explain the Western European transition equally well, local adopters of agropastoralism appear to have contributed to the process far more strongly than migrating farmers (Lemmen, Gronenborn, and Wirtz 2011).

Several other models have been developed to study the dynamics of the Neolithic and the emergence of early societies. One recent agent-based model focused on the function of warfare among early complex, hierarchically structured polities and demonstrated the importance of wealth, power, and well-defined means of succession and internally specialized control mechanisms in predicting the outcome of conflict (Gavrilets, Anderson, and Turchin 2014). Another recent article reported on the construction of a quantitative model of evolution in a patch-structured population that explored the conditions for the transition from egalitarianism to leadership and despotism in the Neolithic and beyond. This model predicts that the transition to a despotic system will occur when surplus resources lead to demographic expansion of groups and dispersal costs limit people's ability to escape the system (Powers and Lehmann 2014).

As we explain in more detail below, our systems-dynamics model has several distinctive features, including its attention to the empirical data derived from the excavations at Çatalhöyük and other sites in the region and its inclusion of "religious" variables (cognitive, moral, ritual, and social) in its causal architecture.

RELIGIOUS ENTANGLEMENT

In this context, our use of the term *entanglement* follows that of Ian Hodder, for whom it refers to the co-evolutionary reciprocal dynamics between humans and (other) things whereby they are increasingly bound together through complex dynamics of (mutual) production, enablement, constraint, and limitation. Hodder (2012) spells out the philosophical sense of this term in most detail in *Entangled: An Archaeology of the Relationships between Humans and Things*, but he has been using it to describe the evolutionary co-dependence of human beings and other material things for many years. In discussing the Neolithic engagement of "spirits" (animal spirits or ancestor ghosts) in *Çatalhöyük: The Leopard's Tale*, for example, Hodder suggests that "as people, society and crafted materials increasingly became entangled and codependent, so the codependent material agents were further enlisted and engaged in a social world in which spirits were involved" (Hodder 2006, 195).

Leading up to and throughout the life cycle of Çatalhöyük, human beings became increasingly entangled with the natural objects of their world, including plants, animals, and building materials, as well as their own social productions, including agricultural technologies, figurines, and houses. Hodder's use of the term *entanglement* has been adapted and adopted by several other scholars associated with the Çatalhöyük project in their attempts to interpret various aspects of the Neolithic. For example, in several places, Dorian Fuller and colleagues have explored the way humans and plants were entangled or mutually "entrapped" in the long pathway to domestication (Fuller, Allaby, and Stevens 2010; Fuller, Asouti, and Purugganan 2012).

Our simulation includes variables related to the domestication of plants and animals, but our primary interest is in teasing out the way things (and processes) deemed "religious" became entangled in the human-thing co-dependent evolutionary nexus during the Neolithic. Hodder emphasized the role of religion in the introduction to the volume reporting on Phase II of the project. He noted that religion, as an integral part of life at Çatalhöyük, played varying roles in instigating and producing transformative change throughout the site sequence from 7400 to 6500 BCE. Religion was central to the complex world of the community, which was constituted by sodalities akin to mystery cults and dominated by symbols such as the leopard and the bear. As Hodder pointed out, however, it was the ancestors and the wild bull that served as the primary foci around which social groups formed and developed relations with each other (Hodder 2014, 3).

In their reflections on the "magical" relation to material artifacts at Çatalhöyük in the same volume, Carolyn Nakamura and Peter Pels also stressed the importance of religion (in a general sense) in the ever-complexifying web of people-thing relations. They noted that "various archaeological accounts of the site point to a fairly distinctive symbolic sphere at Çatalhöyük that was deeply entangled in the daily lives of people and the beings (both living and dead) with whom they shared their world." Nakamura and Pels refer to this sphere of interaction as "the transcendent nonhuman world of Çatalhöyük" (Nakamura and Pels 2014, 195).

The term *religion* is highly contested, and not only in archaeology. Even—or especially—in departments of "religious studies," it is hard to find a consensus on its meaning. For our purposes here, we will use the term in the way it commonly functions in the discourse among scholars interested in studying religion from a cognitive-science perspective. In this context, "religion" is used in a broad sense to designate modes of shared imaginative engagement with the supernatural agents culturally postulated by an in-group, which promote cooperation, commitment, and cohesion in the face of out-group threats and environmental challenges (cf. Shults 2014b, 2018). With reference to Çatalhöyük, this means ritual interaction with culturally postulated, putatively disembodied intentional forces like auroch spirits and ancestor ghosts.

It is important to emphasize that what we are calling "religion" is itself an always evolving nexus of entangled variable forces (e.g., cognitive, moral, ritual, social). This is why describing it well requires the integration of multiple disciplinary perspectives. The model we describe below is in fact based on an *entanglement of theories*, a causal architecture that integrates insights from a wide variety of fields, including cognitive science, psychology, sociology, political theory, and economics—as well as the scientific study of religion.

In the conclusion to the volume reporting on Phase I of the project, Hodder cited the first author of this chapter, who had teasingly admitted his initial annoyance and impatience with scientists' obsession with details about "petrified poop" and relative inattention to the far more interesting philosophical implications of the site (Hodder 2010, 351). In the conclusion to the volume on

Phase II of the project, Hodder noted the danger that interdisciplinary scholars might bring their pet theories and impose them on Çatalhöyük. He wondered whether the "big data" from the site really caused theorists to change their minds (Hodder 2014, 346). Now, in the wake of Phase III of the project, the chapter's first author can speak for himself (still, apparently, only in the third person) and confirm that the whole process has led him to a new appreciation of the importance of both big data and big theory—and to a new tool for experimentally linking them.

THE NEOLITHIC SOCIAL INVESTMENT MODEL (NSIM)

The conceptual framework for our simulation of the Neolithic transition is *lifestyle conversion* within a constrained geographic area (such as the Konya plain). The low-investment (LI) lifestyle encompasses hunter-gatherer groups and people who used farming techniques without embracing the intensification of sociality and human-thing entanglement that we see in Çatalhöyük. They could delay gratification and plan for the future but were not as intensely entangled with people, places, and things. The high-investment (HI) lifestyle involves agricultural settlements with higher levels of entanglement with the vital forces of their environment, including people, animals, plants, objects, and culturally postulated spirits. People in HI lifestyles were far less likely to move around than people in LI lifestyles because of this difference in entanglement with people and place. These two lifestyles are treated as alternatives with transition paths between them. The six variables needed to determine the ratio of LI to HI people are birth rates for each type, rates of exposure to the other type of lifestyle, and rates of conversion to the alternative lifestyle. The stock-and-flow diagram in figure 1.1 illustrates this conceptual framework.

The model presents babies born into LI (top left) and HI (bottom left) lifestyles according to the corresponding birth rates (LI_BirthRate and HI_BirthRate). The LI people are exposed to the HI lifestyle alternative at the rate dictated by the matching variable (LI_ExposureRateToHI). The LI people who are so exposed then convert to the HI lifestyle at a rate depending on the conversion variable (LI_ConversionRateToHI). A parallel conversion path exists to move from the HI lifestyle over to the LI lifestyle (with variables HI_ExposureRateToLI and HI_ConversionRateToLI). Of course, everyone dies in the end. The model also includes a Carrying Capacity parameter that contributes, with total population size, to resource scarcity, which in turn modifies birth rates.

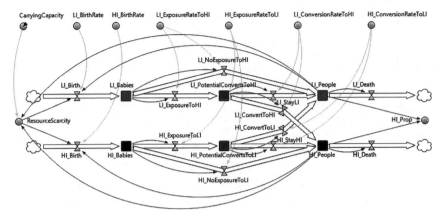

FIGURE 1.1. *Stock-and-flow diagram for conversion model between low-investment (LI) and high-investment (HI) lifestyles.*

The links indicate causal connections between elements of the diagram. For example, the ResourceScarcity variable is defined intuitively in mathematical terms by (LI_People + HI_People) / CarryingCapacity. The stocks (boxes) are defined by first-order differential equations. For instance, the HI_PotentialConvertsToLI stock is defined by d(HI_PotentialConvertsToLI)/dt = HI_ExposireToLI—HI_ConvertToLI—HI_StayHI (the right-hand side is just inputs minus outputs). The flows (broad arrows) are defined by simple mathematical formulas for flow rates (in the center of each flow arrow). For instance, the LI_ConvertToHI flow rate is defined as follows: LI_ConversionRateToHI * HI_PotentialConvertsToLI. These formulas are natural and, we believe, non-controversial.

A conversion model of this kind can be applied to many different scenarios. To apply it to the Neolithic transition, we need a theoretically persuasive way to produce the six key variables (the two birth rates, two exposure rates, and two conversion rates). Think of this "theoretically persuasive way" for now as a "Black Box." We will look into this part of the causal architecture, which is a collection of fully specified and carefully clustered mechanisms, in more detail below. We need this Black Box to be responsive to the HI_Prop variable, which expresses the proportion of HI people in the entire population in the geographic area of interest. We also need the Black Box to produce six variables as output. And we need the Black Box to have a range of parameters as inputs, along with the HI_Prop variable input, so we can explore the model's behavior in a variety of ecologically meaningful conditions. So, what's in the Black Box?

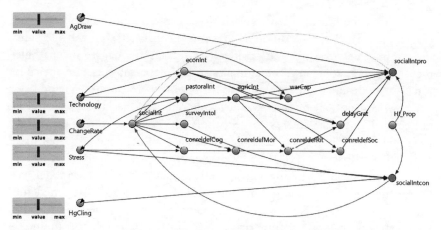

FIGURE 1.2. *Core of the Black Box: causal architecture of the Neolithic transition.*

THE BLACK BOX

The theoretical core of the Black Box is presented in figure 1.2. This is the heart of what we consider to be the causal architecture of the Neolithic transition, at least within the limitations of the NSIM model. The five input parameters on the left side of the diagram define the beginnings of a parameter space whose characteristics we will describe later. The HI_Prop variable (derived from the conversion model explained below) is presented as an input on the right side of the diagram. The remaining thirteen variables and the causal connections between them express the theoretical elements of the causal architecture. After explaining this architecture, we will show how we extract from it the six input variables (i.e., the two birth rates, two exposure rates, and two conversion rates) of the conversion model.

This causal architecture is consistent with Hodder's entanglement hypothesis. However, it also incorporates insights from several other theories derived from empirical research during the last couple of decades within the cognitive science of religion and related disciplines. Unfortunately, there is no space to survey this literature here (see Wildman 2009; Shults 2014b).

The Social Intensity variable (socialInt) toward the left side of the diagram is a quantitative proxy for the social aspects of the degree of entanglement present in the region in question. As social relationships become more complex as a component of the way people's connections to objects, animals, plants, places, and processes become richer, the Social Intensity variable increases, which in turn influences a variety of theoretically motivated intervening variables.

Note that the Social Intensity variable (socialInt) and the proportion of HI people (HI_Prop) are different conceptual constructs. On the one hand, the contrast between HI and LI lifestyles reflects the degree of entanglement between people and the animals, plants, places, objects, and invisible forces of their environment. High social intensity, on the other hand, is a narrower concept—it represents a key component of the HI lifestyle but is not equivalent to it. While we expect these two variables to rise and fall in the same direction, it is conceivable that they could change at quite different rates. For instance, HI_Prop might begin to increase early and change slowly, while socialInt might increase quickly but only after the model has been running for a while; or the opposite might occur. We designed the causal architecture of NSIM to reflect the possibility that the aspect of the HI lifestyle we call Social Intensity, which is particularly relevant to understanding religion, might develop partially independently of the embrace of HI lifestyles. Yet we also expect that HI_Prop could only reach the highest levels if socialInt were also high.

Jumping to the right side of the diagram, we can see that some of these intermediate variables tend to increase Social Intensity (the degree of increase represented by the socialIntpro variable) while others tend to decrease Social Intensity (the degree of decrease represented by the socialIntcon variable). These two variables on the far right increment and decrement, respectively, the Social Intensity variable, with their effects weighted by the proportion of HI people and the proportion of LI people, respectively. The ChangeRate parameter determines how quickly both incremental variables change the Social Intensity variable, thereby speeding up or slowing down the model. The loop established in this way means that Social Intensity both impacts and is impacted by all of the intermediate variables.

Within this fundamental feedback loop, the theoretical action takes place in the interactions between the intervening variables and between those variables and the four parameters that influence them (AgDraw, Technology, Stress, and HgCling). We explain these variables one by one in what follows, spelling out the theoretical underpinnings of the causal architecture along the way. Table 1.1 lists these parameters and variables with their abbreviations and definitions.

First, we describe the five parameters that function as inputs to the Black Box.

Agricultural Draw (AgDraw parameter) is the tendency to want to embrace the HI lifestyle associated with settled agriculture, domesticated animals, and town life. AgDraw impacts socialIntpro: an increased tendency in the population to embrace the HI lifestyle will ratchet up HI growth.

TABLE 1.1. Black Box parameters and variables constituting the causal architecture of the model

Abbreviation	Name	Definition
Parameters		
AgDraw	Agricultural Draw	Varies between 0.01 and 1.00
Technology	Technology	Varies between 0.00 and 1.00
ChangeRate	Change Rate	Varies between 0.01 and 1.00
Stress	Stress	Varies between 0.00 and 1.00
HgCling	Hunter-Gatherer Cling	Varies between 0.01 and 1.00
Variables that Co-Vary with Social Intensity		
econInt	Economic Intensity	(socialInt+Technology)/2
pastoralInt	Pastoral Intensity	Technology*socialInt*(4*Stress*(1−Stress))
agricInt	Agricultural Intensity	(pastoralInt+socialInt)/2
warCap	War-making Capacity	(Technology+agricInt+econInt)/3
surveyIntol	Surveyance Intolerance	socialInt
delayGrat	Delay of Gratification	(econInt+agricInt+conreldefRit)/3
Variables Related to Contesting Evolutionarily Stabilized Religion–Relevant Defaults		
conreldefCog	Contest Religious Defaults: Cognitive	socialInt*4*Stress*(1−Stress)
conreldefMor	Contest Religious Defaults: Moral	(socialInt+conreldefCog)/2
conreldefRit	Contest Religious Defaults: Ritual	(agricInt+conreldefMor)/2
conreldefSoc	Contest Religious Defaults: Social	(conreldefRit+econInt)/2
Variables Structuring the Major Feedback Loop		
socialIntpro	Social Intensity Pro	((econInt+agricInt+warCap+delayGrat+conreldefSoc)/5)*AgDraw*HI_Prop
socialIntcon	Social Intensity Con	((surveyIntol+(1−4*Stress*(1−Stress)))/2)*HgCling*(1−HI_Prop)
socialInt	Social Intensity	min(max(socialInt+ChangeRate*(socialIntpro−socialIntcon),0.01),0.99)

Technology (Technology parameter) is the level (complexity, quantity, quality, entanglement) of technological advancement within the regional population. Technology includes farming know-how, tool design and manufacture, knowledge of medicinal plants, house-building techniques, and so on. Technology impacts:

- *warCap*: Increased technological advancement raises a population's capacity to wage war.
- *econInt*: Increased technological advancement improves a population's capacity for managing more complex economic exchange and increases the diversity and value of traded products.
- *pastoralInt*: Increased technological advancement improves a population's capacity to maintain and breed its domesticated animals.

Change Rate (ChangeRate parameter) sets the rate at which socialIntpro increments and socialIntcon decrements the socialInt variable.

Stress (Stress parameter) is the level of anxiety-producing or survival-threatening conditions in the population's natural environment. Stress impacts other variables in the Black Box in an inverted parabolic way, as the formula $4*Stress*(1-Stress)$ indicates. This means that the greatest impact occurs when Stress is middling and that both low and high Stress have less impact. Stress affects:

- *pastoralInt*: Moderate stress puts pressure on individuals to grow crops and domesticate animals, if they know how. If conditions are less stressful, the LI hunting-and-gathering lifestyle is easier, decreasing pastoralInt. If conditions are more stressful, crops won't grow and domesticated animals die, decreasing pastoralInt.
- *conreldefCog*: Moderate stress puts pressure on individuals to contest their religious-cognitive defaults (e.g., they resist their tribal tendencies and begin to explore new ways to cooperate and share resources with people beyond their immediate kith/kin group). High levels of stress reinstate religious cognitive defaults (terror-management theory), while low levels of stress do not induce people to challenge those defaults.
- *socialIntcon*: High stress puts too much pressure on individuals, overriding their capacity to contest defaults or delay gratification, tempting them to adopt LI lifestyles. Low stress yields no incentive to cooperate or increase Social Intensity. Moderate stress produces the least resistance to Social Intensity. In this case only, the formula regarding stress takes the form $(1-4*Stress*(1-Stress))$ rather than $4*Stress*(1-Stress)$, inverting the parabola.

Hunter-Gatherer Cling (HgCling parameter) is the tendency to want to conserve the LI lifestyle associated with hunter-gatherers. HgCling impacts socialIntcon because an increased tendency in the population to want to conserve the LI lifestyle will ratchet down HI growth and thus Social Intensity.

Second, let's consider the intermediate variables that co-vary with Social Intensity.

Economic Intensity (econInt variable) is the level (complexity, quantity) of economic exchange within and across groups. Economic Intensity increases with both social intensity and technology. The econInt variable impacts:

- *socialIntpro*: Increased economic exchange ratchets up the need to keep investing in an HI sedentary-domestication lifestyle.
- *warCap*: Increased economic exchange improves social networking and cooperation skills and imports technology and know-how, which raise the capacity to wage war.
- *delayGrat*: Increased economic exchange contributes to social entrainment of individuals such that they learn to delay gratification.
- *conreldefSoc*: Increased economic exchange contributes to social entrainment of individuals such that they learn to be less vigilant of non-kith/kin, reciprocators, and cultural others.

Pastoral Intensity (pastoralInt variable) is the level (complexity, extent) of entanglement between a group of humans and the animals they (at least partially) domesticate. Pastoral intensity increases with both the social intensity required to cooperate and the technology required for domesticating animals, and both elements are important simultaneously (this is why the formula uses multiplication rather than averaging). The pastoralInt variable impacts agricInt in the same way because increased pastoral (animal) entanglement requires increased agricultural (plant) entanglement to feed the animals.

Agricultural Intensity (agricInt variable) is the level (complexity, extent) of entanglement between a group of humans and the plants they (at least partially) domesticate. Agricultural intensity rises as pastoral and social intensity rise; both factors are equally important. Of course, agricultural intensity also drives pastoral intensity, a subordinate causal link expressed through the mediation of the socialIntpro variable. The agricInt variable impacts:

- *socialIntpro*: As more of the population becomes agriculturally entangled, the general willingness to invest in HI lifestyles increases.
- *warCap*: The capacity to wage war is improved when agricultural entanglement increases because of the way the latter improves general social cooperation skills and makes it easier to provide regular food for training and using troops.
- *delayGrat*: As more of the population becomes agriculturally entangled, it becomes easier to trust the farming process and delay one's need to gratify desires quickly.

- *conreldefRit*: As more of the population becomes agriculturally entangled, participation in imagistic, high-impact rituals declines in favor of low-impact, oft-repeated rituals that bind people more closely together in socially stable and religiously reinforced configurations.

War-Making Capacity (warCap variable) is the overall capacity of groups to wage war on other groups. The capacity and willingness to wage war increase with technological sophistication, economic intensity, and agricultural intensity. WarCap impacts socialIntpro: increased capacity to wage war ratchets up the general tendency toward investing in HI lifestyles.

Surveyance Intolerance (surveyIntol variable) is the level of annoyance at being watched by others. The more intense sociality becomes, the more everyone knows your business, and the more tension there is around surveyance. The surveyIntol variable impacts socialIntcon: annoyance at being watched can have a ratcheting-down effect, leading to a higher percentage of the population resisting the HI lifestyle.

Delay of Gratification (delayGrat variable) is the willingness to delay immediate gratification of desires. Experimental studies suggest that religion can increase the ability (Rounding et al. 2012), or at least the motivation (Harrison and McKay 2013), to delay gratification, which lends warrant to the idea that it may have played some adaptive role by enhancing self-control and in-group cooperation. The ability to delay gratification is increased to about the same degree by economic intensity, agricultural intensity, and the willingness to contest both religious ritual defaults. The delayGrat variable impacts socialIntpro: the willingness to invest in HI sedentary-domestication lifestyles is ratcheted up by an increase in the population's willingness and ability to delay gratification.

Third, we come to the variables pertaining to contesting evolutionarily stabilized cognitive defaults that are more directly relevant to religion.

Contest Religious Defaults: Cognitive (conreldefCog variable) is the extent to which the evolved religious cognitive defaults (e.g., terror-management–motivated engagement with narratively immediate, small-scale idiosyncratic spirits) are contested. The willingness to contest religious-cognitive defaults increases so long as two other factors both increase (thus multiplication rather than averaging): when social intensity increases and when moderate stress forces people to become more aware of cognitive defaults. Extremely low or extremely high stress has the opposite effect. The conreldefCog variable impacts conreldefMor: preference for and willingness to think outside the scope of a small group calls for contesting default moral intuitions that support small-group lifestyles.

Contest Religious Defaults: Moral (conreldefMor variable) is the extent to which evolved religious moral defaults (e.g., high concern for purity, parochial care for in-group) are contested. The willingness to contest religious moral defaults increases as social intensity increases and when some capacity to contest cognitive defaults is in place. conreldefMor impacts conreldefRit: preference for and willingness to engage in less imagistic, divine agent rituals increases as freedom from small-group moral defaults increases.

Contest Religious Defaults: Ritual (conreldefRit variable) is the extent to which the evolved religious ritual defaults (high-intensity, infrequent, "imagistic" rituals engaging spirit agents) are contested and transformed in favor of low-intensity, more frequent, social-bonding, "doctrinal" rituals that imaginatively engage supernatural agents relevant to larger groups. The willingness to contest religious ritual defaults in this way increases with agricultural intensity and rising willingness to contest moral defaults. The ritualInt variable impacts:

- *delayGrat*: Increased openness and willingness to engage in non-imagistic rituals improves one's ability and willingness to delay gratification because the engaged spirits are concerned with the larger group's welfare.
- *conreldefSoc*: Increased openness and willingness to engage in non-imagistic rituals makes one accustomed to being around and trusting non-kith/kin, reciprocators, and cultural others.

Contest Religious Defaults: Social (conreldefSoc variable) is the extent to which the evolved religious social defaults (vigilance toward non-kith/kin, reciprocators, and cultural others) are contested and transformed in favor of openness to outsiders, to trading goods, and to learning about strangers. The willingness to contest religious social defaults is driven up by increasing willingness to contest religious ritual defaults and by economic intensity that forces mixing with strangers. The conreldefSoc variable impacts socialIntpro: willingness to invest in sedentary-domestication lifestyles is ratcheted up by an increase in the population's trust in non-kith/kin, reciprocators, and cultural others.

Fourth, what about the variables defining the fundamental feedback mechanism of the Black Box?

Social Intensity Pro (socialIntpro variable) is an incrementing mechanism by which SocialInt is increased. Five factors contribute to the increase of social intensity: economic intensity, agricultural intensity, war-making capacity, the willingness to delay gratification, and the willingness to contest religious social defaults. This effect is amplified by both the proportion of HI people and the AgDraw parameter.

Social Intensity Con (socialIntcon variable) is a decrementing mechanism by which socialInt is decreased. Two factors contribute to the decrease of social intensity: allergy to surveyance and either low or high stress (but not moderate stress). This effect is amplified both by the tendency to cling to LI hunter-gatherer lifestyles and by the proportion of LI people (i.e., 1–HI_Prop).

Social Intensity (socialInt variable) has been discussed as an important component of entanglement in HI lifestyles. The socialInt variable is the only one that increments (via socialIntpro) or decrements (via socialIntcon) its own old value. Thus all of the model's reinforcing loops involve the link from socialIntpro to socialInt (+), and all balancing loops involve the link from socialIntcon to socialInt (−). The socialInt variable impacts six intermediate variables:

- *econInt*: Economic exchange within and across groups will become more complex as willingness to invest in an HI sedentary-domestication lifestyle grows.
- *pastoralInt*: Pastoral entanglement will become stronger as people become more willing to invest in HI sedentary-domestication lifestyles.
- *agricInt*: Agricultural entanglement will become stronger as people become more willing to invest in HI sedentary-domestication lifestyles.
- *surveyIntol*: The more people live close to each other in the same place, doing the same things repeatedly, the higher the susceptibility to being annoyed by having people watching them.
- *conreldefCog*: Participation in an HI sedentary-domestication lifestyle forces the contestation of defaults toward focusing on narratively immediate, terror-managing engagement with small-scale idiosyncratic spirits.
- *conreldefMor*: Participation in an HI sedentary-domestication lifestyle forces the contestation of defaults toward overriding purity concerns related to in-group care.

GENERATING OUTPUTS FROM THE BLACK BOX

The final challenge on the way to a complete causal architecture is to make the Black Box we have described generate the six key output variables that function as inputs to the conversion model. To reiterate, those output variables are two birth rates (LI_BirthRate and HI_BirthRate), two exposure rates (LI_ExposureRateToHI and HI_ExposureRateToLI), and two conversion rates (LI_ConversionRateToHI and HI_ConversionRateToLI).

Figure 1.3 depicts the links between the various components of the Black Box and these six output variables. Note that we have suppressed the display

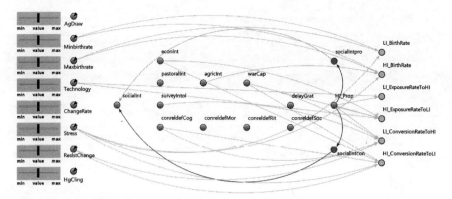

FIGURE 1.3. *Six outputs from the Black Box.*

of all links already discussed, save for the fundamental feedback loop, to clarify the diagram. Also, we now display three more parameters; these are important for generating some of the output variables, but they do not impact variables within the Black Box itself. All three are biologically and environmentally determined aspects of human beings: the minimum birth rate, the maximum birth rate, and the average psychological tendency to resist change.

Table 1.2 lists those parameters, along with the definitions of the six output variables.

We now explain the mathematical formulas for the six *output* variables.

- *LI_BirthRate*: In LI societies, increasing stress drives birth rates down from the maximum toward the minimum. Note that there is a linear rather than a parabolic effect of the Stress parameter here.
- *HI_BirthRate*: In HI societies, either increasing agricultural intensity or decreasing stress (without averages or products; thus: min((agricInt+(1−stress)),1)) will drive birth rates upward from the minimum to the maximum.
- *LI_ExposureRateToHI*: In LI societies, exposure to HI alternatives occurs through either trade or war, amplified by the prevalence of HI people and the overall level of technology.
- *HI_ExposureRateToLI*: In HI societies, exposure to LI alternatives depends on both low technology and the prevalence of LI people (1-HI_Prop).
- *LI_ConversionRateToHI*: In LI societies, conversion to HI alternatives becomes more frequent when economic intensity, war-making capacity, and the willingness to contest religious-cognitive defaults rise. Middling stress increases this conversion rate while low stress or high stress inhibits it. All

TABLE 1.2. Additional parameters impacting output variables but not Black Box variables

Abbreviation	Name	Definition
Parameters		
Minbirthrate	Minimum Birth Rate	Varies between 0.50 and 2.00 (children per person)
Maxbirthrate	Maximum Birth Rate	Varies between 1.00 and 5.00 (children per person)
ResistChange	Resistance to Change	Varies between 0.00 and 1.00
Six Output Variables from the Black Box That Function as Inputs to the Conversion Model		
LI_BirthRate	LI Birth Rate	Minbirthrate+(Maxbirthrate−Minbirthrate)*(1−Stress)
HI_BirthRate	HI Birth Rate	Minbirthrate+(Maxbirthrate−Minbirthrate)*min((agricInt+(1−stress)),1)
LI_ExposureRateToHI	LI Exposure Rate to HI	((econInt+warCap)/2)*Technology*HI_prop
HI_ExposureRateToLI	HI Exposure Rate to LI	(1−Technology)*(1−HI_Prop)
LI_ConversionRateToHI	LI Conversion Rate to HI	((econInt+warCap+conreldefCog+(4*Stress*(1−Stress)))/4)*ResistChange
HI_ConversionRateToLI	HI Conversion Rate to LI	((surveyIntol+HI_Prop+max(0,2*warCap−1)+(1−conreldefCog)+(1−(4*Stress*(1−Stress))))/5)*ResistChange

factors are roughly equal in importance, and their average effect is moderated by resistance to change.

- *HI_ConversionRateToLI*: In HI societies, conversion to LI alternatives occurs under four roughly equally weighted conditions. Surveyance intolerance drives up this conversion rate. So does war-making capacity, but only when it becomes extreme so that it interferes with the benefits of township life (thus the formula max(0,2*warCap−1)). The willingness to contest religious-cognitive defaults has the opposite effect, driving this conversion rate down. And either low stress or high stress makes this type of conversion more likely, while middling stress makes it less likely. All factors are roughly equal in importance, and their average effect is moderated by resistance to change.

Finally, there is the question of time in the model. We needed to explore the model dynamics before we could be sure about what real-world time to associate with one cycle of the model. After running a few experiments, we concluded that we could interpret *one cycle of the model as one year*.

EXPLORING THE MODEL

While we are also working on agent-based models that engage the empirical data at Çatalhöyük even more directly (see below), the systems-dynamics model we have described above is at least indirectly connected to the data in two important ways. On the one hand, NSIM has been able to model the growth of the percentage of high-investment people in a Neolithic population in a way that conforms to widely accepted interpretations of the archaeological evidence from the site and surrounding areas. At the very least, we have constructed a theoretically grounded causal architecture that simulates the emergence of a population pattern that plausibly represents the growth of Çatalhöyük during the first millennium of its existence.

On the other hand, the primary virtue of theory-centric models like NSIM is the way they facilitate the construction and clarification of causal architectures that are based on scientific hypotheses rooted in empirical data and provide a means for experimentally exploring the validity and implications of more general theories about, for example, the transformation of civilizational forms. Although the quantitative aspects of the model are not explicitly testable in the traditional scientific sense, they were rigorously informed by the data from Çatalhöyük and contribute to the confirmation of several important theories. The most prominent among these is Hodder's theory of entanglement, which is grounded in a mass of archaeological data. However, NSIM also helps to validate and explore the consequences of a host of theoretical developments within the cognitive science of religion and related fields, all of which are rooted in current-day experimental data (Wildman 2009). The cognitive-science theories are relevant to Çatalhöyük on the almost universally accepted assumption that intervening genetic changes do not materially impact the relevant aspects of human cognition.

The first step in exploring a model such as this one is to validate it in a variety of ways. At the theoretical level, validation consists in evaluating the plausibility of the various mathematical formulas and the decisions about where the major causal links belong. The mathematical formulas themselves are rather straightforward, given the causal reasoning they reflect. Thus, given that everything affects everything else, the leading theoretical-validation issue is which links (thin gray arrows) in figure 1.2 should be presented as expressions of the strongest lines of causation. In this venue, we cannot pursue this type of analysis beyond what we have already explained, but in the model development process we debated the most important causal links in great detail.

At a technical level, an important validation question is whether NSIM yields stable equilibrium states in between the extremes of 0 and 1 for the proportion of HI people in the wider population (HI_Prop variable). It turns out that the model does produce such stable equilibria intermediate between extremes, which is a sign that NSIM contains interesting behavior worth analyzing in more detail. Another technical validation question is whether the evolution of the simulation within a single run produces an interpretable model of life at Çatalhöyük and within the surrounding Konya plain. Most of our analysis was dedicated to this kind of validation, as we now explain.

The challenge in analyzing this model's parameter space is to understand how the model represents the transition pathways from 100 percent LI people to a high percentage of HI people. For the sake of concrete analysis and manageable communication, we stipulated a specific transition pathway and asked about the conditions under which the model's development approximates that pathway. The pathway we specified had the percentage of HI people (HI_Prop) increasing from zero steadily up to 50 percent at the rate of 10 percent per century, staying around that 50 percent mark for about 200 model cycles (about two centuries), then jumping up to 80 percent in a single century, and then climbing to 100 percent at the rate of 10 percent per century. We think this is a plausible approximation of the growth of Çatalhöyük over the first millennium of its existence: growing steadily, perhaps with a few slowdowns and leaps, and eventually drawing in most of the people from the Konya plain.

We call this pathway the Target transition pathway in view of the fact that we asked our model to approximate it. Figure 1.4 plots NSIM's best solution along with the Target transition pathway. Table 1.3 displays the parameter settings producing this best approximation to the stipulated ideal.

Taking this best approximation to the Target transition pathway as a starting point, we varied parameters around the solution to evaluate the behavior of the model. We were particularly interested in the Stress and Technology parameters, which reflect ecological conditions and accumulating knowledge and know-how. While we were focused on the HI_Prop variable, we also kept track of the Social Intensity variable (socialInt) because this furnishes a convenient way to judge the extent to which changing model dynamics depend on Social Intensity understood as a key component of HI lifestyles. The results of our investigation can be expressed as answers to a series of questions.

First, what happens to Social Intensity (socialInt) as the model moves through the pathway that best approximates the stipulated Target transition pathway? Figure 1.5 gives the answer. As HI_Prop increases relatively smoothly, socialInt stays low, moves through a period of rapid changes, stabilizes at a high value, and finally cycles close to its high-value equilibrium. We had tentatively hypothesized that socialInt would increase smoothly like the

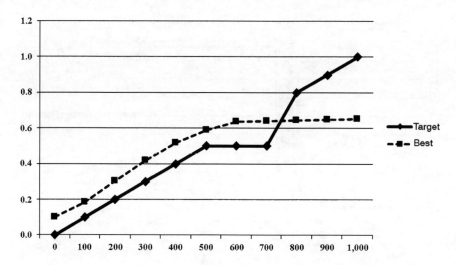

Figure 1.4. *Closest solution to the Target transition pathway for the growing population of Çatalhöyük.*

Table 1.3. Parameter settings yielding the closest approximation to the Target transition pathway through the model's parameter space from a low to a high proportion of HI people

Parameter	Setting
CarryingCapacity	10,000
ChangeRate	0.051
Technology	0.679
ResistChange	0.99
Maxbirthrate	2.5
Minbirthrate	0.5
Stress	0.34
HgCling	1.0
CivDraw	0.34

HI_Prop variable does, though perhaps at a different rate. Instead, the model describes a phase transition between low and high Social Intensity, with the transition involving a lot of ups and downs over the course of about a century. Upon reflection, that pattern probably does make more sense as a description of the rocky process of civilizational transformation. The cyclical behavior of socialInt in the last 150 years of the model's time span is intriguing. It suggests

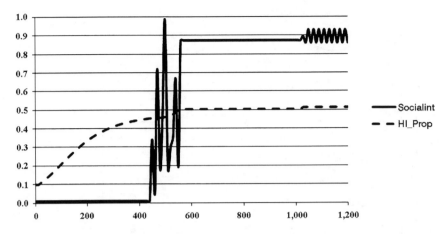

FIGURE 1.5. *Behavior of Social Intensity (socialInt) as the simulation runs through the best approximation to the Target transition pathway.*

periodic upheavals in the social order, but of a relatively minor sort and without much overall effect on the achieved equilibrium.

Second, what is the impact on HI_Prop of varying the Stress parameter while holding other parameters at the values they have in the best approximation to the Target transition pathway? Figure 2.6 tells us that, under low stress, relatively opportunistic agricultural and pastoral labor (mixed with whatever hunting and gathering remained in the Konya plain at the start of the Çatalhöyük settlement) is relatively easy, and the delayed gratification of HI lifestyles is unnecessary and less appealing than it might otherwise be; HI_Prop does not reach its maximum value under these circumstances. Under medium stress, food production and energy capture are difficult enough to induce people to invest in cooperative agricultural and pastoral lifestyles; this is when HI_Prop attains its maximum value. Under high stress, the crops won't grow and the domesticated animals die, so the civilizational transformation to HI settled farming lifestyles is thwarted; HI_Prop stays low in such conditions.

Third, what is the impact on HI_Prop of varying the Technology parameter while holding other parameters at the values they have in the best approximation to the Target transition pathway? Unsurprisingly, figure 1.7 indicates that there is a close link between Technology and the proportion of high-investment people. One cannot run a farming settlement unless one has enough know-how to handle domesticated crops and animals and to defend against opportunistic marauders.

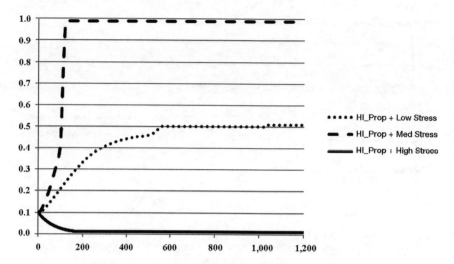

FIGURE 1.6. *Impact on HI_Prop of varying the Stress parameter while holding other parameters at the values they have in the best approximation to the Target transition pathway: low stress = 0.34; medium stress = 0.50; high stress = 0.90.*

Fourth and finally for the purposes of this presentation of NSIM, how often do we get particular ranges of values of socialInt when parameters are varied around their values in the solution to the calibration experiment? From answers to the previous two questions, we know that Stress is particularly sensitive in the interval between 0.3 and 0.4 and that Technology is particularly sensitive in the interval between 0.6 and 0.7. We ran a Monte Carlo experiment varying Stress and Technology within those limits in steps of 0.001, making enough runs to be 95 percent confident that socialInt would be within 5 percent of the reported number. This yielded 10,100 distinct runs with various combinations of parameter settings, from which we generated a probability distribution function and a cumulative distribution function for Social Intensity (the socialInt variable; see figure 1.8).

The graph in figure 1.8 expresses the stochastic character of NSIM. There are some parameter settings that produce very high values for socialInt, but fewer than 13 percent of parameter combinations yield more than 80 percent of people in the population adopting the high degree of social intensity that appears to have been present in Çatalhöyük, which is probably what happened in the Konya plain (these cases are represented in the last two bins on the right side of figure 1.8).

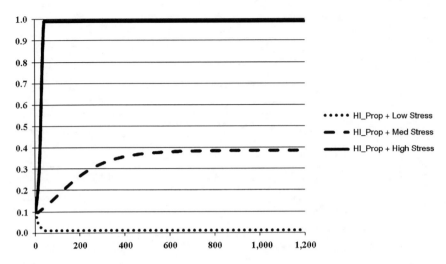

FIGURE 1.7. *Impact on HI_Prop of varying the Technology parameter while holding other parameters at the values they have in the best approximation to the Target transition pathway: low technology = 0.34; medium technology = 0.68; high technology = 0.90.*

Other parameter settings lead nowhere or yield a smaller proportion of the population in HI lifestyles than occurred at Çatalhöyük. By the standards of this model, therefore, Çatalhöyük was a special place, and a lot had to go right to make its phenomenal growth possible.

NSIM articulates a version of the entanglement hypothesis that explicitly incorporates *religious* beliefs and practices as part of the environment within which human beings are entangled. Moreover, NSIM indirectly supports this version of the entanglement hypothesis because disabling the four key religious variables in the Black Box destroys NSIM's causal architecture and renders the model virtually meaningless.

The incorporation of religious variables is perhaps the most important contribution of NSIM to understanding the rise of HI lifestyles in the Konya plain and possibly elsewhere as well. Referring back to NSIM's causal architecture in figure 1.2, for a given value of the ecological Stress parameter, Social Intensity (socialInt) drives the willingness to contest the evolutionarily stabilized cognitive defaults underlying religious belief and practice. This is a plausible causal hypothesis grounded on leading empirically validated theories within the bio-cultural study of religion. As human groups became larger and more entangled, a stronger type of social glue was required to motivate people to cooperate with and commit to individuals beyond the limits of their kith

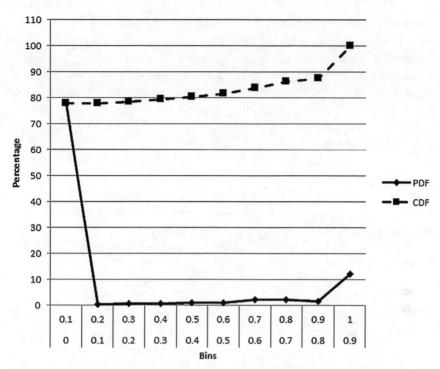

FIGURE 1.8. *Probability distribution function (PDF) and cumulative distribution function (CDF) socialInt. Percentage of cases is on the vertical axis while binned ranges of the socialInt variable are along the horizontel axis. The Çatalhöyük case involves very high Social Intensity so it is located on the right-most bins, which is an indication of how rare that type of situation is in the context of NSIM.*

and kin. Shared imaginative engagement with morally relevant (and potentially punitive) invisible beings that were interested in the behavior of larger groups would have helped.

The fundamental connection between Social Intensity and the willingness to contest religious cognitive defaults impacts the rest of the causal architecture, driving Social Intensity higher in a reinforcing cycle of social complexity and religious entanglement. NSIM suggests that this process would have had several ripple effects. Default moral approaches to outsiders, laced with religiously loaded intuitions about purity, would be supplanted by a wider moral vision, with all moral actors monitored by supernatural watchers. Ritually, people would contest default tendencies to high-arousal, infrequent ritual acts and embrace more regular rituals that bind larger groups more closely together. Socially, clannish ingroup-outgroup monitoring would yield to acceptance of a larger social identity with a wider conception of the in-group.

All of these effects of increased Social Intensity would have played a role in the shift from primarily house-based to primarily sodality-based modes of "history making," as described by Ian Hodder in his introduction to this volume.

NSIM suggests that Social Intensity is a necessary condition for the dominance of HI lifestyles, in a very particular sense. As we saw in figure 1.5 and as seems to have occurred in the Konya plain, HI lifestyles can grow in prominence as people consolidate agricultural and pastoral know-how. At a certain point in the embrace of HI lifestyles, however, Social Intensity becomes a major factor in leveraging a new kind of religiously potentiated entanglement, an increasingly intense form of sociality, and greater numbers of people with shared civilizational projects and imaginative worlds. NSIM suggests that this is what may have happened in Çatalhöyük.

LIMITATIONS AND FUTURE RESEARCH

The main limitation of NSIM is one it shares with other computer simulations: it cannot "prove" the correctness of an archaeological (or any other) theory. It only provides added warrant for accepting the plausibility of Hodder's entanglement hypothesis by demonstrating that a social transition of the sort that did actually occur at Çatalhöyük can emerge based on the interactions among the variables included within its causal architecture. Like all systems-dynamics models, NSIM is also limited by the coherence and constraints expressed in the theoretical integration behind this architecture. Perhaps better—or at least different—theoretical integration and algorithms could be constructed. This is precisely where detractors ought to start their critical analysis.

Another important limitation of this model is that there are fewer explicit links to hard data than we would have liked. Our calibration efforts were based on our review of the empirical literature and long discussions with experts at Çatalhöyük, but the nature of the systems-dynamics M&S technique employed here precluded us from getting down to the level of "petrified poop." We are working on other models (see below) that are more closely tied to the material at the site, which ought to get us (at least) to the "mudbrick" level. It is important to note that NSIM is also limited to the emergence and sustenance of the civilizational form that characterized Çatalhöyük. It does not help us understand why the site was eventually deserted—at least not directly. One might tease out hypotheses about its desertion from the causal architecture by asking questions like: might the abandonment of the site have been related to the disappearance (or weakening) of one or more of the causal links embedded within the "Black Box"?

Despite these limitations, we believe NSIM helpfully shows the plausibility of the religious entanglement theory widely embraced by Çatalhöyük experts. It also serves the function of pressing scientists toward conceptual clarity in their formulations and critiques of theory. Other scholars may not like our general proposal or may disagree with particular aspects of it, but by setting out our assumptions about the relevant factors in such detail and outlining our theoretical integration so openly, we hope to have made it easier for them to criticize us and to generate or reformulate their own proposals. For our part, we plan to continue validating and experimenting with models describing the agricultural transition and to publish more technical articles in M&S journals.

We also hope to develop at least one more model of the transformation of civilizational form that characterized the Neolithic. This simulation is tentatively called the Çatalhöyük House Entanglement Model (CHEM). Unlike NSIM, CHEM is an agent-based model (ABM). Instead of people, however, its "agents" are settlements and houses. While the idea of settlements or houses as agents might initially seem odd, this is a relatively normal approach within M&S. For example, it is not uncommon for a business to develop an ABM in which the agents are factories, which have to "decide" where to get built based on population distribution, market opportunities, and traffic patterns. In the case of CHEM, settlements and houses will "decide" where to get built based on several variables, such as availability of material resources (mud, clay, water) and social resources (builders, people with know-how and motivation). In this way, CHEM enables us to develop even more explicit tie-downs to the data at Çatalhöyük. We aim to run CHEM in an environment that incorporates Geographic Information Systems (GIS) data on the Konya plain as it likely existed between 7400 and 6000 BCE.

We expect that the scientific community studying Çatalhöyük will not have too much difficulty accepting the notion of houses as agents. Hodder's philosophy of entanglement itself is based on the idea that "things" have their own sort of agency or causal efficacy (albeit not intentionality in the usual sense) within an evolutionary nexus. Several other researchers have made similar observations. Anthropologist Mary Weismantel, for example, has referred to the Çatalhöyük house as defined "by the processes of give and take between it and other entities, which keep it constantly in a state of reinvention, reiteration, and reproduction." She encourages us to "see the house in action. Meals were centrifugal forces that gathered people together . . . equally important were the centripetal movements of things between households ... the whole entity changed form as it pulsed to overlapping cycles, frequent and infrequent, predictable and sporadic, festive and workaday" (Weismantel 2014, 271).

As we mentioned in the first section of this chapter, the efforts on which we report here are part of the larger Modeling Religion Project at the Center for Mind and Culture. Set within this broader context, we hope that NSIM and eventually CHEM will contribute not only to a better understanding of the transitional dynamics of the Neolithic but also to the forces of civilizational transformation at work in the contemporary world.

ACKNOWLEDGMENTS

We wish to thank Saikou Diallo, Chris Lynch, and Ross Gore of the Virginia Modeling, Analysis and Simulation Center for their assistance with validating and analyzing NSIM. We are also grateful for funding from the John Templeton Foundation and the Center for Mind and Culture.

REFERENCES

Allaby, Robin, Terence Brown, and Dorian Fuller. 2010. "A Simulation of the Effect of Inbreeding on Crop Domestication Genetics with Comments on the Integration of Archaeobotany and Genetics: A Reply to Honne and Heun." *Vegetation History and Archaeobotany* 19 (2): 151–58. https://doi.org/10.1007/s00334-009-0232-8.

Allaby, Robin, Dorian Fuller, and Terence Brown. 2008. "The Genetic Expectations of a Protracted Model for the Origins of Domesticated Crops." *Proceedings of the National Academy of Sciences of the United States of America* 105 (37): 13982–86. https://doi.org/10.1073/pnas.0803780105.

Axtell, Robert L., Joshua M. Epstein, Jeffrey S. Dean, George J. Gumerman, Alan C. Swedlund, Jason Harburger, Shubha Chakravarty, Ross Hammond, Jon Parker, and Miles Parker. 2002. "Population Growth and Collapse in a Multiagent Model of the Kayenta Anasazi in Long House Valley." *Proceedings of the National Academy of Sciences of the United States of America* 99 (3, Supplement 3): 7275–79. https://doi.org/10.1073/pnas.092080799.

Bell, James A. 1987. "Simulation Modelling in Archaeology: Reflections and Trends." *European Journal of Operational Research* 30 (3): 243–45. https://doi.org/10.1016/0377-2217(87)90065-8.

Costopoulos, Andre, and Mark W. Lake. 2010. *Simulating Change: Archaeology into the Twenty-First Century*. Salt Lake City: University of Utah Press.

Crabtree, Stefani A., and Timothy A. Kohler. 2012. "Modelling across Millennia: Interdisciplinary Paths to Ancient Socio-Ecological Systems." *Ecological Modelling* 241 (August): 2–4. https://doi.org/10.1016/j.ecolmodel.2012.02.023.

Doran, Jim, and Mike Palmer. 1995. "The EOS Project: Integrating Two Models of Palaeolithic Social Change." In *Artificial Societies: The Computer Simulation of Social Life*, ed. Nigel Gilbert, 103–25. London: University College London Press.

Doran, Jim, Mike Palmer, Nigel Gilbert, and Paul Mellars. 1994. "The EOS Project: Modelling Upper Palaeolithic Social Change." In *Simulating Societies: The Computer Simulation of Social Phenomena*, ed. Nigel Gilbert and Jim Doran, 195–221. London: University College London Press.

Fuller, Dorian Q., Robin G. Allaby, and Chris Stevens. 2010. "Domestication as Innovation: The Entanglement of Techniques, Technology, and Chance in the Domestication of Cereal Crops." *World Archaeology* 42 (1): 13–28. https://doi.org/10.1080/00438240903429680.

Fuller, Dorian Q., Eleni Asouti, and Michael D. Purugganan. 2012. "Cultivation as Slow Evolutionary Entanglement: Comparative Data on Rate and Sequence of Domestication." *Vegetation History and Archaeobotany* 21 (2): 131–45. https://doi.org/10.1007/s00334-011-0329-8.

Gavrilets, Sergey, David G. Anderson, and Peter Turchin. 2014. "Cycling in the Complexity of Early Societies." In *History and Mathematics: Trends and Cycles*, ed. Leonid E. Grinin and Andrey V. Korotayev, 136–58. Volgograd, Russia: Uchitel.

Gerbault, Pascale, Robin G. Allaby, Nicole Boivin, Anna Rudzinski, Ilaria M. Grimaldi, J. Chris Pires, Cynthia Climer Vigueira et al. 2014. "Storytelling and Story Testing in Domestication." *Proceedings of the National Academy of Sciences of the United States of America* 111 (17): 6159–64. https://doi.org/10.1073/pnas.1400425111.

Harrison, Justin Marc David, and Ryan Thomas McKay. 2013. "Do Religious and Moral Concepts Influence the Ability to Delay Gratification? A Priming Study." *Journal of Articles in Support of the Null Hypothesis* 10 (1): 25.

Hodder, Ian, ed. 1978. *Simulation Studies in Archaeology*. Cambridge: Cambridge University Press.

Hodder, Ian. 2006. *Çatalhöyük: The Leopard's Tale*. London: Thames and Hudson.

Hodder, Ian, ed. 2010. *Religion in the Emergence of Civilization: Çatalhöyük as a Case Study*. Cambridge: Cambridge University Press. https://doi.org/10.1017/CBO9780511761416.

Hodder, Ian. 2012. *Entangled: An Archaeology of the Relationships between Humans and Things*. New York: John Wiley and Sons. https://doi.org/10.1002/9781118241912.

Hodder, Ian, ed. 2014. *Religion at Work in a Neolithic Society: Vital Matters*. Cambridge: Cambridge University Press.

Jayyousi, Thaer W., and Robert G. Reynolds. 2014. "Exploiting the Synergy between Micro, Meso, and Macro Levels in a Complex System: Bringing to Life an Ancient Urban Center." In *Complexity and the Human Experience: Modeling Complexity in the Humanities and Social Sciences*, ed. Paul A. Youngman and Mirsad Hadzikadic, 241–69. Stanford, CA: Pan Stanford Publishing. https://doi.org/10.1201/b16877-16.

Kohler, Timothy A., R. Kyle Bocinsky, Denton Cockburn, Stefani A. Crabtree, Mark D. Varien, Kenneth E. Kolm, Schaun Smith, Scott G. Ortman, and Ziad Kobti. 2012. "Modelling Prehispanic Pueblo Societies in Their Ecosystems." *Ecological Modelling* 241 (August): 30–41. https://doi.org/10.1016/j.ecolmodel.2012.01.002.

Lake, Mark W. 2014. "Trends in Archaeological Simulation." *Journal of Archaeological Method and Theory* 21 (2): 258–87. https://doi.org/10.1007/s10816-013-9188-1.

Lemmen, Carsten, Detlef Gronenborn, and Kai W. Wirtz. 2011. "A Simulation of the Neolithic Transition in Western Eurasia." *Journal of Archaeological Science* 38 (12): 3459–70. https://doi.org/10.1016/j.jas.2011.08.008.

Lemmen, Carsten, and Aurangzeb Khan. 2012. "A Simulation of the Neolithic Transition in the Indus Valley." In *Climates, Landscapes, and Civilizations*, ed. Liviu Giosan, Dorian Q. Fuller, and Kathleen Nicoll, 107–14. Washington, DC: Wiley. https://doi.org/10.1029/2012GM001217.

Lowe, John W.G. 1985. *The Dynamics of Apocalypse: A Systems Simulation of the Classic Maya Collapse*. Albuquerque: University of New Mexico Press.

Mithen, Steven, and Melissa Reed. 2002. "Stepping Out: A Computer Simulation of Hominid Dispersal from Africa." *Journal of Human Evolution* 43 (4): 433–62. https://doi.org/10.1016/S0047-2484(02)90584-1.

Nakamura, Carolyn, and Peter Pels. 2014. "Using 'Magic' to Think from the Material: Tracing Distributed Agency, Revelation, and Concealment at Çatalhöyük." In *Religion at Work in a Neolithic Society: Vital Matters*, ed. Ian Hodder, 187–224. Cambridge: Cambridge University Press.

Nikitas, Panos, and Efthymia Nikita. 2005. "A Study of Hominin Dispersal Out of Africa Using Computer Simulations." *Journal of Human Evolution* 49 (5): 602–17. https://doi.org/10.1016/j.jhevol.2005.07.001.

Powers, Simon T., and Laurent Lehmann. 2014. "An Evolutionary Model Explaining the Neolithic Transition from Egalitarianism to Leadership and Despotism." *Proceedings of the Royal Society of London: Series B, Biological Sciences* 281 (1791). https://doi.org/10.1098/rspb.2014.1349.

Rounding, Kevin, Albert Lee, Jill A. Jacobson, and Li-Jun Ji. 2012. "Religion Replenishes Self-Control." *Psychological Science* 23 (6): 635–42. https://doi.org/10.1177/0956797611431987.

Shults, F. LeRon. 2010. "Spiritual Entanglement: Transforming Religious Symbols at Çatalhöyük." In *Religion in the Emergence of Civilization: Çatalhöyük as a Case Study*, ed. Ian Hodder, 73–98. Cambridge: Cambridge University Press. https://doi.org/10.1017/CBO9780511761416.003.

Shults, F. LeRon. 2014a. "Excavating Theogonies: Anthropomorphic Promiscuity and Sociographic Prudery in the Neolithic and Now." In *Religion at Work in a Neolithic Society: Vital Matters*, ed. Ian Hodder, 58–85. Cambridge: Cambridge University Press.

Shults, F. LeRon. 2014b. *Theology after the Birth of God: Atheist Conceptions in Cognition and Culture*. New York: Palgrave Macmillan. https://doi.org/10.1057/9781137358035.

Shults, F. LeRon. 2018. *Practicing Safe Sects: Religious Reproduction in Scientific and Philosophical Perspective*. Leiden: Brill Academic.

Squazzoni, Flaminio, Wander Jager, and Bruce Edmonds. 2014. "Social Simulation in the Social Sciences." *Social Science Computer Review* 32 (3): 279–94. https://doi.org/10.1177/0894439313512975.

Weismantel, Mary J. 2014. "The Hau of the House." In *Religion at Work in a Neolithic Society: Vital Matters*, ed. Ian Hodder, 259–79. Cambridge: Cambridge University Press.

Wildman, Wesley J. 2009. *Science and Religious Anthropology: A Spiritually Evocative Naturalist Interpretation of Human Life*. Aldershot, UK: Ashgate.

Yilmaz, Levent, ed. 2015. *Concepts and Methodologies for Modeling and Simulation*. New York: Springer. https://doi.org/10.1007/978-3-319-15096-3.

2

Creating Settled Life

Micro-Histories of Community, Ritual, and Place—the Central Zagros and Çatalhöyük

WENDY MATTHEWS

New research on the origins and nature of early agriculture and settled life has identified both widespread similarities as well as local variations in the options selected in these lifeways across the Middle East (Willcox 2005; Zeder 2009; Asouti and Fuller 2013). To investigate context and cause in the creation of settled life, I examine similarities and differences in the options selected in two geographically distant regions of the Middle East that are not frequently compared. The focus of this research is on early settled life in the east of this region, in the central Zagros of Iraq and Iran, as this was a key heartland of change but has until recently been overlooked in many studies, with selective comparison to settled life in the west at Çatalhöyük in central Turkey. Previous formative studies on the origins of agriculture and settled life in the Zagros (Braidwood et al. 1983, 10) identified significant patterning in the layout of settlements and buildings but concluded "we have no really good evidence that might specify the use to which the various rooms in the house were put" as few artifacts were left on floors, as at many archaeological sites across the Middle East and many regions of the world. One objective in this investigation, therefore, is to evaluate how new forensic-scale micro-archaeological analyses can inform on the origins of settled life by providing new high-resolution micro-historical data on the specific associations between people and place and the role of ritual practices in creating this.

DOI: 10.5876/9781607327370.c002

This research examines three key issues and questions in studies of the adoption of agriculture and settled ways of life that are explored throughout this volume, with particular regard to the new evidence emerging from the Zagros and selective comparison to Çatalhöyük.

What is the timing of the emergence of history making at particular places? Can any correlations be used to suggest causal processes such as agricultural intensification, population increase, and social competition? In particular, was there a 1,500-year gap in occupation in the Zagros ca. 10,000–8500 cal BCE, as previous research suggests (Hole 1996)?

How widespread and how significant were repetitive building and cosmological patterning in the layout of houses in creating and sustaining historical ties to place and to ancestors and long-term social relationships characteristic of delayed-return agricultural systems?

Is there less evidence of Neolithic ritual practices in the Zagros than in other regions of the Near East, as earlier studies suggest (Hole 1996; Bernbeck 2004)?

This study begins with an appraisal of the increasing body of research on early settled life in the Zagros. It critically reviews approaches and theories in macro- and micro-archaeological analyses of histories of place and ritual. It then applies these analyses and theories to examine new and existing macro- and micro-scale evidence for history-making, repetitive building, and cosmological patterning and ritual practices in highland and lowland central Zagros of Iran and Iraq, with selective comparisons to Çatalhöyük, to study local and regional variations in the role of these in creating settled life. The concluding section critically examines how these new integrated macro- and micro-contextual approaches, theories and evidence inform on context and cause in the creation of settled life. All date estimations are calibrated BCE.

CASE STUDIES: NEW AND EXISTING RESEARCH IN THE ZAGROS AND AT ÇATALHÖYÜK

The Zagros highland and piedmont regions of Iraq and Iran were key native habitats of wild plants and animals that were domesticated in the early Holocene, notably barley, goat, sheep, and pig (Zeder 2009). This review reexamines data from pioneering investigations on the origins of agriculture that were conducted in this region in the 1950–70s (figure 2.1). This includes the sites of Asiab and Ganj Dareh in highland Iran and Ali Kosh and Jarmo in piedmont zones of Iran and Iraq, which span ca. 9000/8500—6000 BCE (Braidwood et al. 1983; Smith 1990; Hole, Flannery, and Neely 1969; Zeder 2008, table 3). New excavations at a wide range of Neolithic sites in the Zagros are currently being conducted in both

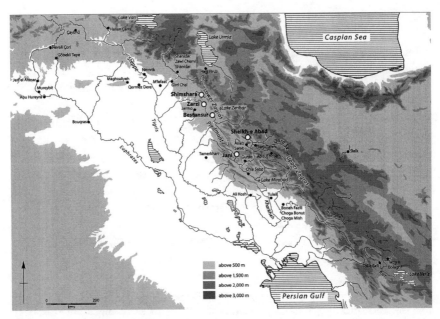

FIGURE 2.1. *Location of Neolithic sites in the Zagros*

Iran (Matthews and Fazeli Nashli 2013; Riehl, Zeidi, and Conard 2013) and Iraq (Matthews, Shillito, and Elliott 2014). The original new micro-archaeological and interdisciplinary data examined in this study are from four sites investigated by the Central Zagros Archaeological Project (CZAP). They comprise two sites in the highland Zagros in Iran (Matthews, Matthews, and Mohammadifar 2013). Sheikh-e Abad, ca. 9800–7590 BCE, at 1,400 m asl, ca. 1 hectare (ha), and 10 m high, was investigated by excavation of three trenches at the base, middle, and top of the mound. Jani, > ca. 8000–7500 BCE, 1,200 m asl, ca. 1 ha, and 8 m high, was investigated by recording and sampling a > 60 m long extant cross-section through the northwest edge of the mound. In the piedmont Zagros in Iraq, Bestansur, ca. 7600–7000 BCE, 553 m asl, 2.5 ha, and ca. 3–4 m high in the Neolithic, was investigated by excavation of thirteen trenches in different sectors of the site; and Shimshara, ca. 7400–6000 BCE, 493 m asl, currently < ca. 1 ha, was explored by rescue excavations and recording of a > 25 m section ca. 2–2.5 m high (Matthews et al. in prep). The principal source of ecological data analyzed here for this region is from Lake Zeribar core (ca. 40,000 BP; Wasylikowa and Witkowski 2008; Matthews 2013). The selected micro-archaeological case studies from Çatalhöyük, ca. 7100–6000 BCE, are drawn from research conducted at the site by the author since 1993 (Matthews 2005a, 2005b).

APPROACHES AND THEORIES IN HISTORIES OF PLACE AND RITUAL: THE POTENTIAL OF MICRO-ARCHAEOLOGY

Micro-Stratigraphic and Micromorphological Analyses

Many archaeological investigations, including those on histories of place and ritual, involve theoretical consideration of the timescales of the everyday and specific events, seasonal, life-cycle, and longer-term time spans, and the dialectic intersections and entanglements of these in decision-making and lived worlds (Lucas 2005; Hodder 2012). Due to methodological constraints, however, many analytical studies are obliged to examine the incremental constructs of building phases and levels (e.g., Kohn and Dawdy 2016, 132–33), which represent aggregations of practices over several years or generations and thereby restrict study of the specificity and temporality of particular actions and events (Foxhall 2000, 496; Robb 2010). Micromorphological analyses provide the opportunity to study intact sequences of surfaces and accumulated activity residues as evidence of micro-histories of place and settled life at multiple timescales. This includes temporal resolutions that are not otherwise recoverable by routine excavation and sampling, which unavoidably lump together many fine strata, often < 1–5 mm, and also disperse and only selectively recover the diverse micro-artifactual, bioarchaeological, and minerogenic materials that, importantly, are present in archaeological deposits and detectable in thin section (Matthews et al. 1997; Matthews 2010).

To study how communities created early settled life and the role of history making and place making and ritual in this, this chapter, therefore, examines new high-resolution data from micro-stratigraphic and micromorphological analyses of intact sequences of surfaces and accumulated deposits in the field and in large resin-impregnated thin sections, 14 cm × 7 cm, 25–30 μm thick. Each micro-stratigraphic section-profile was photographed, drawn, correlated with excavation units, and block-sampled for micromorphological, geochemical, and phytolith analysis. More than 100 thin sections from the Central Zagros Archaeological Project and 350 from Çatalhöyük have been analyzed, described, and interpreted using internationally standardized methodologies and reference materials (Bullock et al. 1985; Courty, Goldberg, and Macphail 1989; Matthews 2005a). Many samples were collected at 1–2 m intervals to study spatial variation within buildings and open areas within settlements and from sequences that span the entire history of particular places. An estimation of the temporality represented by particular strata within micro-stratigraphic sequences is calculated here by analysis of the number of surfaces within ^{14}C estimates of building time spans, as well as geoarchaeological data on deposit formation (Matthews 2005b).

Theories in the Study of Micro-Histories of Place

A range of archaeological, anthropological, and architectural theories and observations highlights ways in which surfaces and the residues on them can be investigated as highly resolved indicators of histories of place and the ecological and social roles and relations of the communities that created them. Ethnoarchaeological and experimental observations attest that, first, surfaces are key indicators of histories of place and ritual actions, as surface materials are often expressly chosen for specific places and intended events and bear impact traces of subsequent actions and events (Kramer 1979; Matthews, Hastorf, and Ergenekon 2000; Boivin 2000). Second, it is the smaller artifactual and bioarchaeological remains present within buildings that are more likely to become primary refuse and indicators of actual use, as they are more likely to remain in context, even after periodic cleaning (McKellar in Schiffer 1987). Even in environments where there are limited materials and choices and resources, architectural materials and surfaces are attributed with and gain significance from the specific circumstances of their selection, arrangement, and placement (Leatherbarrow and Mostafavi 2002, 23) and are key media in cultural place making. The particular significance of materials, although culturally specific, is also context-dependent and as such may be investigated by comparative geographical and chronological contextual analysis of the properties, potential performance characteristics, and macro- and micro-contextual associations of surface materials and accumulated residues (Matthews 2005a, 2005b; see Robb 2010, 506–7).

With regard to interpretation of repetitive actions and building and continuity or change in these, Bloch (2010, 154) argues that a temporal dialectic is developed between actions and the places in which they are conducted. By their durable nature, settings, features, and buildings may represent continuity in social concepts, roles, and relations that transcend the flux of daily, seasonal, and life-cycle changes. They enable roles to be passed on from person to person and across communities, both contemporaneously as well as through time, and by example as well as through any direct social or genetic links, which may or may not be present and are not implicit in history making or in memorialization.

The choice to continue or change materials, actions, and relations includes simultaneous consideration of material attributes and concepts that Robb (2010, 506–7) argues are:

1. shared more widely across different materials and contexts
2. associated with particular fields of action and expected norms for that context type
3. related to a specific event, actor(s), audience, and moments in time.

Study of the biographies and boundaries of architectural materials and surfaces in buildings and settlements and the impacts and residues of actions on them, therefore, have the potential to inform on how individuals and communities construct, articulate, and change particular fields of action and roles and relationships in specific situations, both in the short as well as the longer term, and thereby how they engaged in creating and changing histories.

The capacity for places to restrict or enable social relations and performance can be measured as metric distance, based on experiments in interpersonal perception (Fisher 2009). The measures used in this research are drawn from Moore (1996, 153–67; Bowser and Patton 2004, figure 7), whereby it is calculated that distances of < 1 m enable intimate interactions (soft voice, close visibility), 1–4 m permit social "styles of interaction" (relaxed voice, whole body movement visible), and 4–8 m allow public-near interactions and require a loud voice but still enable perception of general facial gestures. Facial gestures, however, become difficult to discern beyond 8 m at public-far distances, at which point communication requires more accentuated and "stylized" interactions.

Theories in Micro-Contextual Analyses of Cosmology and Ritual

With regard to approaches and theory in the study of cosmology and ritual, conceptual distinctions between "domestic" and "ritual" and "sacred" and "profane" have been widely and robustly critiqued. In many societies, daily, periodic tasks are part of the re-creation of the cosmos, and houses themselves may be both shelter and "temple" or "shrine" (Dean and Kojan 2001, 126). The house, normally a locus for food preparation, cooking, sleeping, and shelter, may be transformed during sacred ceremonies into the cosmos and a world of myth (Carsten and Hugh-Jones 1995, 234) by creating ritual place through modulation of the location, time, and particular sequences of events and associated materials (Hastorf 2001, 1, 9).

Bradley (2005, 33–34) argues for a definition of ritual and ritualization as "a kind of practice—a performance which is defined by its own conventions" in which certain aspects of life are selected and provided with added emphasis (see also Bell 1997). Such performances may have been "composed out of elements that had wider resonance, for this is how they would have gained their social significance and why they could have been understood" (Bradley 2005, 119). Ritual in this definition is not restricted to communicating religious beliefs and reproduces and is reproduced by everyday life and wider social practice (Hunt 2005). Ceremony and ritual are frequently associated with reproduction of identity, including stages in the life course, gender, ethnicity, and profession; emotion and catharsis; symbolic values, especially of death and rebirth; as well as mechanisms for bringing the individual into the community (Bell 1997, 89). Hunt (2005, 26) argues that it is possible in the social sciences generally to move toward a synthesis of perspectives that considers substantive (ritual as belief system), functionalist (ritual as underscoring social values), phenomenological (ritual as construction of reality and making of the world), and interpretative (historically and individually contingent) viewpoints.

In light of this, analysis of finely stratified depositional sequences provides new avenues for study of ritualized practice by examination of the precise spatial and temporal context and associations of materials and thereby how they may have been used to add emphasis to specific events and places. Micromorphological analysis of surfaces and residues in thin section provides a robust high-resolution technique for teasing apart different temporal and contextual strands in examination of ritual in ways similar to that explored by Insoll (2004, 125) and Hastorf (2001, 3). It enables us to approach ritual and religion by analysis of highly specific associations of materials and traces of actions and by looking for them "as existing in multiple contexts . . . shrines, houses, funerary practices, diet, agricultural practice, technology, landscape alteration and perception and by looking at the overall context," as suggested by Insoll (2004, 151).

ORIGINS OF SETTLED LIFE AND EMERGENCE OF HISTORY MAKING AT PARTICULAR PLACES

Infrequent Visits/Low-Density Activities

Previous excavations indicate that many sites in the Zagros began as places of infrequent but repeated temporary activities associated with communal food processing and cooking, attested by clusters of hearths and pits with fire-cracked stones. Sites such as Ganj Dareh Level E and Asiab were clearly used for hunting, collecting, and perhaps managing seasonally abundant resources—including wild birds, land snails, and clams—and social gatherings and perhaps feasting (figure 2.2a; Smith 1990; Braidwood et al. 1983). Ganj Dareh E is likely to have been occupied during the months of May–October to collect pistachio nuts and harvest and sow wild and domesticated barley, attested in these levels.

New excavations at Sheikh-e Abad indicate that such repeated associations with place began as early as ca. 9810 BCE in the high Zagros and now

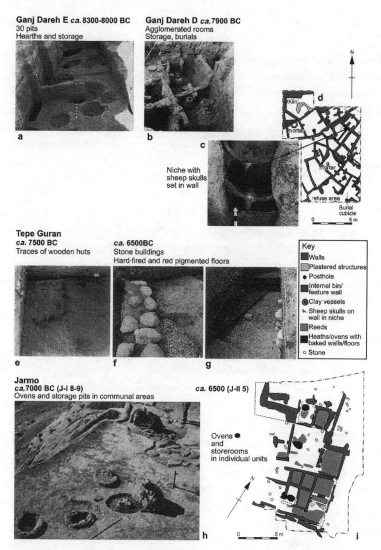

FIGURE 2.2. *Origins of early settled life. Ganj Dareh: (a) communal probable storage pits and hearths with fire-cracked stones, Level E (after Smith 1975, plate IIIa); (b) and (d) agglomerated rooms with enclosed storage facilities, Level D (after Smith 1971, 9; Smith 1990, figure 1); (c) sheep skulls in niche (after Smith 1972, plate Ib). Tepe Guran: (e) traces of wooden huts, G I Level N; (f) stone wall buildings with hard-fired and red-pigmented floors, G I Level J, and (g) Level E (after Mortensen 2014, 26, 31, 44). Jarmo: (h) ovens and storage pits in communal areas, J-I 8–9; (i) ovens and storerooms within architectural units, J-II 5 (after Braidwood et al. 1983, figures 74, 51, courtesy of the Oriental Institute of the University of Chicago).*

fill an apparent gap in occupation in the Zagros, together with research at Chogha Golan and East Chia Sabz (Matthews and Fazeli Nashli 2013). At Sheikh-e-Abad, this early occupation is represented by repeated surfaces of gravel and mud and accumulations of ash with burned aggregates, bone, and omnivore coprolites from ca. 9810–200 BCE in Trench 1, collections of wild large-seed grasses, and emerging traces of dung burned as fuel (Matthews, Shillito, and Elliott 2013, figure 7.2; Whitlam et al. 2013; Matthews et al. in prep). These repeated associations with place correspond directly with a rapid increase in the spread of grasses at the end of the Younger Dryas, as attested by the Zeribar core (Wasylikowa and Witkowski 2008), which were used during these early visits to the site.

Similar evidence of initially infrequent visits to particular sites and places is attested at Jani prior to 8000 BCE, in Phase 1 (Matthews, Shillito, and Elliott 2013, figure 7.1). Here, the first activities at the site are represented by sparse traces of burned aggregates, bone, charred wood, and fire-cracked stones from food processing and cooking in accumulations predominantly of natural sediment (70%–80%). These traces resemble those from ethnoarchaeological and micro-archaeological observations of residues from temporary hunter-gatherer hearths that have been reworked by natural agencies after abandonment (Mallol et al. 2007). These residues also resemble basal and edge of settlement deposits at Çatalhöyük, pre-ca. 7100 BCE, which similarly indicate infrequent/low-density activities early in the history of this site/area (Matthews 2005a, 384).

Repeated Activities and Increased Intensity of Occupation

A marked increase in the intensity of activities at both of these highland Iran sites is attested by at least ca. 8000 BCE. At Sheikh-e Abad this is marked by repeated sequences of finer surfaces and in situ burning rich in ash from nut shells, burned animal dung, and a figurine, ca. 7960 BCE, in Trench 2, 6 m above the base of the mound (Matthews, Shillito, and Elliott 2013, figure 7.3). At Jani it is marked by repeatedly laid surfaces and accumulations of ash- and phytolith-rich deposits with scatters of red pigment, ca. 8000 BCE, Phase 2 (Matthews, Shillito, and Elliott 2013, figure 7.8). At both sites the thinness (< 2 mm–2 cm) and repeated frequency and preservation of deposits with fragile articulated plant remains are indicative of rapid deposition and burial and thereby represent longer-term, more intensive occupation and associations with place, as also observed in ethnoarchaeological research by Mallol and colleagues (2007).

The presence of construction material aggregates in these deposits at both sites may indicate adjacent buildings pending further excavation. At Sheikh-e Abad certainly, these deposits in Trench 2 are directly overlaid by buildings and include house-mouse droppings that attest to year-round occupation, diverse charred plant remains from flotation that include domesticated barley and wheat, and evidence for storage of large-seeded pest-infected legumes, as well as almond and pistachio nuts (Whitlam et al. 2013).

At both Sheikh-e Abad and Jani, this increase in intensity of repeated activities and occupation corresponds with early evidence for use of herbivore dung as fuel. This correlation suggests that proximity and management of ruminant herds and their ready supplies of meat, leather, sinews, marrow, and dung fuel was a key factor in enabling greater sedentism and associations with place, in addition to use of diverse wild and newly domesticated plant resources.

These rapid and repeatedly accumulating deposits also correspond with an increase in the number of sites across the Zagros ca. 8500–8000 BCE (Matthews and Fazeli Nashli 2013; Matthews 2012) and a peak in the extent of grasslands and evidence of vegetation disturbance attested by *Plantago lanceolata* pollen in the Lake Zeribar core (Wasylikowa and Witkowski 2008; Matthews 2013), which suggest increasing human management and impact on plant, animals, and landscapes.

At Çatalhöyük, the earliest deposits with sparse traces of activity, ca. 30–45 cm thick, are also overlain by deposits that attest a marked increase in intensity of activity ca. 7000 BCE in Space 181 in the Deep Sounding (figure 2.3b; Cessford 2007a; Matthews 2005a, 384). Here the > 2.5 m deep sequence of midden deposits also includes construction materials, most notably aggregates of fired lime with red and yellow ochre surfaces, that must derive from elaborate adjacent buildings, pending future excavation (figure 2.3c). Like similar deposits and relative site levels in the Zagros, these deposits include micromorphological indicators of use of dung fuel from the earliest increase in intensity of activity, notably aggregates that are compacted and laminated and attest penning of herbivores in the vicinity of this area or site, which would have contributed to the viability of settled life. A large walled animal pen, Space 199–98, > 9 m × 5 m, with repeated accumulations of compacted dung, was rebuilt twice in the same place at the end of this sequence (figure 2.3b, e–g) and was directly overlain by two elaborate buildings with wall painting and sculpture (figure 2.3a, Buildings 23 (X.1) and 18 (X.8); Matthews 2005a, 389–91). The size of the pen and the intensity and duration of its use indicate significant management of animal herds and their resources immediately prior to a phase of settlement expansion.

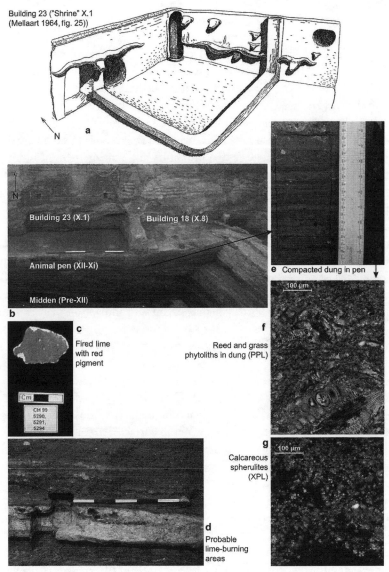

FIGURE 2.3. Çatalhöyük, Turkey. Transition in S area to (a) "history house" B23 (X.1) from (b) midden and probable lime-burning areas, Space 181, pre-XII then animal pens, Spaces 199–198 XII-XI (scale = 2 m); (c) fired lime plaster with red-pigment pre-XII midden; (d) probable lime-burning area (after Matthews 2005a, figure 19.24); (e) compacted dung in pen micromorphology sample (scale in cm); (f) reed and grass phytoliths (PPL); (g) calcareous dung spherulites 10–20 μm in size (XPL).

Repeated Building

Some of the most striking examples of repeated building on the same place and plot were constructed in central Anatolia, not only at Çatalhöyük but at the early adjacent site of Boncuklu (Baird et al. 2012; although with greater re-cutting into earlier buildings), as well as at Aşıklı Höyük in the second half of the ninth millennium BCE (Özbaşaran 2012). As Bloch (2010) and Hodder and Pels (2010) argue, this repeated building represents remarkable continuity in social roles and relations as well as evidence of history making.

From previous excavations, some of the earliest evidence of repeated building in the Zagros is from Zawi Chemi Shanidar ca. 11,150–10,400 BCE (Solecki 1981; Matthews and Fazeli Nashli 2013), where a round house, Structure 1, was repeatedly constructed three times in the same place and associated with a deposit of ritual paraphernalia in an adjacent open area, discussed below. At later sites with rectilinear architecture (figure 2.2), building complexes were agglutinative and included shared party walls, as at Ganj Dareh Level D ca. 7900 BCE, suggesting more communal or extended social and political relations during the foundation and early history of settled communities, as at Jarmo ca. 6900–6500 BCE (figure 2.2i) and in Anatolia at Çatalhöyük ca. 7000 BCE, Levels XII–X. There is some evidence of repeated building at Jarmo, but as Braidwood and colleagues (1983, 10) noted, there was little macro-evidence to specify whether there was continuity in use of these rooms, as few artifacts were left on floors. At Tepe Guran, the early construction of wooden huts in the Aceramic Neolithic, from ca. 7050 BCE, continued with and was then replaced by mudbrick architecture, some with hard-fired and red-pigmented walls and floors and terrazzo, with evidence of repeated building in the latest levels, K–E (figure 2.2e–g; Mortensen 2014, figures 26, 44, 31). At both sites, increasing investment and marking of place and social groups corresponds with increasing use of managed and domesticated plants and animals, which would have both encouraged and supported more settled life. At Ganj Dareh, the construction of substantial architecture in Level D correlates with the choice to invest in early barley cultivation and management of goats, based on kill-off patterns (Zeder 2009). At Tepe Guran, current evidence suggests that the transition to mud-brick architecture correlates with cultivation of founder crops and domesticated animals (Meldgaard, Mortensen, and Thrane 1963; Mortensen 2014).

New excavations have identified further evidence of repeated building in the piedmont Zagros at Bestansur ca. 7660 BCE. Here, a comparatively large, elaborate building was constructed at least twice on the same plot, within a complex of abutting buildings in Trench 10 (figure 2.6). The aggrandizement of the building from Building 8 to Building 5, which housed > seventy-two human burials, however, suggests that there was an increase in the range and complexity of social roles and relations of the communities associated with these buildings and in history making, discussed below in the section on cosmological layout and ritual.

In the highland Zagros, distinctly separate rather than agglomerated abutting buildings were constructed ca. 7590 cal BCE, as revealed at Sheikh-e Abad in Trench 3 at the top of the mound (figure 2.4). Here, surrounding open areas indicate that external, more public spaces remained important loci for activities. Building 1 was constructed with small rooms, 2 m × 2 m, like many at Ganj Dareh (figure 2.2d; Smith 1990), although the latter may have comprised sub-floor cubicles. At least one burial, that of an infant, was placed below the floors of a living room (figure 2.4j). An overlying building was probably rebuilt on the same location based on the corresponding alignment of later burials dug into this building from a now eroded level (Cole 2013, figure 14.1). It is significant that apparent historical continuity in the creation of links between people and place is associated with acts of sub-floor intramural burial in many highland and piedmont Zagros sites, discussed further below.

Micro-stratigraphic and micromorphological analyses of the surfaces and residues within these buildings and settlements have enabled identification of remarkable continuity in the use of specific areas for particular activities, as attested by the remarkable repetition in the type and thickness and frequency of surfaces and activity residues within particular rooms and areas of buildings and the settlement. This repetition strongly suggests that within specific building phases in the central Zagros, there was remarkable stability in social roles and relations (see Bloch 2010). At Sheikh-e Abad ca. 7590 BCE, this included continuity in the creation of separate places for living and for penning of managed animals within the small rooms in Building 1 (figure 2.4). This is attested by evidence, respectively, of repeatedly well-prepared and maintained plaster floors with overlying accumulations of occupation debris with sparse fragments of pottery and burned plant remains, as well as sub-floor burial (figure 2.4i–m), and of compacted dung (figure 2.4c–e; Matthews, Shillito, and Elliott 2013). Outside buildings, open areas were used for access and for discard of working tools, as well as being shared in places with animals—as attested by concentrations of dung (figure 2.4a, f–h). Similar continuity in use of particular areas within buildings as well as open areas and, thereby, in the social roles and relations associated with these areas is also attested in the large-scale excavations at Bestansur in the piedmont zone (Matthews et al. 2016, in prep).

At Jani, continuity and stability in micro-histories of place is marked by repeated re-plastering of constructed hearths/ovens in Phase 3, in the eighth

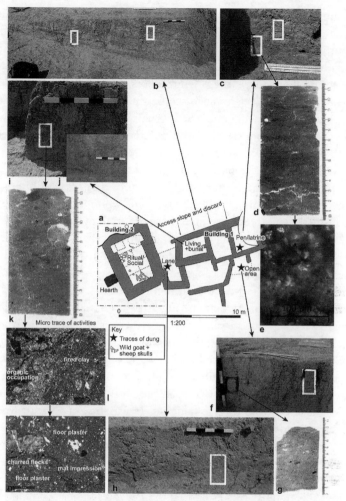

FIGURE 2.4. *Sheikh-e Abad, Iran: (a) plan of Trench 3 cosmological patterning of wild animal skulls in Building 2 and animal penning, living and infant burial in Building 1. Repeated sequences of associated surfaces and activity residues: (b) laid packing and ashy discard in sloping open area; (c) lenses of ruminant and omnivore coprolites in a small room, Space 8, (d) in thin section with (e) abundant calcareous dung spherulites 10–20 μm in size (XPL); (f) lenses of charred flecks and trampled sediments and dung in an open area (g) in thin section and (h) a lane; (i) silt loam floors, plasters, and lenses of occupation deposits in living room, Space 15, with (j) an infant burial, (k) in thin section; (l) occupation deposits < 1–5 mm thick with phytoliths, charred flecks, and fired clay (probable pottery fragments) (XPL); (m) silt loam floor plasters with mat impressions and lenses of occupation (PPL) (adapted from Matthews, Shillito, and Elliott 2013, figures 7.5, 7.6, 7.9, 7.10).*

millennium BCE. The following Phase 4 was marked by extensive leveling and settlement expansion, with the construction of large rectilinear buildings with abutting mudbrick walls and remarkable repetitive renewal of fine white plaster floors in sequences more than 25–30 cm thick (figure 2.5; Matthews, Shillito, and Elliott 2013). These white floors, as well as the rectilinear and plano-convex mudbricks, are macroscopically similar to those at Çatalhöyük and antecedents at Boncuklu in central Anatolia. Microscopically, however, it is evident that they were created from different local materials and technologies, with the addition of dung ash to white floors, for example, at Jani (figure 2.5a–f). Other similarities between the central Zagros and central Anatolia include emphasis on the whiteness and cleanliness of areas of interior space. At Shimshara, fragments of multiple layers of white wall plaster/wash have been identified that closely resemble those of Çatalhöyük and may have been applied in the order of annually or intra-annually (figure 2.5g–h; Matthews 2005b). The possible ritual significance of this is discussed in the final section below. A range of technological and cultural concepts was shared across different regions during the Neolithic, as highlighted by Özdoğan (1999, 232) and supported by analyses of artifacts and materials from Bestansur and Shimshara by Richardson (Matthews et al. 2016, in prep). Her analyses attest that these two sites were engaged in extensive networks, whether directly or indirectly, that include indications of shared networks by communities in central Zagros and central Anatolia. One striking example is of carved stone bracelets of similar "rare supra-regional" form that span the ninth–eighth millennia BCE and were manufactured from diverse materials (see Astruc et al. 2011; Kozlowski and Aurenche 2005).

COSMOLOGY AND RITUAL

Previous research has suggested that there was less focus on ritual practices in the Zagros than in other regions of the Middle East in the Neolithic (Hole 1996; Bernbeck 2004). Reexamination of evidence from previous excavations as well as new excavations and micromorphological analyses, however, indicate that ritual was important in the creation of early settled life and communities in the Zagros. Here I examine the evidence for ritual practices in the creation of communal buildings and cosmological patterning, placement and treatment of animal skulls, human burial and figurines, as well as use of pigment, wall paintings, and surfaces to mark specific places and events in the central Zagros, all of which are attested in other regions of the Middle East. This section begins with analysis of the nature and significance of each

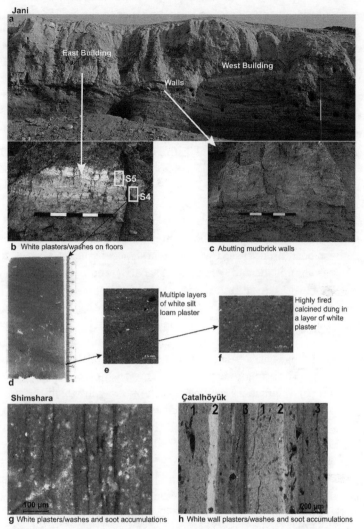

FIGURE 2.5. *Jani, Iran: transition from (a) repeated sequence of midden deposits (Phase 2) to construction of abutting buildings (Phase 4) (scale = 1.5 m), with (c) mudbrick walls and (b) repeated white floor plasters/washes > 25 cm deep (scale = 50 cm), (d) < 0.5–1 mm thick in thin section, (e) white silt loam plasters, (f) some with highly fired calcined dung ash (adapted from Matthews, Shillito, and Elliott 2013, figure 7.16). Multiple lenses of white silt loam plasters/washes and soot accumulations at (g) Shimshara, Iraq, and (h) Çatalhöyük ca. annual wall plaster sequences with thick foundation layer [1], white dolomite rich silty clay initial wash [2], and white marl washes with increasing accumulations of soot [3].*

of these categories of ritual practice, with reference to the earlier discussion of theories and approaches to cosmology and ritual. It then assesses how these different practices and cosmological patterning were employed and articulated in the Zagros.

Ritual Practice in Early Settled Communities in the Middle East

Communal buildings have been distinguished in studies of early settled life in the Middle East by their size, enabling public-scale interactions over distances greater than 4–8 m (Moore 1996); separation from other buildings; distinctive and specially prepared construction materials, floors, and surfaces; and fixtures such as benches (Keane 2010). Traces of ritual have been distinguished by identification of specific fixtures, materials, and artifacts that are likely to have been used to separate and mark specific places, people, actions, and events as significant and often transformative and transcendent (Keane 2010; see also Hastorf 2001; Bradley 2005).

Many researchers argue that mortuary practices were central in creating and maintaining identities and social relationships in early settled communities in the Middle East (Croucher 2012; Hodder 2010; Kuijt 2000). Baird (2005) argues that buildings with burials were key social arenas, not only during burial ceremonies but also throughout their life history. Kuijt (2000, 143) argues that secondary mortuary practices may have been one way in which the timing of ceremonies was controlled and reenacted to coincide with other social gatherings and ritual events or delayed until the deceased or other members of the household or community were returned to the settlement. The placement of burials within buildings provides some indication of the associations of individuals and groups with buildings and houses at particular points in time, although those interred may only represent a proportion of the population associated with a particular place and may not have been resident (Kuijt 2000).

Figurines, as miniature representations of aspects of being, are intrinsically symbolic. Their creation, use, and discard are likely to have varied according to context and may have included use in a wide range of rites, "wish magic," and representations of identity and affiliation (Daems 2008; Nakamura and Meskell 2013).

The repeated use of whitewashes and plasters and red pigment in many regions of the Middle East to mark burials, wall paintings, and molded sculpture indicates that these materials in particular were highly significant and charged with meanings that included associations with related ceremonies and events. Study of their occurrence in other contexts, therefore, enables investigations of traces of possible rituals in other domains (Insoll 2004; Hunt 2005). Cross-culturally, white as a bright luminous material is potently associated with purity (Bachelard 1994) and red, when mixed with water, with blood and life (Taçon 2004).

Previous Zagros Case Studies

The early settled communities identified in the Zagros to date were smaller than those represented by mega-sites in the Middle East such as Çatalhöyük, up to 13 hectares in size and 20 m high. It remains possible, however, that some large early settlements lie buried below much later urban mounds in the Zagros region. Many Neolithic Zagros sites were 1–3 hectares in size, including those excavated by CZAP, similar to many current village sites in the Middle East (Baird 2005).

From previous excavations in the Zagros, it is evident that some open areas and buildings were large enough to host public-scale communal gatherings and include indications of possible ritual practice. At Zawi Chemi Shanidar ca. 11,150–10,400/9700 BCE, a hoard of ritual paraphernalia that included fifteen wild goat/sheep skulls and the articulated wing bones of seventeen large birds of prey was deposited in an open area and is associated, paradoxically, with indications of possible sheep herd management (Zeder 2009; Solecki 1981). This paraphernalia was deposited next to Structure 1, which was rebuilt at least three times, attesting strong links between ritual performance and history making. At Asiab ca. 9000–8500 BCE, a large semi-subterranean round structure more than 10 m in diameter was constructed that included at least two human burials with traces of red pigment and many figurines, in addition to its later uses for a diverse range of food processing and cooking activities (Braidwood 1960, figure 3). This structure regrettably remains incompletely published but is being reinvestigated in new excavations at Asiab (Darabi, Richter and Mortensen 2018).

Other known constructions with indications of possible ritual in the central Zagros were smaller and only permitted social or intimate interactions. At Ganj Dareh ca. 7900 BCE, two wild sheep skulls with horns were set into the wall of a small room/sub-floor cubicle, 1.3 m × 1 m (figure 2.2c–d). More than sixty-nine individuals were buried across the settlement in a remarkably diverse range of practices that included burial in a sarcophagus and below floors and placement of the remains of body parts on floors (Smith 1990). The placement of these individuals within a complex of other rooms/cubicles with features for storage and food preparation indicates that ritual practices were incorporated into places of everyday life, as at Çatalhöyük.

Previous excavations, however, reveal that more separate and distinctive elaborate, and possibly ritual, buildings were constructed at other sites and suggest that in these cases there was a greater cosmological separation between ritual practices and everyday life. Examples of this include two buildings with terrazzo and red-pigmented floors at Tepe Guran (J, E) ca. 7000–6600 BCE (figure 2.2f–g; Mortensen 2014) and Haji Firuz Tepe Structure VI ca. 6300 BCE, which has a podium and human and animal bone deposits (Voigt 1983, 314–16). New examples are discussed below.

A range of primary and secondary burial practices were conducted at many Zagros sites, suggesting that there was a range of networks and interactions between different communities. There is evidence of skull modification (Croucher 2012) and an urgent need for new scientific analyses of human burial remains. Figurines were widely used at many sites, and more than 5,500 were found at Jarmo (Morales 1990). Red pigment was also used from the earliest levels of many sites to mark materials and individuals in burials (e.g., at Asiab), to color floors (e.g., Karim Shahir, Tepe Guran J, E), and in residues scattered on floors (e.g., at Karim Shahir, Asiab, Chogha Bonut, and Jarmo); and it is found on groundstones (e.g., at Ali Kosh).

In sum, therefore, these examples suggest that ritual intersections in early transitions to settled life were diverse and widespread in the Zagros, contra Hole (1996) and Bernbeck (2004).

New Zagros Case Studies: The Central Zagros Archaeological Project

New excavations are also contributing significantly to our knowledge of ritual and burial practices in early settled life in the Zagros.

At Sheikh-e Abad in the highlands ca. 7590 BCE, there was a clear cosmological separation between larger social-scale interactions in a gathering place associated with reification of skulls from four large wild male goats, one with red ochre, and a wild sheep in a distinctive single-room building with a T-shaped interior, Building 2, on the one hand; and on the other more intimate to social-scale interactions in a building with smaller rooms used for living, infant burial, and penning of animals, Building 1 (Matthews, Shillito, and Elliott 2013; see Bowser and Patton 2004, figure 7). More public-near and public-far interactions were possible in open areas, associated with widespread traces of dung as well as discard from craft activities. The walls of the probable ritual/more communal Building 2 were constructed from distinctive miniature mudbricks, and the floor was clean, with only sparse traces of red pigment and charred plant remains. The wild goat and sheep skulls had been deliberately placed in rows in the south of the room, and a pestle and a large bird bone, perhaps also relating to ritual paraphernalia, were placed in the northeast corner. The placement of both these wild animal skulls and those at Ganj Dareh in the south of rooms may speculatively suggest some cosmological patterning, but a sample size of two is too small to affirm this. The clear differentiation of Building 2 from Building 1 in materials, layout, features, and traces of use points to a marked reification and separation of the wild from the more domestic at precisely the time when humans and animals were living in greater proximity, as discussed by Hodder (1990) for the Neolithic of Europe more widely. Speculatively, this highly ritualized cultural place making may represent both myth making as well as history making. Pending further excavation at this site, it is not yet known whether this cosmological separation is a repetition of earlier patterns.

At the 2.5 ha site of Bestansur in the piedmont ca. 7660 BCE, there is some evidence for community-wide collaboration in the ordering of place, as all of the architecture identified in seven of thirteen trenches is oriented NW-SE (figure 2.6). This orientation was common in later settlements and may have included environmental as well as cosmological considerations, as explored by Shepperson (2009) for Mesopotamia more widely. The site was located close to a large perennial spring, and the community had access to a wide range of wild and domesticated resources (Matthews et al. 2016, in prep).

This study focuses on the cosmological layout and evidence for ritual in two successive buildings at Bestansur, Buildings 8 and 5, in a neighborhood of abutting buildings in Trench 10 (figure 2.6). These buildings are large and multi-roomed with accentuated portico entrances, and > seventy-two individuals were buried in the main room of Building 5. The closest parallels for the layout of these buildings are from Tell Halula, Syria, where buildings of similar date ca. 7500 BCE were also constructed with a portico entrance, large main room, smaller rooms, and up to five–fifteen burials placed within them (Molist 2013).

At Bestansur, the earlier building, Building 8, has not yet been excavated, but it was clearly elaborate as the walls are ca. 70 cm thick and coated in multiple layers of brown, green, and white plaster and traces of red-pigmented wall paintings that were partially exposed in plan in the field and analyzed in micromorphological thin section (Godleman, Almond, and Matthews 2016). The later building, Building 5, was also painted, in this case with indistinct traces of red pigment on the base of some walls. It was enlarged, with additional rooms added. The use of Building 5 for ritual activities is indicated by

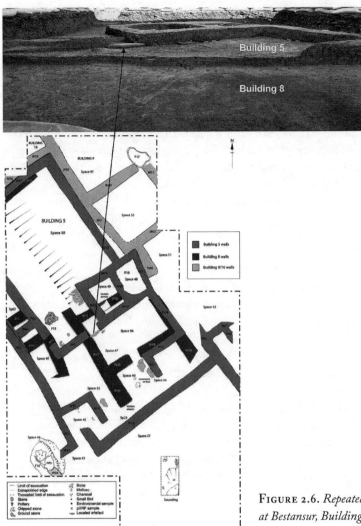

FIGURE 2.6. *Repeated building at Bestansur, Buildings 8 and 5, ca. 7660 BC.*

the placement of more than seventy-two primary, secondary, and skull burials—some with beads, cowrie shells, pigment, and traces of molded plaster—below and on the floors of the main room, which at ca. 7.7 m × 5 m in size was large enough to host public-scale gatherings. A range of evidence indicates that this building was primarily used for reception (including perhaps sleeping), burial, and ritual practices. The entrances to the building and to the main room were both marked by the placement of large stones, one of which was incised and would have emphasized the passage into these interior places. The artifacts on the latest floors and surfaces of the entrance room include a polished pierced stone disc, figurine, amulet, and muller. The interior floors of the main room, Space 50, in particular were repeatedly plastered and kept immaculately clean, with extensive and repeated traces of floor coverings and some red pigment. The few finds in the deep sequence of reed and dung ash fuel in the large rectilinear oven include a large bird wing bone and a neonatal human femur, as well as a mace head and a carved stone mortar placed upside down in an act of closure at the end of this installation's use. Of particular significance to questions of the role of history making and ritual in the creation of settled life is the evidence for repeated human burial throughout this building's life history as well as cosmological patterning in the placement of some individual remains below the floors according to life-cycle stage. There was some tendency to place infants and juveniles close to the entrance—some with their skulls facing the large stone threshold—and adults further into the room, as at Tell Halula (Molist 2013; Matthews et al. 2016, in prep). Occasionally, they were placed together in specific groupings.

Cosmological Patterning and Burial Practices at Çatalhöyük

Large-scale excavations at Çatalhöyük have identified remarkable shared community-wide concepts in the cosmological layout of buildings from ca. 6900–6200 BCE. The northern half of the main rooms was most frequently used for burials, wall paintings, and molded sculpture; and the walls and sitting platforms were marked by multiple applications of whitewashes and plaster. The southern half and side rooms were often used for food preparation and cooking and storage and were endowed with fire installations and bins and surfaces more frequently coated with thicker non-white plasters and accumulated residues (Matthews 2005a). Within this shared conceptual framework, however, the layout and use of buildings were historically contingent and did change with shifts in social roles and relations, as attested by changes in locations of some key features including ovens and hearths, bins, and crawl-hole entrances, for example, and in associated surfaces and residues (Matthews 2005b; Farid 2007; Hodder and Pels 2010). The final case study considers one example of patterning from Building 1 for comparison to the preceding examples from the Zagros on cosmological layout and ritual and the role of intramural burial in particular in these early settled communities.

There is some evidence of cosmological patterning in the placement of the dead, not only according to life stage, as at Bestansur, but also commensality at Çatalhöyük in Building 1, where the remains of sixty-two individuals were buried. This correlation is based on micro-stratigraphic analyses of floors and a micro-contextual reanalysis by the author of the human isotope analyses conducted by Jessica Pearson (figure 2.7; Richards and Pearson 2005; Richards et al. 2003), analyses of the human remains (Andrews, Molleson, and Boz 2005; Molleson, Andrews, and Boz 2005), and excavation data (Cessford 2007b). As at Bestansur, juveniles and adults tended to be buried separately or occasionally together in specific groupings. In Building 1 at Çatalhöyük, 86 percent of the burials in the northwest (NW) platform are juveniles (19/22); 93 percent in the central-east (CE) platform were adult (13/14). Most of these are primary burials and were placed in raised platforms perhaps used for sitting, arguably by these respective social groups. The height and repeated whitewashing of these platforms may have been used to mark the status and separation of these groups (figure 2.7b, e). Isotopic analysis indicates that these two groups had different but related diets and suggests that they were closely related commensal groups (figure 2.7d NW, CE). A separate group of individuals, however, represented by secondary (66%) or double female and infant burials, perhaps related to childbirth, was placed not in platforms but below floor level and covered only with brown mud plasters in the north-center (figure 2.7c–d NC). The diet of this group differed from that of the others and may suggest that these individuals were less frequently commensal and, speculatively, that they may not have lived in this building. The eight individuals buried as secondary remains may have died away from the site, as suggested by Kuijt (2000) for at least some of the individuals who were subject to secondary mortuary practices more widely in the Neolithic. Notably, when the building was destroyed by fire and reconfigured as a smaller unit (Phase B1.4), burials were no longer placed in the NC area (figure 2.7d). This correlation suggests that a reduction in the size of the house and social arenas was coincident with a reduction in the extended networks of the group associated with Building 1.

CONCLUSION

In conclusion, therefore, it is clear that early Neolithic communities in the Zagros engaged more widely in ritual activities than previously suggested and that integration of forensic-scale micro-archaeological methodologies and theoretical approaches can provide new informative insights into ritual, history, and place at temporal resolutions much closer to lived worlds in the past.

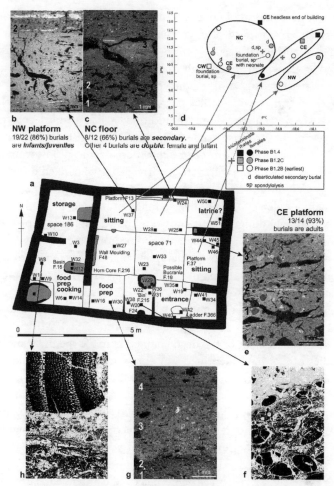

FIGURE 2.7. *Placing the dead (a) Çatalhöyük, Building 1. Repeated surfaces and residues: (b) NW platform, white silty clay plasters [1] and washes [2] (oblique incident light (OIL)); (c) NC floor, brown mud plasters [1], thin lens of "trample" [2], occasional orange silty clay plaster with aggregates of reworked wall plaster [3] (OIL); (d) (see below); (e) CE platform, white silty clay plasters [1, 3] and washes with rare traces of accumulated sediment [2] (OIL); (f) "bin" containing charred and calcitic ash remains of lentils (PPL, height = 11 mm); (g) SW platform, brown silt loam plasters [1], lenses with charred plant remains [2], pale brown/whitish sandy silt loam plasters [3], mat impression [4] (OIL); (h) burned floor plaster (after Matthews 2005a, figure 19.13), retted grass floor covering, and charred wood (ulm); (d) $\delta^{13}C$ and $\delta^{15}N$ stable isotope analysis of human bone in B1 burials by placement and phase (data drawn from Richards and Pearson 2005; Richards et al. 2003; Andrews, Molleson, and Boz 2005; Molleson, Andrews, and Boz 2005; Cessford 2007b).*

In particular, micromorphology has enabled in-depth characterization of the materiality and histories of place at the scale of individual surfaces, impact traces, and depositional events *within* as well as at the scales of building phases and levels, which are difficult to resolve and separate during excavation. It has permitted identification of the composition, technology, and impacts on surfaces, together with other micro-analytical techniques, not evident from macroscopic analyses alone. In addition, it has enabled detection and characterization of a diverse range of mineral, bioarchaeological, and micro-artifactual residues and their precise depositional associations, which are not routinely detectable or are aggregated or irreversibly separated during routine excavations and sampling. Crucially, particularly for the origins of agriculture, this has included detection of a diverse range of burned and unburned plant remains and animal dung that are providing new insights into plant use and animal management, prior to detectable zoomorphological changes.

With regard to ritual, micro-stratigraphic analyses have enabled detection of the context of traces of red pigment, both in highly ritualized contexts such as burials and placement of animal skulls as well as in more "everyday" contexts such as food preparation and cooking areas and its use as a key marker of reified events (Hastorf 2001; Taçon 2004: Hunt 2005). Micro-contextual analyses of the placement of the dead coupled here for the first time with isotopic studies of diet have identified ways in which commensal and extended non-commensal networks may have played a role in the social and ritual creation of Neolithic communities and settled life and continuity and change in those communities within the history of a building, Building 1 at Çatalhöyük.

Anthropological, ethnographic, and architectural observations and theory (Hastorf 2001; Leatherbarrow and Mostafavi 2002; Bloch 2010; Robb 2010; Hodder 2012) have provided supporting insight into how surface materials are often expressly chosen for specific places and intended events and bear the subsequent impact and residual traces of continuity or change (Kramer 1979; Matthews, Hastorf, and Ergenekon 2000; Boivin 2000; Leatherbarrow and Mostafavi (2002). These interdisciplinary theories and approaches have also provided important rationale, first, for comparative micro-contextual approaches to the study of the significance of surfaces and residues, with particular reference to Robb (2010). Second, with regard to interpretation of repetitive actions and building, Bloch (2010) has established a basis for analysis of settings, features, and buildings as representations of social concepts, roles, and relations and the study of continuity and change in these.

The research here has also highlighted the benefits of reexamining data from old excavations and integrating this with analysis and interpretation of new data through the lenses of new excavations, analytical techniques, and theory. As a result, this research has brought the Zagros region back into focus in studies of the origins of agriculture, the creation of settled life, and the importance of history and ritual in the making of these, in agreement with new genetic insights (Lazaridis et al. 2016).

In response to the questions raised at the beginning, the following observations are proposed for the central Zagros, which until recently has not been as frequently included in wider debates on early settled life since the pioneering investigations in this region in the 1950–70s. The contributions of micro-analytical and micro-historical approaches are also highlighted.

Concerning the timing of associations with particular places, there was little or no gap in occupation in the Zagros in the early Holocene, as newly discovered sites were founded from ca. 9810 BCE, including Sheikh-e Abad and the slightly later sites of East Chia Sabz and Chogha Golan (Matthews, Shillito, and Elliott 2013; Matthews and Fazeli Nashli 2013). Many Zagros sites throughout the origins of early settled life, from ca. 9800–6000 BCE, attest increasing engagement in history making in the construction, renewal, and elaboration of buildings and placement of the dead. Micro-analyses in particular demonstrated that the earliest visits to some sites, such as Jani pre-8000 BCE, were periodic, with low-intensity/density activities resembling those from shifting modern hunter-gatherer sites (Mallol et al. 2007) and basal areas of Çatalhöyük. When communities became more settled, deposition was much more rapid and repeated, indicating greater investment in place making and history making.

With regard to the context and causes for early settled life and history making, correlations suggest that they include climatic amelioration, use of locally biodiverse resources, greater proximity to and management of animals, population increase and settlement expansion, and access to wider networks. Micromorphological identification of hitherto undetected unburned dung has established that animals were penned and lived among communities within settlements and selected buildings or rooms by 8000–7000 BCE, as at Sheikh-e Abad and Çatalhöyük. Macro- and micro-stratigraphic analyses also indicate that periods of transformation from periodic to more settled life and subsequent settlement expansion at particular sites were often rapid and abrupt and a comparative step-change.

In response to discussion of how widespread and significant repetitive building and cosmological patterning were in creating and sustaining historical ties and social relations, new micro-stratigraphic and micromorphological evidence suggests that there was considerable continuity in social roles and relations at the generational timescales of individual building levels at sites in the Zagros. This is based on the remarkable repetition and continuity in sequences of floors and residues within particular open areas and buildings. The emphasis on repeated rendering of reception, burial, and ritual areas in particular with specific distinctive materials highlights the new central focus on these interior places as key social arenas in creating cultural and historical ties. There is less evidence of repeated construction of buildings to date in the central Zagros than in central Anatolia, although there are notable exceptions, from very early evidence of more settled life at Zawi Chemi Shanidar, Structure 1, and later evidence from Bestansur, ca. 7660, where elaborate buildings were rebuilt on the same plot.

The layout of settlements, buildings, and placement of the dead attests some cosmological patterning in specific circumstances such as separation according to life-cycle stage of juveniles and adults as well as separation of wild and domestic spheres. This patterning and separation are likely to have served to highlight the significance of particular roles, relations, events, and ways of living among these communities.

Lastly, with regard to the geographic extent of ritual practices in the Neolithic, there is increasing evidence of the diversity and significance of ritual practices in the Zagros and of similarities with practices across the Middle East. Shared practices include reification of wild animal skulls and large birds, intramural sub-floor primary and secondary mortuary practices, and use of red pigments in wall paintings and a wide range of contexts to mark particular events and associations. Micro-contextual analyses of the placement of the dead within buildings have identified long historical ties between a supra-household community of > seventy-two individuals at Bestansur in Building 5 ca. 7660 BCE and some cosmological separation according to life stage as well as group, as at sites such as Halula (Molist 2013) and explored above for Çatalhöyük.

It is hoped that this research has established some of the ways in which integration of macro- and micro-histories and cross-regional approaches can contribute to our understanding of the creation and development of settled life in key early agricultural heartland communities and to studies of history, religion, and place more widely.

ACKNOWLEDGMENTS

I am very grateful to Iran's Culture, Heritage, Handicrafts and Tourism Organisation, the Iranian Centre for Archaeological Research, the State

Organisation for Antiquities and Heritage Iraq, Baghdad; the Erbil Directorate of Antiquities and Heritage and its director, Abubakir O. Zainadin; the Sulaimaniyah Directorate of Antiquities and Heritage and its director, Kamal Rasheed Raheem; the Turkish Ministry of Culture and Tourism, and the General Directorate of Cultural Heritage and Museums for their kind permission and extremely helpful support in conducting excavations and export of samples for scientific analysis. I would also like to thank the Central Zagros Archaeological Project co-directors Roger Matthews, Yaghoub Mohammadifar, and Abbass Motarjem and all team members; the Sulaimaniyah Directorate of Antiquities and Heritage representative, Kamal Rouf Aziz; and the director of the Çatalhöyük Research Project, Ian Hodder, and team members for their collaboration in fieldwork and research. I thank John Jack and Julie Boreham for manufacturing thin sections and Sarah Lambert-Gates and Margaret Mathews for help with the illustrations. This research was supported by the British Academy (BARDA-48993), the Arts and Humanities Research Council (AH/H034315/1), the British Institute of Persian Studies, and the Department of Archaeology, School of Archaeology, Geography and Environmental Science, University of Reading, to which I am extremely grateful. I am very grateful also to the following for permission to reproduce images in this article: British Institute of Persian Studies (figures 2.2a, 2.2c); Routledge Taylor & Francis Group (figure 2.2d); Professor Peder Mortensen (figures 2.2e–g); Oriental Institute of the University of Chicago (figures 2.2h–i); and British Institute at Ankara (figure 2.3a).

REFERENCES

Andrews, Peter, Theya Molleson, and Basak Boz. 2005. "The Human Burials at Catalhoyuk." In *Inhabiting Catalhoyuk: Reports from the 1995–99 Seasons*, edited by Ian Hodder, 261–79. Cambridge: McDonald Institute for Archaeological Research and British Institute of Archaeology at Ankara.

Asouti, Eleni, and Dorian Q. Fuller. 2013. "A Contextual Approach to the Emergence of Agriculture in Southwest Asia: Reconstructing Early Neolithic Plant-Food Production." *Current Anthropology* 54 (3): 299–345. https://doi.org/10.1086/670679.

Astruc, Laurence, M. Roberto Vargiolu, M. Ben Tkaya, Nur Balkan-Atlı, Mihriban Özbaşaran, and Hassan Zahouani. 2011. "Multi-Scale Tribological Analysis of the Technique of Manufacture of an Obsidian Bracelet from Aşıklı Höyük (Aceramic Neolithic, Central Anatolia)." *Journal of Archaeological Science* 38 (12): 3415–24. https://doi.org/10.1016/j.jas.2011.07.028.

Bachelard, Gaston. 1994. *The Poetics of Space*. Translated from the French "La Poetique de L'espace" (Presses Universitaires de France, 1958) by Maria Jolas (Orion, 1964), with a new foreword by John R. Stilgoe. Boston: Beacon.

Baird, Douglas. 2005. "The History of Settlement and Social Landscapes in the Early Holocene in the Çatalhöyük Area." In *Çatalhöyük Perspectives. Themes from the 1995–99 Seasons*, ed. Ian Hodder, 55–74. Cambridge: McDonald Institute for Archaeological Research and British Institute of Archaeology at Ankara.

Baird, Douglas, Andrew Fairbairn, Louise Martin, and Caroline Middleton. 2012. "The Boncuklu Project: The Origins of Sedentism, Cultivation, and Herding in Central Anatolia." In *The Neolithic in Turkey: The Cradle of Civilization. New Discoveries*, ed. Mehmet Özdoğan, Nezih Başgelen, and Peter Kuniholm, 219–44. Istanbul: Arkeoloji ve Sanat Yayınları.

Bell, Catherine. 1997. *Ritual Perspectives and Dimension*. Oxford: Oxford University Press.

Bernbeck, Reinhard. 2004. "Iran in the Neolithic." In *Persiens Antike Pracht*, ed. Thomas Stöllner, 140–47. Bochum, Germany: Deutsches Bergau-Museum Bochum.

Bloch, Maurice. 2010. "Is There Religion at Çatalhöyük ... Or Are There Just Houses?" In *Religion in the Emergence of Civilization: Çatalhöyük as a Case Study*, ed. Ian Hodder, 146–62. Cambridge: Cambridge University Press. https://doi.org/10.1017/CBO9780511761416.006.

Boivin, Nicole L. 2000. "Life Rhythms and Floor Sequences: Excavating Time in Rural Rajasthan and Neolithic Çatalhöyük." *World Archaeology* 31 (3): 367–88. https://doi.org/10.1080/00438240009696927.

Bowser, Brenda J., and John Q. Patton. 2004. "Domestic Spaces as Public Places: An Ethnoarchaeological Case Study of Houses, Gender, and Politics in the Ecuadorian Amazon." *Journal of Archaeological Method and Theory* 11 (2): 157–81. https://doi.org/10.1023/B:JARM.0000038065.43689.75.

Bradley, Richard. 2005. *Ritual and Domestic Life in Prehistoric Europe*. London: Routledge.

Braidwood, Linda, Robert J. Braidwood, Bruce Howe, Charles A. Reed, and Patty Jo Watson, eds. 1983. *Prehistoric Archaeology along the Zagros Flanks*. Chicago: University of Chicago Oriental Institute.

Braidwood, Robert J. 1960. "Seeking the World's First Farmers in Persian Kurdistan." *Illustrated London News* 237: 695–97.

Bullock, Peter, Nicolas Fedoroff, A. Jongerius, Georges Stoops, and Tatiana Tursina. 1985. *Handbook for Soil Thin Section Description*. Wolverhampton, UK: Waine Research.

Carsten, Janet, and Stephen Hugh-Jones. 1995. *About the House: Levi Strauss and Beyond.* Cambridge: Cambridge University Press. https://doi.org/10.1017/CBO9780511607653.

Cessford, Craig. 2007a. "Level Pre-XII.E-A and Levels XII and XI, Spaces 181, 199, and 198." In *Excavating Çatalhöyük: South, North, and KOPAL Area Reports from the 1995–99 Seasons*, ed. Ian Hodder, 59–101. Cambridge: McDonald Institute for Archaeological Research and the British Institute at Ankara.

Cessford, Craig. 2007b. "Building 1." In *Excavating Çatalhöyük: South, North, and KOPAL Area Reports from the 1995–99 Seasons*, ed. Ian Hodder, 405–549. Cambridge: McDonald Institute for Archaeological Research and the British Institute at Ankara.

Cole, Garrard. 2013. "Human Burials." In *The Earliest Neolithic of Iran: 2008 Excavations at Sheikh-e Abad and Jani*. Central Zagros Archaeological Project CZAP Reports, vol. 1, ed. Roger Matthews, Wendy Matthews, and Yaghoub Mohammadifar, 163–74. Oxford: British Institute of Persian Studies and Oxbow Books.

Courty, Marie Agnes, Paul Goldberg, and Richard I. Macphail. 1989. *Soils and Micromorphology in Archaeology*. Cambridge: Cambridge University Press.

Croucher, Karina. 2012. *Death and Dying in the Neolithic of the Near East*. Oxford: Oxford University Press. https://doi.org/10.1093/acprof:osobl/9780199693955.001.0001.

Daems, Aurelie. 2008. "Evaluating Patterns of Gender through Mesopotamian and Iranian Human Figurines: A Reassessment of the Neolithic and Chalcolithic Period Industries." In *Gender through Time through the Ancient Near East*, ed. Diane Bolger, 77–118. Los Angeles: Altamira.

Darabi, Hojjat, Tobias Richter, and Peder Mortensen. 2018. "New Excavations at Tappeh Asiab, Kermanshah Province, Iran." *Antiquity* 92 (361): 1–6. https://doi.org/10.15184/aqy.2018.3.

Dean, Emily, and David Kojan. 2001. "Ceremonial Households and Domestic Temples: 'Fuzzy' Definitions in the Andean Formative." In *Past Ritual and the Everyday*. Kroeber Anthropological Society Papers 85, ed. Christine A. Hastorf, 109–33. Berkeley: Kroeber Anthropological Society.

Farid, Shahina. 2007. "Neolithic Excavations in the South Area, East Mound, Çatalhöyük 1995–99." In *Excavating Çatalhöyük: South, North, and KOPAL Area Reports from the 1995–99 Seasons*, ed. Ian Hodder, 41–344. Cambridge: McDonald Institute for Archaeological Research and the British Institute at Ankara.

Fisher, Kevin D. 2009. "Placing Social Interaction: An Integrative Approach to Analyzing Past Built Environments." *Journal of Anthropological Archaeology* 28 (4): 439–57. https://doi.org/10.1016/j.jaa.2009.09.001.

Foxhall, Lin. 2000. "The Running Sands of Time: Archaeology and the Short-Term." *World Archaeology* 31 (3): 484–98. https://doi.org/10.1080/00438240009696934.

Godleman, Jessica, Matthew J. Almond, and Wendy Matthews. 2016. "An Infrared Microspectroscopic Study of Plasters and Pigments from the Neolithic Site of Bestansur, Iraq." *Journal of Archaeological Science: Reports* 7: 195–204. https://doi.org/10.1016/j.jasrep.2016.04.013.

Hastorf, Christine A. 2001. "Studying Ritual in the Past." In *Past Ritual and the Everyday*. Kroeber Anthropology Society Papers 85, ed. Christine A. Hastorf, 1–15. Berkeley: Kroeber Anthropological Society.

Hodder, Ian. 1990. *The Domestication of Europe*. Oxford: Blackwell.

Hodder, Ian. 2012. *Entangled: An Archaeology of the Relationships between Humans and Things*. New York: John Wiley and Sons. https://doi.org/10.1002/9781118241912.

Hodder, Ian, ed. 2010. *Religion in the Emergence of Civilization: Çatalhöyük as a Case Study*. Cambridge: Cambridge University Press. https://doi.org/10.1017/CBO9780511761416.

Hodder, Ian, and Peter Pels. 2010. "History Houses: A New Interpretation of Architectural Elaboration at Çatalhöyük." In *Religion in the Emergence of Civilization: Çatalhöyük as a Case Study*, ed. Ian Hodder, 163–86. Cambridge: Cambridge University Press. https://doi.org/10.1017/CBO9780511761416.007.

Hole, Frank. 1996. "The Context of Caprine Domestication in the Zagros Region." In *The Origins and Spread of Agriculture and Pastoralism in Eurasia*, ed. David Harris, 263–81. Washington, DC: Smithsonian Institution Press.

Hole, Frank, Kent V. Flannery, and James A. Neely. 1969. *Prehistory and Human Ecology of the Deh Luran Plain: An Early Village Sequence from Khuzistan, Iran*. Memoirs of the Museum of Anthropology. Ann Arbor: University of Michigan.

Hunt, Stephen. 2005. *Religion and Everyday Life*. London: Routledge.

Insoll, Timothy. 2004. *Archaeology, Ritual, Religion*. London: Routledge.

Keane, Webb. 2010. "Marked, Absent, Habitual: Approaches to Neolithic Religion at Çatalhöyük." In *Religion in the Emergence of Civilization: Çatalhöyük as a Case Study*, ed. Ian Hodder, 187–219. Cambridge: Cambridge University Press. https://doi.org/10.1017/CBO9780511761416.008.

Kohn, Alison, and Shannon Lee Dawdy. 2016. "Archaeologies of an Informal City: Temporal Dimensions of Contemporary Andean Urbanism." In *Elements of Architecture: Assembling Archaeology, Atmosphere, and the Performance of Building Spaces*, ed. Mikkel Bille and Tim Flohr Sørensen, 121–40. London: Routledge.

Kozlowski, Stefan Karol, and Olivier Aurenche. 2005. *Territories, Boundaries, and Cultures in the Neolithic Near East*, Maison de l'Orient et de la Mediterranee. Lyon: Jean Pouilloux.

Kramer, Carol. 1979. "An Archaeological View of a Contemporary Kurdish Village: Domestic Architecture, Household Size, and Wealth." In *Ethnoarchaeology: Implications of Ethnography for Archaeology*, ed. Carol Kramer, 139–63. New York: Columbia University Press.

Kuijt, Ian. 2000. "Keeping the Peace: Ritual, Skull Caching, and Community Integration in the Levantine Neolithic." In *Life in Neolithic Farming Communities: Social Organization, Identity, and Differentiation*, ed. Ian Kuijt, 137–64. New York: Kluwer Academic/Plenum.

Lazaridis, Iosif, Dani Nadel, Gary Rollefson, Deborah C. Merrett, Nadin Rohland, Swapan Mallick, Daniel Fernandes, Mario Novak, Beatriz Gamarra, Kendra Sirak et al. 2016. "Genomic Insights into the Origin of Farming in the Ancient Near East." *Nature* 536 (7617): 419–24. https://doi.org/10.1038/nature19310.

Leatherbarrow, David, and Mohsen Mostafavi. 2002. *Surface Architecture*. London: MIT Press.

Lucas, Gavin. 2005. *The Archaeology of Time*. London: Routledge.

Mallol, Carolina, Frank W. Marlowe, Brian M. Wood, and Claire C. Porter. 2007. "Earth, Wind, and Fire: Ethnoarchaeological Signals of Hadza Fires." *Journal of Archaeological Science* 34 (12): 2035–52. https://doi.org/10.1016/j.jas.2007.02.002.

Matthews, Roger, and Hassan Fazeli Nashli, eds. 2013. *The Neolithisation of Iran: The Formation of New Societies*. Oxford: Oxbow Books.

Matthews, Roger, Wendy Matthews, and Yaghoub Mohammadifar, eds. 2013. *The Earliest Neolithic of Iran: 2008 Excavations at Sheikh-e Abad and Jani*. Central Zagros Archaeological Project CZAP Reports, vol. 1. Oxford: British Institute of Persian Studies and Oxbow Books.

Matthews, Roger, Wendy Matthews, Kamal Rasheed Raheem, and Kamal Rauf Aziz. 2016. "Current Investigations into the Early Neolithic of the Zagros Foothills of Iraqi Kurdistan." In *The Archaeology of the Kudistan Region of Iraq and Adjacent Regions*, ed. Konstantinos Kopanias and John MacGinnis, 219–28. Oxford: Archaeopress.

Matthews, Roger, Wendy Matthews, Amy Richardson, and Kamal Rasheed Raheem, eds. In prep. *Sedentism and Resource Management in the Neolithic of the Central Zagros*. Central Zagros Archaeological Project CZAP Reports, vol. 2. Oxford: Oxbow Books.

Matthews, Wendy. 2005a. "Micromorphological and Microstratigraphic Traces of Uses of Space." In *Inhabiting Çatalhöyük: Reports from the 1995–99 Seasons*, ed. Ian Hodder, 355–98, 553–72. Cambridge: McDonald Institute for Archaeological Research and British Institute of Archaeology at Ankara.

Matthews, Wendy. 2005b. "Life-Cycle and Life-Course of Buildings." In *Çatalhöyük Perspectives: Themes from the 1995–99 Seasons*, ed. Ian Hodder, 125–51. Cambridge:

McDonald Institute for Archaeological Research and British Institute of Archaeology at Ankara.

Matthews, Wendy. 2010. "Geoarchaeology and Taphonomy of Plant Remains and Microarchaeological Residues in Early Urban Environments in the Ancient Near East." *Quaternary International* 214 (1–2): 98–113. https://doi.org/10.1016/j.quaint.2009.10.019.

Matthews, Wendy. 2012. "Defining Households: Micro-Contextual Analysis of Early Neolithic Households in the Zagros, Iran." In *New Perspectives on Household Archaeology*, ed. Bradley J. Parker and Catherine P. Foster, 183–218. Winona Lake, IN: Eisenbrauns.

Matthews, Wendy. 2013. "Contexts of Neolithic Interaction: Geography, Palaeoclimate, and Palaeoenvironment of the Central Zagros." In *The Earliest Neolithic of Iran: 2008 Excavations at Sheikh-e Abad and Jani*. Central Zagros Archaeological Project CZAP Reports, vol. 1, ed. Roger Matthews, Wendy Matthews, and Yaghoub Mohammadifar, 13–20. Oxford: British Institute of Persian Studies and Oxbow Books.

Matthews, Wendy, Charles A.I. French, Thomas Lawrence, David F. Cutler, and Martin K. Jones. 1997. "Microstratigraphic Traces of Site Formation Processes and Human Activities." *World Archaeology* 29 (2): 281–308. https://doi.org/10.1080/00438243.1997.9980378.

Matthews, Wendy, Christine A. Hastorf, and Begumsen Ergenekon. 2000. "Ethnoarchaeology: Studies in Local Villages Aimed at Understanding Aspects of the Neolithic Site." In *Towards Reflexive Method in Archaeology: The Example at Çatalhöyük*, ed. Ian Hodder, 177–89. Cambridge: McDonald Institute of Archaeological Research and British Institute of Archaeology at Ankara.

Matthews, Wendy, with contributions from Lisa-Marie Shillito and Sarah Elliott. 2013. "Investigating Early Neolithic Materials, Ecology, and Sedentism: Micromorphology and Microstratigraphy." In *The Earliest Neolithic of Iran: The Central Zagros Archaeological Project 2008 Excavations at Sheikh-e Abad and Jani*. Central Zagros Archaeological Project CZAP Reports, vol. 1, ed. Roger Matthews, Wendy Matthews, and Yaghoub Mohammadifar, 67–105. Oxford: British Institute of Persian Studies and Oxbow Books.

Matthews, Wendy, Lisa-Marie Shillito, and Sarah Elliott. 2014. "Neolithic Lifeways: Microstratigraphic Traces within Houses, Animal Pens, and Settlements." In *Early Farmers: The View from Archaeology and Science*, ed. Alasdair Whittle and Penny Bickle, 251–79. London: British Academy. https://doi.org/10.5871/bacad/9780197265758.003.0014.

Meldgaard, Jorgen, Peder Mortensen, and Henrik Thrane. 1963. "Excavations at Tepe Guran, Luristan: Preliminary Report of the Danish Archaeological Excavation to Iran 1963." *Acta Archaeologica* 34: 7–133.

Molist, Miquel, ed. 2013. *Tell Halula: Un Poblado de los Pirmeros Agricutores en el Valle del Eufrates, Siria*, vol. 1. Barcelona: Ministero de Educacion, Cultura y Deporte.

Molleson, Theya, Peter Andrews, and Başak Boz. 2005. "Reconstruction of the Neolithic people of Catalhoyuk." In *Inhabiting Catalhoyuk: Reports from the 1995–99 Seasons*, edited by Ian Hodder, 279–300, 521–32. Cambridge: McDonald Institute for Archaeological Research and British Institute of Archaeology at Ankara.

Moore, Jerry. 1996. *Architecture and Power in the Ancient Andes*. Cambridge: Cambridge University Press. https://doi.org/10.1017/CBO9780511521201.

Morales, Vivian Broman. 1990. *Figurines and Other Clay Objects from Sarab and Cayonu*. Chicago: Oriental Institute of the University of Chicago.

Mortensen, Peder, ed. 2014. *Excavations at Tepe Guran: The Neolithic Period*. Leuven, Belgium: Peeters.

Nakamura, Carolyn, and Lynn Meskell. 2013. "Figurine Worlds at Çatalhöyük." In *Substantive Technologies at Çatalhöyük: Reports from the 2000–2008 Seasons*, ed. Ian Hodder, 201–34. Los Angeles: Cotsen Institute of Archaeology at UCLA.

Özbaşaran, Mihriban. 2012. "Aşıklı Höyük." In *The Neolithic in Turkey: The Cradle of Civilization, New Discoveries*, ed. Mehmet Özdoğan, Nezih Başgelen, and Peter Kuniholm, 135–58. Istanbul: Arkeoloji ve Sanat Yayınları.

Richards, Michael P., Jessica A. Pearson, Theya I. Molleson, Nerissa Russell, and Louise Martin. 2003. "Stable Isotope Evidence of Diet at Neolithic Çatalhöyük, Turkey." *Journal of Archaeological Science* 30 (1): 67–76. https://doi.org/10.1006/jasc.2001.0825.

Riehl, Simone, Mohsen Zeidi, and Nicholas J. Conard. 2013. "Emergence of Agriculture in the Foothills of the Zagros Mountains of Iran." *Science* 341 (6141): 65–67. https://doi.org/10.1126/science.1236743.

Richards, Michael P., and Jessica A. Pearson. 2005. "Stable-Isotope Evidence of Diet at Catalhoyuk." In *Inhabiting Catalhoyuk: Reports from the 1995–99 Seasons*, edited by Ian Hodder, 313–22. Cambridge: McDonald Institute for Archaeological Research and British Institute of Archaeology at Ankara.

Robb, John. 2010. "Beyond Agency." *World Archaeology* 42 (4): 493–520. https://doi.org/10.1080/00438243.2010.520856.

Schiffer, Michael B. 1987. *Formation Processes of the Archaeological Record*. Albuquerque: University of New Mexico Press.

Shepperson, Mary. 2009. "Planning for the Sun: Urban Forms as a Mesopotamian Response to the Sun." *World Archaeology* 41 (3): 363–78. https://doi.org/10.1080/00438240903112229.

Smith, Philip E.L. 1971. "Iran, 9000–4000 B.C.: The Neolithic." *Expedition* (Spring–Summer 1971): 6–13.

Smith, Philip E.L. 1972. "Ganj Dareh: Survey of Excavations in Iran during 1970–71." *Iran* 10: 165–68.

Smith, Philip E.L. 1975. "Ganj Dareh: Survey of Excavations in Iran, 1973–74." *Iran* 13: 178–80.

Smith, Philip E.L. 1990. "Architectural Innovation and Experimentation at Ganj Dareh, Iran." *World Archaeology* 21 (3): 323–35. https://doi.org/10.1080/00438243.1990.9980111.

Solecki, Rose L. 1981. *An Early Village Site at Zawi Chemi Shanidar*. Malibu: Undena.

Taçon, Paul S.C. 2004. "Ochre, Clay, Stone, and Art: The Symbolic Importance of Minerals as Life-Force among Aboriginal Peoples of Northern and Central Australia." In *Soils, Stones, and Symbols: Cultural Perceptions of the Mineral World*, ed. Nicole Boivin and Mary Ann Owoc, 31–42. London: UCL Press.

Voigt, Mary. 1983. *Haji Firuz Tepe, Iran: The Neolithic Settlement*. Philadelphia: University Museum, University of Pennsylvania.

Wasylikowa, Krystyna, and Andrzej Witkowski, eds. 2008. *The Palaeoecology of Lake Zeribar and Surrounding Areas, Western Iran, during the Last 48,000 Years*. Ruggell, Liechtenstein: A.R.G. Gantner Verlag KG.

Whitlam, Jade, Hengameh Ilkhani, Amy Bogaard, and Michael Charles. 2013. "The Plant Macrofossil Evidence from Sheikh-e Abad: First Impressions." In *The Earliest Neolithic of Iran: 2008 Excavations at Sheikh-e Abad and Jani*. Central Zagros Archaeological Project CZAP Reports, vol. 1, ed. Roger Matthews, Wendy Matthews, and Yaghoub Mohammadifar, 175–84. Oxford: Oxbow Books and British Institute of Persian Studies and Oxbow Books.

Willcox, George. 2005. "The Distribution, Natural Habitats, and Availability of Wild Cereals in Relation to Their Domestication in the Near East: Multiple Events, Multiple Centres." *Vegetation History and Archaeobotany* 14 (4): 534–41. https://doi.org/10.1007/s00334-005-0075-x.

Zeder, Melinda. 2008. "Animal Domestication in the Zagros: An Update and Directions for Future Research." In *Archaeozoology of the Near East VIII*, ed. Emmanuelle Vila, Lionel Gourichon, Alice M. Choyke, and Hijlke Buitenhuis, 243–77. Lyon: Travaux de la Maison de l'Orient et de la Méditerranée.

Zeder, Melinda. 2009. "The Neolithic Macro-(R)evolution: Macroevolutionary Theory and the Study of Culture Change." *Journal of Archaeological Research* 17 (1): 1–63. https://doi.org/10.1007/s10814-008-9025-3.

3

Long-Term Memory and the Community in the Later Prehistory of the Levant

Nigel Goring-Morris
and Anna Belfer-Cohen

The prehistoric record provides ample evidence in general (herewith pertaining to the Levant in particular) to indicate that "shared history," created and sustained through cultivating and keeping long-term memories, predates sedentism and agricultural lifeways. "Group identity," "group cohesion" and "belonging," among others, were reinforced through sustaining and adhering to common history and shared long-term memories. They played a pivotal role in supporting viable mating networks throughout the Paleolithic, long before the need arose for social cohesion as a means to alleviate the pressures inherent in a delayed-return economy (see discussion below).

"History making" may result from habituated behavior, and there could have been a continuity of practices based on "it had always been done this way"; still, it seems that much earlier than the appearance of a delayed-return economy, "shared history" was intentionally cultivated and retained through the creation of long-term memories affecting customs and behaviors. The evidence for this differs when discussing mobile versus sedentary societies. For example, long-term memory is reflected in Çatalhöyük mostly through observations of the architectural domain. However, with respect to pre-sedentary or semi-sedentary societies, one needs to examine other, different domains of existence to identify evidence for "shared history" and the preservation/retention of long-term memory in groups at large, sodalities, or even extended kin.

DOI: 10.5876/9781607327370.c003

During the Paleolithic one can observe "long-term memory" expressed by ties to localities, that is, repetitive intra-site patterning through all the stages of a site occupation going back to the Lower and Middle Paleolithic, at sites such as Acheulian Gesher Benot Ya'aqov and Mousterian Kebara (Alperson-Afil et al. 2009; Bar-Yosef and Meignen 2007). This intentional repetitive patterning is most obvious in open-air sites, since the spatial evidence from caves and rock shelters can be biased; the very same repetitious use may stem from physical constraints, such as within the spatially restricted Upper Paleolithic Levantine Aurignacian occupations in Hayonim Cave (Belfer-Cohen 1980; Belfer-Cohen and Bar-Yosef 1981). Examples from open-air occupations include the repetitious use of precisely the same location in the Initial Upper Paleolithic sequence at Boqer Tachtit. Upper Paleolithic Ahmarian occupations at Boqer and Qadesh Barnea 601 (Gilead and Bar-Yosef 1993; Marks 1983) likely derive from specific, tactical considerations of the occupants within the landscape as part of wider strategic subsistence decisions associated with the annual rounds of mobile foragers (see Binford 1983a, 1983b). Yet it is clear that there was a need for retaining "long-term memories" of a social "shared history" for Paleolithic bands to survive as a biological entity, as they were mostly living in small and dispersed groups, too small to sustain a viable mating network. The discussion of the mechanisms employed during the Paleolithic is beyond the scope of this chapter. Here we discuss the archaeological evidence of "shared history construction" and long-term memories from the Epipaleolithic cultures as documented by the material record in the southern Levant, preceding the onset of the Neolithic and the shift to a "delayed-return economy," that is, farming.

EPIPALEOLITHIC LONG-TERM MEMORY AND THE COMMUNITY

Without going back too far in time, a case in point is the archaeological evidence predating the Levantine Neolithic by at least 12,500 years, at the beginning of the Epipaleolithic (ca. 24,000 cal BP). The presently available evidence indicates that already, with the beginning of the early Epipaleolithic in the region, a significant shift in adaptations and settlement patterns may be identified, with the emergence of pan-regional large-scale aggregation sites in the Azraq basin, eastern Jordan (Byrd, Garrard, and Brandy 2016; Jones et al. 2016; Maher et al. 2016; Martin, Edwards, and Garrard 2010; Richter et al. 2011, 2013). These localities would have facilitated the disparate bands from far and wide, comprising a particular mating network to congregate on a seasonal basis during times of plenty for social purposes, as reflected by the evidence for

FIGURE 3.1. *Section of Locus 15 at Early Epipaleolithic Ohalo II. Note superimposed occupation layers of darker organic material. Courtesy, D. Nadel.*

feasting and exchange of desirable commodities—for example, marine mollusks from both the Mediterranean and Red Seas—not to mention mates (Belfer-Cohen and Goring-Morris 2011). Indeed, Kharaneh IV and Jilat 6 appear to have been founded as a consequence of the exceptional convergence of particularistic local environmental conditions with the onset of the Last Glacial Maximum (LGM) and the large-scale herd behaviors of ungulates in the flattish, open steppic terrain and plains of eastern Transjordan and Syria. This unique combination of factors—cooler conditions and decreased evaporation—during the Early and Middle Epipaleolithic (ca. 24,000–15,000 cal BP) provided ideal foraging conditions within an open steppe landscape interspersed with local ponding.

With regard to long-term intra-site patterning, the open-air Masraqan site of Ohalo II on the shore of the Sea of Galilee, dated to ca. 23,000–22,000 cal BP, displays the presence of repeated occupations and the construction of brush huts in the very same spot for decades, if not centuries—at least according to the extensive range of 14C dates available (Nadel, Carmi, and Segal 1995; Nadel et al. 2004). Here, returning to *precisely* the same locality and location had nothing to do with space limitations; this was an open-air site at the edge of the lake, with possibilities of shifting the camp and huts in either direction along the shore. The huts had been used, abandoned, and then remodeled and reused following a period of abandonment of unknown duration (figure 3.1). Yet the inner arrangement of activities within the respective huts was retained on each sequential floor. This is well illustrated by the recent analysis of Hut 1, with three successive floors and the intra-site patterning of activities, reflected by the chipped stone, faunal, and botanical remains (Snir, Nadel, and Weiss 2015). A similar series of superimposed hut floors was documented also at early Epipaleolithic Ein Gev I, on the other side of the Sea of Galilee, although no spatial analysis of activities within the *fond de cabane* there has been published to date (Bar-Yosef 1970; Stekelis and Bar-Yosef 1965).

Yet, it is with the emergence of the later Epipaleolithic Natufian complex, ca. 15,000–11,600 cal BP (i.e., a duration of 3,400 calendric years and thus representing at least 140 generations), that the retention of long-term memory, of spatial arrangements mundane and/or symbolic, is most evident in open-air semi-sedentary or sedentary settlements.

The open-air hamlet of Eynan (Ain Mallaha), in the upper Jordan Valley adjacent to Lake Hula, provides clear evidence for the long-lasting use of the very same location within the site for successive huts. The early Natufian Structure 131 architectural complex (including Structure 51—see figure 3.2) comprises multiple occupation phases (Perrot 1966; Perrot and Ladiray 1988; Valla 1988, 1991; see Haklay and Gopher 2015 for a recent reassessment of the structure). Here, concentrations of colored pebbles and stone tools on two successive floors separated by a sterile fill in the very same location within Structure 131 indicate the retention of memory as regards the specific placing of symbolic items (Valla 1984, fig. 11). It is difficult to evaluate whether this represents habituated, unconscious behavior, but a certain time lapse is indicated—the sterile fill—and the return to the very same spot and creating similar arrangements was clearly intentional.

Indeed, during the Natufian we have other lines of evidence for constructing long-term memory in domains clearly related to social/group identity and cohesion. The Natufian "identity," as a prehistoric cultural entity, is reflected by various shared themes, inherent not only in the lithics—the most common remains of its material culture—but also in more obviously symbolic domains. Thus, one can observe the use of the same motifs continuing from early through the late and final Natufian phases documented throughout the Levant, regardless of the specific ecological setting. Sometimes this takes the form of immobile site furniture, as with the lozenge-shaped engravings on upright limestone and siltstone slabs at Wadi Hammeh 27 (Edwards 2013, figs. 12:3, 12:4) or fragments of a massive basalt mortar, similar to an engraved limestone bowl at Eynan (Perrot 1966, fig. 4:23).

However, it is also possible to recognize and identify expressions of local, community "identity." This is reflected in the presence of distinctive ornaments, with each base camp displaying its own idiosyncratic and particularistic decorative/symbolic motifs (Belfer-Cohen 1988a; 1991a, 1991b). Although these

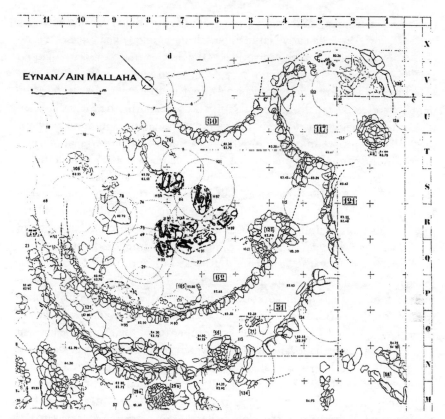

FIGURE 3.2. *Plan of the Loci 51, 62, and 131 architectural complex at Eynan. After Perrot 1966.*

ornaments sometimes appear sporadically in other sites, each locality retains its own distinctive "stamp" through the length of the occupation sequence at that site, sometimes through all of the 3,400-year sequence, as at Eynan (Belfer-Cohen and Goring-Morris 2013, fig. 2).

Community identity within the wider Natufian entity is clearly illustrated by the decorated early Natufian burials. Even the most lavish decorations comprise ornaments typical of the local community, differing from one site to another. An apt example is the comparison between the decorated burials in Hayonim Cave and those in el-Wad (none located more than 75 km from one another). Thus, the early Natufian decorated burials in Hayonim Cave display mostly necklaces, bone spatulas, armbands, and G-strings of bone pendants and fox canines, for example, the burials in Grave VII and Grave XIII

(Belfer-Cohen 1988a, 1988b, 1995; Goring-Morris and Belfer-Cohen 2013). By contrast, at el-Wad the early Natufian burials have unique headdresses but no decorations observed on the torso itself (Belfer-Cohen and Goring-Morris 2013; Garrod 1936–37, 1957; Wright 1978; Wyllie and Hole 2012).

Currently, all of the decorated burials, that is, those with beads and pendants, were recovered from early Natufian contexts. In the later Natufian phases they disappear, and the rare accompanying, seemingly funerary paraphernalia comprise unique items, the meaning of which was apparently known to the community members and is not evident to us. At late Natufian Eynan, of four primary burials placed within the same grave (Locus 10), two—found one on top of the other, though separated by a stone slab—have gazelle horn-core headdresses (Perrot and Ladiray 1988, fig. 32). This again illustrates intentionality of retaining local customs that mark a specific congregation through the mechanism of long-term memory construction. Unusual accoutrements accompany the "shaman's" burial at Hilazon Tachtit Cave—two marten skulls, an ox tail, a raptor's wing, a leopard's pelvis, an extra articulated human foot, and so on, in addition to more than seventy carefully broken tortoise carapaces (Grosman, Munro, and Belfer-Cohen 2008). These accoutrements clearly reflect a code or "language" in terms of their symbolic content, available to at least the local community, which could be used as evidence for the existence and long-term retention of group identity transferred through the construction of long-term, communal ritualistic memories.

During the late Natufian, a rise in the number of secondary burials is observed, while it appears that some sites became primarily graveyards, if not actual cemeteries. When the interred corpse was complete, it was tightly flexed, sometimes with hands crossed on the chest, most probably bundled in a sack, for example, Eynan Locus 20 (Perrot and Ladiray 1988, fig. 33). The most parsimonious explanation appears to be that the deceased were returned to their "home" sites and ancestral localities (see discussion in Goring-Morris and Belfer-Cohen 2013).

Another phenomenon, skull caching, practiced through to and including the Neolithic, also clearly indicates deep memory construction (e.g., Belfer-Cohen 1988a, 1988b, 1995; Goring-Morris and Belfer-Cohen 2013; Bienert 1991; Bonogofsky 2006; Kuijt 2008; Wright 1988). The practice first appears sporadically during the early Natufian but becomes more prevalent in the late/final Natufian; for example, of sixteen individuals in seven late Natufian graves at Hayonim Cave, only two retained their skulls (Belfer-Cohen 1988b). Sometimes the skulls were intentionally cached in pits; in burial grounds that also included primary burials, as at Eynan and Hayonim Cave; or elsewhere, by themselves, as in early Natufian Erq el-Ahmar beneath a slab pavement and Grave II at late Natufian Hayonim Cave (Belfer-Cohen 1988b; Neuville 1951). In the late Natufian layers of Eynan, a cache of eight skulls was placed together in Locus 9, mixed with insufficient postcranial elements for the individuals represented (Perrot and Ladiray 1988, fig. 31).

In the late/final Natufian cemetery on Nahal Oren terrace, the sometimes headless skeletons were often accompanied by breached mortars (also called "stone pipes"); these were interpreted by the excavator as a conduit for pouring libations for the deceased (Stekelis and Yisraely 1963).

In other instances skulls can be found on the surface of the occupation floors, seemingly intentionally placed. Thus, in Structure 131 at Eynan, a carefully cut cranium (*calotte*) had been placed on the floor close to the colored pebble concentration described above (Perrot and Ladiray 1988, Planche XVIII). At early Natufian Wadi Hammeh, twenty-seven fragmented and scattered skull fragments were found within the floor fill of both Structures 1 and 2 (Webb and Edwards 2002). In this instance, it is difficult to ascertain intentionality, but since there was no other human material together with the skull fragments, perhaps one can consider them a foundation deposit, associating the structure with particular ancestors. Since the practice of skull removal is evident from the Natufian all through the pre-Pottery Neolithic and even later, > 6,000 years, it is difficult to ascertain whether this custom retained its original significance and meaning. Still, it undoubtedly followed some specific social rules, since skull removal occurred only in certain instances while the majority of burials retained their heads all through this time period (e.g., Rollefson 2000).

Whether long-term memory is pivotal in the sense of the chronological relationships between burials and overlying structures in Natufian habitation sites remains open to debate, since the intensity of occupation and repeated rebuilding and remodeling of surfaces and features render the finer details of stratigraphy notoriously problematic. Here one can cite the different versions of the early Natufian stratigraphic sequence at Eynan of Structure (Abri) 1 and the burials associated with it (see the successive interpretations presented by Perrot and Ladiray 1988, fig. 9). Similar complex stratigraphic situations are also recognized elsewhere, for example, at el-Wad terrace. There, the early Natufian sequence began with an "ephemeral" initial occupation, followed by burials comprising a "cemetery," and culminating with a series of architectural features including a "terrace" wall and a structure, among others (see plans and photos in Yeshurun, Bar-Oz, and Weinstein-Evron 2014; Yeshurun et al. 2014). The association between the burials and the succeeding architectural remains is mute, and although it is tempting to assume that—just as at the chronologically later instance observed at Hilazon Tachtit Cave (see below)—the burials provided the "raison d'être" for the occupations that followed, there are clear instances indicating more complexity. Thus, for example, while there was an agglomeration of burials (Cemetery B) under the Structure 131 complex in Eynan, a posthole was driven through the burial of H.92, indicating ignorance of its whereabouts (Perrot and Ladiray 1988). The same observation can be made of H.33 (Grave XIII) in Hayonim Cave, which had its legs "cut off" by the wall of the structure (Locus 5) erected immediately above it (Belfer-Cohen 1991b).

Clearly, toward the end of the early Natufian within the Mediterranean zone, society was in the process of transforming into something else. While some early Natufian base camp sites of long duration were deserted, for example, Wadi Hammeh 27 (Edwards 2013), others served mainly as cemeteries mostly for secondary burials, indicating the maintenance of spatial traditions and the retention of long-term memory.

New, small-scale special sites (mostly graveyards) were founded during the late Natufian. They may be interpreted as territorial markers within the landscape, reflecting the preservation of long-term memories and ties to specific localities (Goring-Morris and Belfer-Cohen 2013). Examples include late/final Natufian Raqefet Cave and Nahal Oren terrace, both on Mount Carmel, as well as Hilazon Tachtit Cave, western Galilee (Grosman 2003; Grosman and Munro 2007; Nadel et al. 2013; Noy 1989, 1993; Stekelis and Yisraely 1963).[1] Indeed, the late/final Natufian occupation at the latter site encompassed an area of barely 30 m^2 and was not suitable for habitation, having no terrace (Grosman and Munro 2007). The elaborate "shaman's burial" in Structure A there has been interpreted as representing the earliest burial on-site, subsequently attracting additional, partial secondary burials and associated commemorative feasting (Grosman, Munro, and Belfer-Cohen 2008; Munro and Grosman 2010). Thus, one may consider the "shaman" burial as an indicator of a special meaning to a particular group through a prolonged period of time.

DISCUSSION

Long-term memory as a mechanism for group cohesion and alleviating social stress should have played a role much earlier than with the development and intensification of the delayed-return economy. Small-sized Paleolithic groups required such mechanisms to construct common beliefs and communal history so they could retain and reinforce mating networks with other groups to sustain a viable genetic pool. Indeed, one can consider the appearance of the large-scale aggregation sites in eastern Jordan at the beginning of the Epipaleolithic as a novel means of addressing the need to ensure the existence of long-term mating networks different from those of the Upper Paleolithic. It was the specific combination of the seasonal abundance of food and other resources in particular localities that "enabled" aggregation on a scale not previously possible. Whether this was maintained through habituated behavior as opposed to the conscious building of social memories and historical links to the past is difficult to assess for these early periods.

Still, we believe it was the shift toward sedentism during the late Epipaleolithic and the emergence of Natufian hamlets that introduced yet another, different motivation for the creation of long-term memories/history construction. Indeed, the effects of living in larger numbers for longer periods of time necessitated the generation of new social and ritual means to alleviate and relieve scalar stress. In this, the construction of community identity was paramount as a means for reinforcing the cohesion of kin and non-kin members of the sedentary groups, providing justification for land/resources ownership, and the like (for a more detailed discussion, see Belfer-Cohen and Goring-Morris 2013). Dominant among the various mechanisms that facilitated the construction of community identity was the intentional creation and retention of long-term memories by the group pertaining to places, events, and people. This shift preceded the processes related to delayed-return economy by several millennia.

Indeed, we encounter conscious efforts on the part of specific Natufian communities to construct and retain their individual identities. This is demonstrated in various aspects of their material culture but most clearly in their treatment of the dead. Within this context one can explain skull removal, often considered a particularly prevalent Neolithic practice, as reflecting remembrance and the honoring of chosen individuals (Kuijt 2008). Whether through time the skulls "lost" their individuality as persons and became a group symbol of "the ancestors" or "parent figures" at large merits separate discussion. In the context of the present chapter, suffice it to say that they clearly indicate a continuum, bridging the living members of the group with those who had passed away.

Certainly, many facets of Natufian "practice" and tradition related to treatment of the dead were shared with the pre-Pottery Neolithic B (PPNB), separated by ca. > 1,000 years. Developments during the intervening interval, namely the pre-Pottery Neolithic A (PPNA), are currently rather enigmatic and murky; but, in part at least, it seems likely that this reflects more the history of research than "facts on the ground" (though see, e.g., Makarewicz and Rose 2011). Does such shared "practice" necessarily reflect a common belief system/ideology spanning the Natufian through to the end of the Neolithic? Given the longevity of oral traditions in addition to "practice" in preliterate societies, especially as it relates to the sacred, we believe the (cautious) answer should be in the affirmative.

With the exceptions outlined above, the evidence for direct connections, and hence long-term memory, between preceding cemeteries and subsequent architectural features in (early) Natufian hamlets remains moot and open to debate. So, too, the reuse of previous architectural structures that had originally been dug into slopes—the "matryoshka" (Russian-doll) approach (terminology courtesy of Ian Hodder)—could simply reflect mundane issues concerning the "path of least" effort with regard to the ease of construction (see figure 3.2). Thus, one cannot be precise and point to a direct connection between specific instances of burials and structures. Still, a solid accumulation of data indicates evidence for long-term memory connecting the various intra-site features such as burials and architectural phases (see examples above). Moreover, the very fact that several Natufian hamlets demonstrate evidence for their early use as cemeteries is intriguing and appears to us to indicate long-term memory and the veneration of ancestors. The ways and means of such practices became both more elaborate and clearer with time, as evidenced in Aceramic Neolithic Aşıklı Höyük and later at Çatalhöyük; still, one can say with certainty that those behaviors were grounded in earlier traditions (Hodder 2012; Özbaşaran and Duru 2011). Obviously, the new ways of living in larger communities, which emphasized territorial "ownership" and rights of access, land tenure, and so on at both community and individual levels, were instrumental for long-term nurturing and retaining group and individual memories pertaining to the above.

In concluding, it is interesting to recall that more than fifty years ago Ruth Amiran noted the parallels between early Mesopotamian myths of the creation of man and PPNB plastered skulls and statues, spanning a broadly similar period of time as the one between the Natufian and the PPNB (Amiran 1962). Apparently, social groups in the throes of major changes in lifeways, such as those faced by the Neolithic communities and their immediate predecessors, required connections with the past to provide, among others, social stability; indeed, it may be that such upheavals and changes necessitated the retention of previous frames of reference to afford a psychological "security blanket," beyond the more mundane needs of personal and community land tenure and rights. Adherence to ancient beliefs and rituals during times of fundamental changes indicates intentionality by consciously retaining/constructing "evidence" for shared history and long-term memories.

ACKNOWLEDGMENTS

We are grateful for the invitation by Ian Hodder to participate in the session Religion, History, and Place in the Origins of Settled Life in the Middle East in the framework of the 2015 SAA annual meeting in San Francisco. We appreciate the valuable and constructive criticism from two anonymous readers; needless to say, we take full responsibility for the current chapter. Support for the research presented herein has been provided by generous grants to Goring-Morris from the Israel Science Foundation funded by the Israel Academy of Sciences and Humanities (Grants 840/01, 558/04, 755/07, and 1161/10), the National Geographic Society (Grant #8625/09), and the Irene Levi Sala CARE Foundation.

NOTE

1. Other sites, such as late/final Natufian Shukbah Cave in Samaria, are difficult to interpret given the field methodology used in its excavation (Garrod 1942; Weinstein-Evron 2003). The same may be said of the early Natufian occupation of Kebara cave (Turville-Petre 1932a, 1932b). Throughout the Natufian occupation at Hayonim, the cave was reserved for special activities and burials, as opposed to the occupation area on the terrace (Belfer-Cohen and Bar-Yosef 2012; Valla 2012).

REFERENCES

Alperson-Afil, Nira, Gonen Sharon, Mordechai Kislev, Yoel Melamed, Irit Zohar, Shoshana Ashkenazi, Rivka Rabinovich, Rivka Biton, Ella Werker, Gideon Hartman et al. 2009. "Spatial Organization of Hominin Activities at Gesher Benot Ya'aqov, Israel." *Science* 326 (5960): 1677–80. https://doi.org/10.1126/science.1180695.

Amiran, Ruth. 1962. "Myths of the Creation of Man and the Jericho Statues." *Bulletin of the American Schools of Oriental Research* 167 (167): 23–25. https://doi.org/10.2307/1355683.

Bar-Yosef, Ofer. 1970. "The Epi-Palaeolithic Cultures of Palestine." PhD dissertation, Hebrew University of Jerusalem.

Bar-Yosef, Ofer, and Liliane Meignen, eds. 2007. *Kebara Cave, Mt. Carmel, Israel: The Middle and Upper Paleolithic: Archaeology, Part I*. Cambridge, MA: Peabody Museum of Archaeology and Ethnology, Harvard University.

Belfer-Cohen, Anna. 1980. "The Aurignacian at Hayonim Cave." MA dissertation, Hebrew University of Jerusalem.

Belfer-Cohen, Anna. 1988a. "The Natufian Settlement at Hayonim Cave: A Hunter-Gatherer Band on the Threshold of Agriculture." PhD dissertation, Hebrew University of Jerusalem.

Belfer-Cohen, Anna. 1988b. "The Natufian Graveyard in Hayonim Cave." *Paléorient* 14 (2): 297–308. https://doi.org/10.3406/paleo.1988.4476.

Belfer-Cohen, Anna. 1991a. "Art Items from Layer B, Hayonim Cave: A Case Study of Art in a Natufian Context." In *The Natufian Culture in the Levant*, ed. Ofer Bar-Yosef and François R. Valla, 123–48. Ann Arbor, MI: International Monographs in Prehistory.

Belfer-Cohen, Anna. 1991b. "The Natufian in the Levant." *Annual Review of Anthropology* 20 (1): 167–86. https://doi.org/10.1146/annurev.an.20.100191.001123.

Belfer-Cohen, Anna. 1995. "Rethinking Social Stratification in the Natufian Culture: The Evidence from Burials." In *The Archaeology of Death in the Ancient Near East*, ed. Stuart Campbell and Anthony Green, 9–16. Edinburgh: Oxbow Monographs.

Belfer-Cohen, Anna, and Ofer Bar-Yosef. 1981. "The Aurignacian in Hayonim Cave." *Paléorient* 7 (2): 19–42. https://doi.org/10.3406/paleo.1981.4296.

Belfer-Cohen, Anna, and Ofer Bar-Yosef. 2012. "The Natufian in Hayonim Cave and the Natufian of the Terrace." In *Les Fouilles de la Terrasse d'Hayonim (Israel) 1980–1981 et 1985–1989*, ed. François R. Valla, 471–519. Paris: De Boccard.

Belfer-Cohen, Anna, and Adrian Nigel Goring-Morris. 2011. "Becoming Farmers: The Inside Story." *Current Anthropology* 52 (S4): S209–20. https://doi.org/10.1086/658861.

Belfer-Cohen, Anna, and Adrian Nigel Goring-Morris. 2013. "Breaking the Mold: Phases and Facies in the Natufian of the Mediterranean Zone." In *Natufian Foragers in the Levant: Terminal Pleistocene Social Changes in Western Asia*, ed. Ofer Bar-Yosef and François R. Valla, 543–61. Ann Arbor, MI: Monographs in Prehistory.

Bienert, Hans-Dieter. 1991. "Skull Cult in the Prehistoric Near East." *Journal of Prehistoric Religion* 5: 9–23.

Binford, Lewis R. 1983a. *In Pursuit of the Past: Decoding the Archaeological Record.* London: Thames and Hudson.

Binford, Lewis R. 1983b. *Working at Archaeology.* New York: Academic.

Bonogofsky, Michelle, ed. 2006. *Skull Collection, Modification, and Decoration.* Oxford: BAR International Series.

Byrd, Brian, Andrew N. Garrard, and Paul Brandy. 2016. "Modeling Foraging Ranges and Spatial Organization of Late Pleistocene Hunter-Gatherers in the Southern Levant—a Least-Cost GIS Approach." *Quaternary International* 396: 62–78. https://doi.org/10.1016/j.quaint.2015.07.048.

Edwards, Philip C., ed. 2013. *Wadi Hammeh 27, an Early Natufian Settlement at Pella in Jordan*. Leiden: Brill.

Garrod, Dorothy A.E. 1936–37. "Notes on Some Decorated Skeletons from the Mesolithic of Palestine." *Annual of the British School in Athens* 37: 123–27. https://doi.org/10.1017/S0068245400018037.

Garrod, Dorothy A.E. 1942. "Excavations at the Cave of Shukbah, Palestine, 1928." *Proceedings of the Prehistoric Society* 8: 1–20. https://doi.org/10.1017/S0079497X0002017X.

Garrod, Dorothy A.E. 1957. "The Natufian Culture: The Life and Economy of a Mesolithic People in the Near East." *Proceedings of the British Academy* 43: 211–27.

Gilead, Isaac, and Ofer Bar-Yosef. 1993. "Early Upper Paleolithic Sites in the Kadesh Barnea Area, Northeastern Sinai." *Journal of Field Archaeology* 20: 265–80.

Goring-Morris, Adrian Nigel, and Anna Belfer-Cohen. 2013. "Different Strokes for Different Folks: Near Eastern Neolithic Mortuary Practices in Perspective." In *Religion at Work in a Neolithic Society: Vital Matters*, ed. Ian Hodder, 35–57. Cambridge: Cambridge University Press. https://doi.org/10.1017/CBO9781107239043.004.

Grosman, Leore. 2003. "Preserving Cultural Traditions in a Period of Instability: The Late Natufian of the Hilly Mediterranean Zone." *Current Anthropology* 44 (4): 571–80. https://doi.org/10.1086/377650.

Grosman, Leore, and Natalie D. Munro. 2007. "'The Sacred and the Mundane: Domestic Activities at a Late Natufian Burial Site in the Levant before Farming." *Archaeology and Anthropology of Hunter-Gatherers* 4: 1–14.

Grosman, Leore, Natalie D. Munro, and Anna Belfer-Cohen. 2008. "A 12,000-Year-Old Shaman Burial from the Southern Levant (Israel)." *Proceedings of the National Academy of Sciences of the United States of America* 105 (46): 17665–69. https://doi.org/10.1073/pnas.0806030105.

Haklay, Gil, and Avi Gopher. 2015. "A New Look at Shelter 131/51 in the Natufian Site of Eynan (Ain-Mallaha), Israel." *PLoS One* 10 (7): e0130121. https://doi.org/10.1371/journal.pone.0130121.

Hodder, Ian. 2012. "Renewed Work at Catalhuyuk." In *The Neolithic in Turkey: New Excavations and Research*, ed. Mehmet Özdoğan, Nezih Basgelen, and Peter Kuniholm, 245–77. Istanbul: Archaeology and Art Publications.

Jones, Matthew D., Lisa A. Maher, Danielle A. Macdonald, Conor Ryan, Claire Rambeau, Stuart Black, and Tobias Richter. 2016. "The Environmental Setting of Epipalaeolithic Aggregation Site Kharaneh IV." *Quaternary International* 396: 95–104. https://doi.org/10.1016/j.quaint.2015.08.092.

Kuijt, Ian. 2008. "The Regeneration of Life: Neolithic Structures of Symbolic Remembering and Forgetting." *Current Anthropology* 49 (2): 171–97. https://doi.org/10.1086/526097.

Maher, Lisa A., Danielle A. Macdonald, Adam Allentuck, Louise Martin, Anna Spyrou, and Matthew D. Jones. 2016. "Occupying Wide Open Spaces? Late Pleistocene Hunter-Gatherer Activities in the Eastern Levant." *Quaternary International* 396: 79–94. https://doi.org/10.1016/j.quaint.2015.07.054.

Makarewicz, Cheryl A., and Katherine Rose. 2011. "Early Pre-Pottery Neolithic Settlement at el-Hemmeh: A Survey of the Architecture." *Neo-Lithics* 1 (11): 23–29.

Marks, Anthony E., ed. 1983. *Prehistory and Paleoenvironments in the Central Negev, Israel*, vol. 3: *The Avdat/Aqev Area, Part 3*. Dallas: Southern Methodist University Press.

Martin, Louise, Yvonne Edwards, and Andrew Garrard. 2010. "Hunting Practices at an Eastern Jordanian Epipalaeolithic Aggregation Site: The Case of Kharaneh IV." *Levant* 42 (2): 107–35. https://doi.org/10.1179/175638010X12797237885613.

Munro, Natalie D., and Leore Grosman. 2010. "Early Evidence (ca. 12,000 BP) for Feasting at a Burial Cave in Israel." *Proceedings of the National Academy of Sciences of the United States of America* 107 (35): 15362–66. https://doi.org/10.1073/pnas.1001809107.

Nadel, Dani, Israel Carmi, and Dror Segal. 1995. "Radiocarbon Dating of Ohalo II: Archaeological and Methodological Implications." *Journal of Archaeological Science* 22 (6): 811–22. https://doi.org/10.1016/0305-4403(95)90010-1.

Nadel, Dani, Avinoam Danin, Robert C. Power, Arlene M. Rosen, Fanny Bocquentin, Alexander Tsatskin, Danny Rosenberg, Reuven Yeshurun, Lior Weissbrod, Noemi R. Rebollo et al. 2013. "Earliest Floral Grave Lining from 13,700–11,700-y-Old Natufian Burials at Raqefet Cave, Mt. Carmel, Israel." *Proceedings of the National Academy of Sciences of the United States of America* 110 (29): 11774–78. https://doi.org/10.1073/pnas.1302277110.

Nadel, Dani, Ehud Weiss, Orit Simchoni, Alexander Tsatskin, Avinoam Danin, and Mordechai Kislev. 2004. "Stone Age Hut in Israel Yields World's Oldest Evidence of Bedding." *Proceedings of the National Academy of Sciences of the United States of America* 101 (17): 6821–26. https://doi.org/10.1073/pnas.0308557101.

Neuville, René. 1951. *Le Paléolithique et le Mésolithique du Desert de Judée*. Paris: Archives de l'institut de Paleontologie Humaine.

Noy, Tamar. 1989. "Some Aspects of Natufian Mortuary Behaviour at Nahal Oren." In *People and Culture in Change*, ed. Israel Hershkovitz, 53–57. BAR International Series 508(i). Oxford: British Archaeological Reports.

Noy, Tamar. 1993. "Oren, Nahal." In *The New Encyclopedia of Archaeological Excavations in the Holyland*, ed. Ephraim Stern, 1166–70. Jerusalem: Israel Exploration Society and Carta.

Özbaşaran, Mihriban, and Gunes Duru. 2011. "Asikli Hoyuk: Un Village d'Anatolie Centrale, il y à 10,000 Ans." *Archeologia* 489: 2–13.

Perrot, Jean. 1966. "Le Gisement Natoufien de Mallaha (Eynan), Israel." *L'Anthropologie* 70 (5–6): 437–83.

Perrot, Jean, and Daniel Ladiray. 1988. *Les Hommes de Mallaha (Eynan), Israel*. Paris: Paléorient Association.

Richter, Tobias, Andrew N. Garrard, Samantha Allock, and Lisa A. Maher. 2011. "Interaction before Agriculture: Exchanging Material and Sharing Knowledge in the Final Pleistocene Levant." *Cambridge Archaeological Journal* 21 (1): 95–114. https://doi.org/10.1017/S0959774311000060.

Richter, Tobias, Lisa A. Maher, Andrew N. Garrard, Kevan Edinborough, Matthew D. Jones, and Jay T. Stock. 2013. "Epipalaeolithic Settlement Dynamics in Southwest Asia: New Radiocarbon Evidence from the Azraq Basin." *Journal of Quaternary Science* 28 (5): 467–79. https://doi.org/10.1002/jqs.2629.

Rollefson, Gary. 2000. "Ritual and Social Structure at Neolithic 'Ain Ghazal." In *Life in Neolithic Farming Communities: Social Organization, Identity, and Differentiation*, ed. Ian Kuijt, 165–90. New York: Kluwer Academic/Plenum.

Snir, Ainit, Dani Nadel, and Ehud Weiss. 2015. "Plant-Food Preparation on Two Consecutive Floors at Upper Paleolithic Ohalo II, Israel." *Journal of Archaeological Science* 53: 61–71. https://doi.org/10.1016/j.jas.2014.09.023.

Stekelis, Moshe, and Ofer Bar-Yosef. 1965. "Un Habitat du Paléolithique Superieur a Ein Guev (Israel): Note Préliminaire." *L'Anthropologie* 69 (1–2): 176–83.

Stekelis, Moshe, and Tamar Yisraely. 1963. "Excavations at Nahal Oren: Preliminary Report." *Israel Exploration Journal* 13: 1–12.

Turville-Petre, Francis. 1932a. "Excavations at the Cave Mugharet-el-Kebarah, Near Zichron Jakob, Palestine." *Man* 32 (20): 15.

Turville-Petre, Francis. 1932b. "Excavations at the Mugharet El-Kebarah." *Journal of the Royal Anthropological Institute* 62: 270–76.

Valla, François R. 1984. *Les Industries de Silex de Mallaha (Eynan) et du Natoufien dans le Levant*. Paris: Association Paléorient.

Valla, François R. 1988. "Aspects du Sol de L'abri 131 de Mallaha (Eynan)." *Paléorient* 14 (2): 283–96. https://doi.org/10.3406/paleo.1988.4475.

Valla, François R. 1991. "Les Natoufiens de Mallaha et L'espace." In *The Natufian Culture in the Levant*, ed. Ofer Bar-Yosef and François R. Valla, 111–22. Ann Arbor, MI: International Monographs in Prehistory.

Valla, François R., ed. 2012. *Les Fouilles de la Terrasse d'Hayonim (Israël) 1980–1981 et 1985–1989*. Paris: De Boccard.

Webb, Steven G., and Phillip C. Edwards. 2002. "The Natufian Human Skeletal Remains from Wadi Hammeh 27 (Jordan)." *Paléorient* 28 (1): 103–23. https://doi.org/10.3406/paleo.2002.4741.

Weinstein-Evron, Mina. 2003. "In B or Not in B: A Reappraisal of the Natufian Burials at Shukbah Cave, Judaea, Palestine." *Antiquity* 77 (295): 96–101. https://doi.org/10.1017/S0003598X0006138X.

Wright, Gary A. 1978. "Social Differentiation in the Early Natufian." In *Social Archaeology: Beyond Subsistence and Dating*, ed. Charles L. Redman, Mary Jane Berman, Edward V. Curtin, William T. Langhorne, Nina M. Versaggi, and Jeffery C. Wanser, 201–33. New York: Academic.

Wright, George R.H. 1988. "The Severed Head in Earliest Neolithic Time." *Journal of Prehistoric Religion* 2: 51–56.

Wyllie, Cherra, and Frank Hole. 2012. "Personal Adornment in the Epi-Paleolithic of the Levant." In *Proceedings of the 7th International Congress on the Archaeology of the Ancient Near East, 2010*, vol. 3, ed. Roger Matthews and John Curtis, 707–17. Wiesbaden: Harrassowitz Verlag.

Yeshurun, Reuven, Guy Bar-Oz, Daniel Kaufman, and Mina Weinstein-Evron. 2014. "Purpose, Permanence, and Perception of 14,000-Year-Old Architecture: Contextual Taphonomy of Food Refuse." *Current Anthropology* 55 (5): 591–618. https://doi.org/10.1086/678275.

Yeshurun, Reuven, Guy Bar-Oz, and Mina Weinstein-Evron. 2014. "Intensification and Sedentism in the Terminal Pleistocene Natufian Sequence of el-Wad Terrace (Israel)." *Journal of Human Evolution* 70: 16–35. https://doi.org/10.1016/j.jhevol.2014.02.011.

4

Establishing Identities in the Proto-Neolithic

"History Making" at Göbekli Tepe from the Late Tenth Millennium cal BCE

Lee Clare, Oliver Dietrich, Jens Notroff, and Devrim Sönmez

Göbekli Tepe in southeast Turkey is a long recognized key site for the study of socio-ritual components of transitional Neolithic communities living in Upper Mesopotamia, a core zone of Neolithization, in the late tenth millennium cal BCE. In addition to the construction of the large monumental buildings with their T-shaped monoliths, these groups can be credited with early domestication activities involving wild plant and animal species, which from the mid-ninth millennium cal BCE began to show characteristic morphological changes associated with the emergence of identifiable domesticated forms. Ritual practices and belief systems identified at Göbekli Tepe provide unprecedented insights into the worldview of these "proto-Neolithic" communities at this important juncture in world history. Not only this, the site offers explanations as to how these groups could have overcome various challenges presented by "Neolithization" processes, including demographic growth, increasing competition over biotic and abiotic resources, and a more pronounced vertical social differentiation, with division of labor and craft specialization. In this contribution, it is posited that "history making" at Göbekli Tepe, as reflected, for example, through repetitive building activities at the site, could have been used to encourage group identity and to promote a sense of belonging to a common "cultic community," so important in the face of these challenges. Furthermore, it is proposed that these same "history making events" might also have been harnessed

DOI: 10.5876/9781607327370.c004

FIGURE 4.1. *Aerial view of Göbekli Tepe excavations. The southeast hollow (main excavation area) is clearly visible on the right-hand side from which an extension of several trenches branches off westward onto the southwest mound. Excavations on the northwest mound can be seen at the top left; southwest of these trenches are the (at this time) newly opened trenches in the northwest hollow. Photo by E. Küçük, 2011.*

by individuals and sub-groups in an attempt to legitimize social status and local, perhaps even regional, political influence.

GÖBEKLI TEPE

Göbekli Tepe lies 15 km east of the modern city of Şanlıurfa, on the northern fringe of the Harran plain. The mound has a maximum height of 15 m and a diameter of 300 m and covers an area of 9 hectares (figure 4.1). Göbekli Tepe accumulated over approximately 1,500 years (from the mid-tenth millennium cal BCE) on a prominent limestone plateau, 750 m above sea level and approximately 400 m above the southerly adjacent Harran plain. Two decades of archaeological research have revealed large numbers of architectural structures that have been assigned to the pre-Pottery Neolithic A (PPNA) and early/middle pre-Pottery Neolithic B (E/MPPNB; cf. Schmidt 2000, 51–53, 2006, 2011, 42) (figure 4.2). Among buildings erected in the earliest (PPNA)

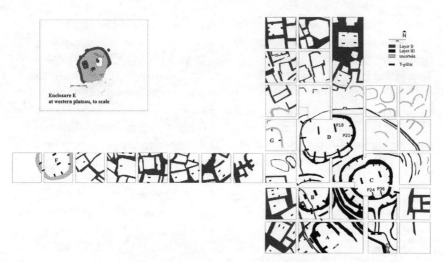

FIGURE 4.2. *Göbekli Tepe: schematic plan of excavations in the southeast hollow or main excavation area (right) and on the southwest mound (left) with the location of pillars (P18, P21, P24, P36) mentioned in the text. Image produced by K. Schmidt and J. Notroff, German Archaeological Institute (DAI).*

phase (Layer III) are nine predominantly round-oval structures, labeled A to H in the order of their discovery. These megalithic, presumably subterranean buildings, frequently referred to as "temples," typically comprise a circular-oval ground plan demarcated by a series of monolithic limestone T-shaped pillars set within encompassing limestone walls. At their center, the buildings feature two taller central T-pillars (up to 5.5 m in height). The T-shape is clearly discernible as an abstract representation of the human form, the enclosures apparently depicting a gathering of these anthropomorphic monoliths. Higher-lying deposits (Layer II) feature multiple rectangular buildings assigned to the EPPNB, some of which also contain T-pillars, though these are less "monumental" in their dimensions than their PPNA predecessors (heights up to 1.5 m). Finally, an uppermost layer of modern soil formation (Layer I) features mixed find materials from PPNA, PPNB, and later phases.

Tell accumulation at Göbekli Tepe was by no means homogeneous, as is testified by its asymmetrical shape, it being composed of a series of small mounds and connecting hollows. While excavations on the flanks and summits of the former have revealed youngest archaeological deposits, that is, rectangular buildings assigned to the EPPNB (Layer II), lower-lying hollows appear only to have accumulated in the PPNA phase (Layer III). This phenomenon was first observed in the southeast hollow where most of the last two decades of fieldwork has been concentrated, and in 2011 it was found repeated in the northwest hollow when round-oval structures with large T-pillars characteristic of the PPNA (Layer III) were revealed directly beneath the surface.

Until recently, a total of fifty-one radiocarbon ages provided the only points of reference for absolute dating at Göbekli Tepe (Dietrich 2011; Dietrich et al. 2013a; Dietrich and Schmidt 2010; Kromer and Schmidt 1998; Pustovoytov 2002, 2006; Pustovoytov, Schmidt, and Taubald 2007). However, no more than half of these ages conform to state-of-the-art knowledge of absolute chronology of the Upper Mesopotamian PPNA/EPPNB, with many samples producing aberrant younger ages. Doubtless, this outcome is related to inherent difficulties associated with absolute dating of the monumental structures at Göbekli Tepe, including the complex nature of archaeological contexts from which most samples have been collected, and the dating of unreliable apatite fractions from bone samples as a result of the poor preservation of collagen in the carbonate-rich sediments of the site. In addition, there is the comparative absence of well-preserved datable (especially short-lived) botanical remains. A radiocarbon dating project currently under way in the frame of the *Our Place: Our Place in the World* project (John Templeton Foundation) promises to provide new insights into the absolute chronology of the site and its PPNA megalithic monuments (Watkins and Schmidt 2012).

GÖBEKLI TEPE AND ÇATALHÖYÜK: A REVIEW

By way of introduction to the topic at hand, "history making" at Göbekli Tepe, we begin with a brief review of socio-ritual contexts at the site and this by comparison with the approximately 3,000 years younger (late Neolithic) settlement of Çatalhöyük East, which also features prominently in this volume (table 4.1).

The most significant difference between the two sites concerns layout and population. Çatalhöyük East features densely clustered contemporaneous houses separated by middens, suggestive of a clear continuation of preceding (PPNB) building traditions, for example, as encountered at Aşıklı Höyük in adjacent Cappadocia (Özbaşaran 2012). In contrast, monumental round-oval structures discovered at Göbekli Tepe appear clustered, though at present it is unclear whether these were coeval, partially coexistent, or even non-contemporaneous; however, it is unlikely that this uncertainty will be resolved by new high-resolution AMS absolute dates alone, and clarity in this matter can only be expected when dedicated stratigraphic studies are undertaken in still unexcavated areas between the different buildings.

TABLE 4.1. Socio-ritual contexts at Göbekli Tepe (PPNA) and Çatalhöyük East (LN).

	Layoutv	Population	Communal Structures	Ritual/Burial	Feasting
Göbekli Tepe (PPNA)	Clustered (?) "communal" buildings; no domestic buildings (?)	No permanent residents (?)	Multiphase "communal" buildings; ceremonial complex (?)	Intermural ritual installations; no burials	Yes, combined with communal building activities
Çatalhöyük East (LN)	Clustered domestic buildings	3,500–8,000 people*	None: domestic buildings only	Intermural ritual installations and burials	Yes, unknown contexts

* For details pertaining to reconstructed population at Çatalhöyük East, cf. Düring 2011, 118.

Further, it remains undetermined whether there was ever a permanent settlement at Göbekli Tepe. Nevertheless, occupation by an at least semi-permanent population should not be ruled out, especially considering the time required for the erection of the monumental buildings (see below) and in light of recent discoveries from a small area of the site that has yielded architecture and finds perhaps indicative of domestic activities. In this context, reference can also be made to a network of channels and cisterns carved into the surrounding limestone plateau (Herrmann and Schmidt 2012); although these cannot all be assigned to the PPN with absolute certainty, collection of precipitation runoff would have provided any population (whether permanent, semi-permanent, or visiting) with essential drinking water. Alternative water sources are otherwise unknown in close proximity to the site.

A second area in which the two sites differ pertains to the function of buildings. At Göbekli Tepe the round-oval structures with their monolithic T-pillars lack evidence for a domestic function; rather, they are interpreted as public, communal, or ritual spaces (Schmidt 2006). As such, the spatial agglomeration of these buildings could be suggestive of a regional ceremonial center (see figure 4.2). This line of interpretation appears to be confirmed by the monumental multiphase plan of the structures; their apparent longevity, which could surpass decades or even centuries; as well as the considerable investment of time and resources that would have been required for their construction (cf. Piesker 2014). In contrast, all buildings at Çatalhöyük East, regardless of the symbolism and ritual they contained, served as domestic houses, albeit that a number of more elaborate buildings at Çatalhöyük have been discerned. These buildings feature larger numbers of intermural burials and multiple rebuild phases that have been referred to as "history houses," domestic buildings in which Çatalhöyük people actively accumulated more extensive transcendent knowledge and symbolic capital (Hodder and Pels 2010, 164).

Ritual/communal buildings at Göbekli Tepe and the much later Çatalhöyük history houses would have been focal points for their communities (or parts thereof); in the case of Çatalhöyük East this "community" may have comprised several generations of a family group, their history house serving the accumulation of "historical memory," as expressed, for example, through the intermural burial of pots, tools, hunting trophies, and human bodies (Hodder and Pels 2010, 182–83). In contrast, at Göbekli Tepe it is suggested that buildings were erected by different (clan) groups who came to the site for this particular purpose. Yet here, too, buildings would have fulfilled a similar function, to preserve historical and mythical memories of these people. Further, associated building activities could have been harnessed by individuals and groups in an attempt to legitimize social status and local, perhaps even regional, political influence.

As we shall see, at the end of their use-lives the monumental buildings at Göbekli Tepe were intentionally buried or interred and their locations physically marked, though perhaps surprisingly and in contrast to the Çatalhöyük history houses, there is still no evidence for intermural human burials in the buildings. Sporadic finds of human bone are limited to fragments recovered from the backfill, though here we could be dealing with unintentional inclusions that had become inadvertently mixed with fill material (cf. Clare et al. n.d.).

At Göbekli Tepe it has been suggested that the interment of the monumental buildings was accompanied by feasting. In addition to gazelle, this included the consumption of large game such as aurochs and onager, as testified by extensive finds of animal bone in the burial fill (Peters and Schmidt 2004; Dietrich et al. 2012). The communal efforts necessary for the realization of such events (building-burial, hunting, and feasting) would have been immense, thus implying the cooperation of several different PPNA communities from the catchment of the site. Feasting at Çatalhöyük, though probably not undertaken in the context of such large-scale communal work, nevertheless shared features of earlier proto-Neolithic practices, especially the clearly defined consumption of large and symbolically charged animals. Be this as it may, the most significant aspect of feasting at both sites would have been the social interactions these events would have provided (Twiss 2012, 2008). In the following we take a closer look at the life cycle of monumental buildings at Göbekli Tepe. It is put forward that each of the structures progressed through three abstract stages of being: erection, modification, and death/burial. All three stages would have represented what are termed here "history-making events."

ENCLOSURE "LIFE CYCLES"

A total of three important stages can be distinguished in the life cycles of buildings, all of which would have brought together the extended community living in the catchment of the site: (1) erection, (2) modification(s), and (3) burial. These three stages likely resulted from more than one, possibly multiple temporally distinct dedicated gatherings by groups at the site.

Erection

Results from experimental studies on the time and labor required for the erection of a monumental building at Göbekli Tepe are still preliminary (cf. Notroff, Dietrich, and Schmidt 2014, 94–95; Beuger n.d.). Nevertheless, unprecedented insights into essential work strategies, techniques, and manpower can be gained from studying living societies that still use "traditional" methods to build megalithic structures. One illuminating series of ethnographic studies has focused on the erection of megalithic tombs in West Sumba, Indonesia (Adams 2010; Adams and Kusumawati 2010). West Sumbanese tombs are typically constructed from four large limestone blocks used for the walls, each about 1.5 m high, and a larger capstone for the roof. The distance from the quarry to the tomb owner's village, where the structures are erected, typically ranges from 500 m to 5,000 m; the lower end of this range corresponds to distances between the tell at Göbekli Tepe and the limestone quarry on the adjacent plateau. In West Sumba the limestone blocks are quarried by a crew of 5–10 men led by a stone-working specialist; subsequently, the blocks are hauled to the village using vine ropes over wooden rollers. A total of 300–1,000 people are needed to pull the largest slab (capstone), and the time required for transportation can vary from one day to one month, depending on the size of the stone and the distance covered. An additional month may be needed for final construction work and the carving of different designs on the stone exteriors (Adams and Kusumawati 2010).

If we assume that the scale of labor required for the construction of West Sumbanese tombs is comparable to efforts required for the construction of Göbekli Tepe enclosures, it follows that approximately 600 individuals were involved in dragging one of the larger (5.5-m-long and ~15 ton) central T-pillars from the quarry to the site, a feat scarcely realizable by one small community. Recently, we posited that Upper Mesopotamian PPNA settlements could have reached sizes of ~150 individuals, a number based on calculations combining horizontal extensions of PPNA sites with ethno-demographic data (Clare et al. n.d.). Provided that ~50 individuals from any one settlement were involved in T-pillar transportation, a cooperative of several communities would have been required to accomplish this task; smaller stones could have been moved by smaller groups. This conclusion is in good accordance with the hypothesis that the erection of an enclosure could have brought together numerous local groups from different territories in the catchment of the site. This conclusion is enhanced by the recent realization that, for example, the innermost of the three concentric dry-stone walls in Enclosure C was not erected piecemeal but following a "uniform" building concept, as indicated, for example, by the accentuation of particular parts of the structure. Unlike the erection of a single T-pillar, it is doubtful that such an ambitious building project was realized in the scope of a single "short-lived" feasting event (Piesker 2014, 23, 28); instead, this finding could imply a prolonged stay of a small building/quarrying crew. In this scenario, the existence of a (semi-)permanent settlement at Göbekli Tepe becomes an increasingly likely prospect.

Equally significant in the context of enclosure construction at Göbekli Tepe would have been the socio-political implications for the communities involved. Again, this aspect is extremely well documented in the West Sumbanese case study where relation building, power acquisition, as well as competition and solidarity within and between clan groups are played out against the backdrop of the construction process (Adams 2010; Adams and Kusumawati 2010). Provision of labor and victuals, including animals (pigs and water buffaloes) for small-scale and large-scale feasting events, is one of the most important factors, especially as a clan's political power and wealth are traditionally expressed in and through the organization of feasts; for example, feasts held by communities in the Torajan highlands of the Indonesian island of Sulawesi can involve the slaughter of large numbers of pigs and water buffaloes, as well as recreational cock and buffalo fighting (Adams and Kusumawati 2010).

Modification

Monumental enclosures erected at Göbekli Tepe were never what can be termed "completed." Rather, buildings might best be described as "dynamic," repeatedly modified by their constructors and possibly by their descendants. Different observations point to the "dynamicity" of buildings, including multiphase construction, the use and incorporation into enclosures of "recycled" T-pillars, and the reworking of pillar shapes and surfaces.

Multiphase building activities are best demonstrated by recent results from architectural investigations in Enclosure C, a building that witnessed a successive reduction in the size of its central enclosed area, as testified by its three temporally successive, concentric dry-stone walls (Piesker 2014, footnote 12) (see figure 4.2). Mention has been made of the innermost wall of this building that appears to have been erected after a "uniform" building concept. Intriguingly, a subsequent building phase can be observed in a section of this wall between Pillar 24 (P24) and Pillar 36 (P36), which appears to mark a former entrance to the center of the structure by way of a "dromos," a U-stone "portal" and stairway (Dietrich et al. 2013b, 2014; Piesker 2014, 31). The sealing of this entrance meant that access was no longer given at ground level, a scenario that prompts the search for alternative means of access.

Further evidence for the modification of buildings stems from the T-pillars themselves. In numerous cases these crucial architectural and monumental elements appear to have been "curated" from other structures. In fact, the reuse of pillars is attested in all investigated PPNA monumental buildings. One example is Pillar 21 (P21) from Enclosure D; the low reliefs carved on its south-facing broad side are partially hidden by the enclosing limestone wall (figure 4.3, left). In many cases the alternative explanation that the wall is a later addition to an already existent circular arrangement of free-standing T-pillars can be ruled out. With the exception of the two central monoliths, the T-pillars at Göbekli Tepe were never free-standing; they lack foundations and were erected either on a layer of soil or atop underlying courses of limestone laid on the plateau surface. This brought the added advantage that smaller pillars (and those broken at the shaft) could be artificially elevated. In addition, there is convincing evidence that narrow gaps were intentionally left in the walls; it was into these gaps that the pillars were heaved or lowered and wedged fast (cf. Piesker 2014, 23).

Another example for pillar reuse was recently discovered in Enclosure H in the northwest hollow (Dietrich et al. 2016) (figure 4.4). Initially, attention was drawn to Pillar 66 (P66) because of its anomalous orientation; instead of one of its front narrow sides, one of its broad sides faces toward the center of the enclosure (figure 4.5). Pillar 66 is not complete but broken at the shaft, thus indicative of its secondary usage. This monolith features a (30 cm) deep cavity at its center, just below the transition from head to shaft. Pillars of this type are also attested in Enclosures B and D. P66 stands on the inner stone bench of the building; its rear broad side is incorporated into the dry-stone wall. Two large worked rectangular limestone blocks placed on the stone bench and to the foot of the pillar draw the observer's gaze toward the cavity. The inner stone bench shows remnants of a mud plaster on its surface. A radiocarbon measurement made on a charcoal sample collected from this plaster has recently produced an age of 9280 ± 30 ^{14}C-BP (UGAMS-21060) (figure

FIGURE 4.3. *Two examples of T-pillars with indications of reuse and alteration. Left—Enclosure D, Pillar 21 (P21): the south-facing broad side of this 2.5-m-tall monolith carries depictions in low relief of (from top to bottom) a gazelle, an onager, and a quadruped. These reliefs are hidden to various extents by the innermost enclosing wall of Enclosure D (cf. Peters and Schmidt 2004, fig. 16). Image from German Archaeological Institute (DAI). Right—Enclosure C, Pillar 36 (P36): pick marks attest to the removal of reliefs from the upper part of its shaft, while depictions of wild boars and cranes on its lower part are preserved. Image from German Archaeological Institute (DAI).*

4.6). Two further dates made on charcoal from the fill of Enclosure H were extracted in close proximity to the stone bench. All three measurements have produced statistically comparable ages (table 4.2). At present, the first mentioned (plaster) date is the most reliable dating evidence for this particular building. The two remaining ages made on charcoal from its backfill attest to the incorporation of older organic remains into the rubble matrix; as such, they are not reliable age estimations for the abandonment and subsequent "interment" of the building, for which they can serve merely as "terminus post quem." In addition to P66, Enclosure H shows evidence for pillar extraction. A gap in the enclosure is situated north of Pillar 56 in Area K10-35 and testifies

FIGURE 4.4. *Enclosure H in the northwest hollow. Detailed plan (left) shows the extent of current excavations in this building and locations of T-pillars mentioned in the text (P51, P54, P56, P66). A gap (left) following the extraction of a pillar from the east side of the enclosing wall (right) is clearly discernible in Area K10-35; view from the north, looking south. Photo by N. Becker, German Archaeological Institute (DAI).*

to the removal of a pillar, perhaps prior to the final interment of the building at the end of its life cycle (figure 4.5, left and right).

The erasure (picking away) of earlier depictions is also a known indicator for building and pillar modification. It can be observed, for example, in Enclosure C where the east-facing broad side of Pillar 36 (P36) testifies to the removal of reliefs from the upper part of its shaft, while the depictions of wild boars and cranes on its lower part were not affected (see figure 4.3, right). It appears that the latter were covered and therefore "protected" by an installation (podest) erected directly adjacent to and east of this monolith (Piesker 2014, 30).

Burial

Although it is beyond any doubt that enclosures were intentionally backfilled at the close of their life cycles, it is open to debate how long they were in use prior to burial; certainly, based on aforementioned architectural observations in Enclosure C, it is not unrealistic that their "life spans" numbered several decades,

FIGURE 4.5. *Pillar 66 (P66; left) and Pillar 54 (P54; right) in Enclosure H. P66 is broken at the shaft and appears to have been "curated" from an earlier building. It stands on the inner stone bench of the enclosure, its broad side facing inward. A total of four animal engravings can be discerned: two aurochs and a bird on the head of the pillar and an unidentifiable quadruped on its shaft. The circular cavity does not perforate the monolith. Photo by N. Becker, German Archaeological Institute (DAI).*

possibly longer. Backfill material comprises tons of fist-sized angular limestone rubble. Unfortunately, it proves extremely difficult to detect any signs relating to deposition speed; however, sections through this material do suggest that it accumulated as a culmination of multiple and temporally distinct deposition episodes (figure 4.7). The origin of the backfill material is unknown, though it is likely that it is the detritus from quarrying activities on the adjacent plateau. Significantly, the rubble is mixed with considerable amounts of flint artifacts and faunal material and sporadic fragments of human bone; as mentioned, faunal remains, including aurochs and Asiatic wild ass but also smaller animals such as gazelle, could stem from large-scale feasts or acts of "theatrical social consumption" held to mark and commemorate the "interment/burial" of the respective building (Peters and Schmidt 2004; Peters, Arbuckle, and Pöllath 2014).

It is at this point that we recall that Göbekli Tepe is composed of a series of mounds and adjacent hollows. Excavations in the latter have so far revealed

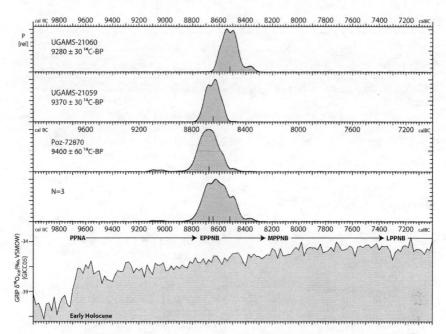

FIGURE 4.6. *Distribution of calibrated radiocarbon ages from Enclosure H at Göbekli Tepe (cf. table 4.2). All three ages correspond to the latest PPNA and earliest EPPNB, a possible indicator for the longevity of PPNA-type, round monumental buildings at the site. Bottom: Greenland ice core stable oxygen $\delta^{18}O$ record (GRIP) with GICC05 age model (Vinther et al. 2006) serves as proxy for North Atlantic air temperature.*

archaeological features assigned to the earliest phase of the site (Layer III), while characteristic rectangular buildings from the subsequent EPPNB (Layer II) are only found on the higher-lying slopes and on the summits of adjacent mounds. The hollows are in fact reminiscent of "amphitheaters" surrounded on three sides by clearly higher mound deposits and open on the fourth. In the case of the southeast hollow (main excavation area), this "amphitheater" opens to the SSE, giving views over the Harran plain. Intriguingly, this "amphitheater" arrangement is underlined by the presence of a dry-stone "terrace wall," 40 m in diameter, that physically separates the lower-lying area (with the buried enclosures) from younger rectangular structures standing on the higher-lying slopes of the surrounding mounds (Schmidt 2010). Therefore, EPPNB observers would have gazed down from this (artificially) higher ground onto the buried enclosures below; their sight was doubtlessly drawn to the earlier monuments by the protruding heads of some of their T-pillars, which may

TABLE 4.2. AMS-radiocarbon ages from Enclosure H in the northwest hollow

Lab Code	¹⁴C-Age [BP]	Material	δ¹³C[‰]	Context	CalAge (68%) [cal BCE]
UGAMS-21060	9280 ± 30	Charcoal (undet.)	−25.7	Area K10-25; Locus 68.1a; Enclosure H; sample from mud plaster (diameter ~30 cm) on stone bench adjacent to (and SW of) Pillar 54 (P54); elevation 773.27 m above NN	8520 ± 60
UGAMS-21059	9370 ± 30	charcoal (Pistacia)	−26.0	Area K10-25; Locus 62.7; Enclosure H; sample from backfill of enclosure, taken several cm south of the stone bench and 1 m SE of Pillar 66 (P66); elevation 773.09 m above NN	8650 ± 50
Poz-72870	9400 ± 60	charcoal (Pistacia)	-	Area K10-25; Locus 65.1; Enclosure H; sample from fill of Enclosure H, taken 1.30 m south of stone bench and 2 m SE of Pillar 66 (P66); elevation 773.08 m above NN	8680 ± 80

have appeared as a series of stone "benches" on the surface. Access to these "benches" was provided by a narrow stone staircase located at a central (northerly) position in the terrace wall (figure 4.8).

In a further turn of events, a large "robber pit" was dug into the burial fill of Enclosure C; this action is linked to the intentional destruction of its two large central pillars. None of the remaining pillars in this or in any of the other enclosures were targeted. This might suggest that (a) these two pillars were the only monoliths visible at the time or (b) all remaining (visible) pillars were deemed an "insufficient threat" to warrant destruction. It is still unclear when this act was carried out; although an EPPNB destruction date is possible, a much later intervention cannot be entirely ruled out, especially as sporadic pottery sherds were recovered from the upper backfill of the intrusive pit. In the case of the eastern pillar (P35), only the stump of the originally ~5-m-tall pillar remained intact; it was subsequently subjected to extreme burning, resulting in its now "flaky" appearance. The western central pillar (P37) was also demolished, though signs of burning are lacking.

FIGURE 4.7. *Excavations in Enclosure D showing the eastern central T-pillar (P18) still partially embedded within the characteristic rubble backfill used to bury the enclosures. Different layers (deposition events?) are visible in the western trench section, just behind P18. Image from German Archaeological Institute (DAI).*

Excavations under way since 2011 in Enclosure H in the northwest hollow have revealed further evidence of potential "post-burial" destruction within a buried enclosure (Dietrich et al. 2014). The eastern central pillar (P51) of Enclosure H carries the depiction of a large pouncing feline (leopard) on one of its broad sides; the monolith lies at a slight angle, and judging from its poor state of preservation its T-shaped head was exposed to the elements for some time. The second central pillar has not yet been discovered, though a large limestone block exposed in Area K10-24 could belong to it (see figure 4.4). Notably, the outline of a large pit can be clearly discerned in a section of balk (separating trenches K10-24 and K10-25). This pit could be linked to the intentional destruction of this second central pillar, thus mirroring the situation attested in Enclosure C in the southeast hollow.

DISCUSSION

If we stand back and take count of the above observations, what does it tell us about how history making was implemented and employed by Göbekli

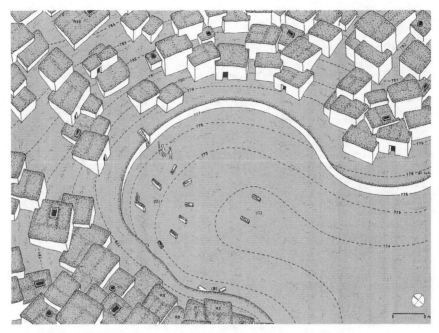

FIGURE 4.8. *Reconstruction of the southeast hollow (main excavation area). Source: Kurapkat 2015, fig. 248.*

Tepe people? Some general points of discussion can be noted:

1. Monumental buildings played a key role in history making at Göbekli Tepe. As a result of their three different stages of being (erection, modification, and burial)—all of which probably resulted from activities of multiple gatherings of groups over an undetermined time span—the buildings functioned as a physical record of historical and mythical memories; in the case of the latter, mention need only be made of some of the most richly adorned monoliths, which could feature scenes from founding myths. Although buried at the close of their respective life cycles, they remained omnipresent: the tallest pillars of interred enclosures were still visible long after this event, and areas in which buildings laid buried appear to have been physically demarcated, in the case of the southeast hollow by means of a terrace wall. The idea that at certain times the histories associated with these monuments were no longer desirable is implied by the documented destruction of central pillars in Enclosures C and H. In other words, attempts were made to actively destroy and eradicate earlier versions of historical and mythical memories.

2. Feasting events would have been central to bringing communities together for communal erection/modification/burial activities at Göbekli Tepe. Although archaeological evidence from the site only attests to feasts at the closure (interment) of buildings, it can be assumed that feasting would also have played an important role during erection and modification projects. Any provision of large quarry (e.g., aurochs) could imply communal hunting activities aimed at large and symbolically charged animals. Evidence from contemporaneous (PPNA) settlements in adjacent regions indicates that feasting events were probably commemorated through the display of skulls and bucrania from hunted animals, frequently in the context of domestic structures (cf. Dietrich et al. 2012). At Hallan Çemi, for example, a complete aurochs skull was found in a circular subterranean (domestic) structure, where it has been suggested that this might have once hung on the northern interior wall, facing the entrance (Rosenberg and Redding 2000, 45). Notably, symbolic elaboration of buildings using bucrania is still observed among megalithic tomb builders of West Sumba, Indonesia (Adams 2005). Here, corporate feasting activities are typically commemorated by the display of bucrania from slaughtered and consumed water buffalo at the front of large ancestral houses, an act interpreted as signaling corporate wealth and success of groups within the larger community; indeed, a parallel function of bucrania was previously postulated for Çatalhöyük East (Adams 2005). In contrast, Göbekli Tepe has so far failed to produce evidence for the display of commemorative bucrania. In contrast, bucrania carved into stone are encountered on several T-pillars at the site, most prominently as a pendant hanging from the "neck" of the western central T-pillar (P31) in Enclosure D and on the northwest-facing narrow side of the eastern central T-pillar (P2) in Enclosure A.

Keeping with the topic of communal hunting of symbolically charged animals, we must return briefly to the aforementioned Pillar 66 (P66) in Enclosure H (see figure 4.5); clearly, this was a special monolith in this structure, judging from its anomalous orientation and the fact that it was marked from above, perhaps subsequent to the backfilling of the enclosure, by a large worked stone with a deep cavity. Most significant, however, are the engravings adorning the inner-facing broad side of the pillar. They depict a scene with two aurochs (one larger, one small), their tongues hanging out and their legs buckled. Without a doubt, the engravings are telling the story of the deaths of these two beasts. Yet the most intriguing and exciting aspect of this depiction is its similarity to a wall painting discovered at Çatalhöyük East (Russell 2012, 79–82, fig. 2). Here, too, we see an aurochs in its final death throes. The similarities are certainly remarkable; and, as we see it, they must attest to continuity in this form of depiction and the act of bull baiting in Anatolia spanning from the PPNA to the late Neolithic.

The P66 engravings also imply that bull baiting was a highly ritualistic or sacred event and that it was doubtlessly related to high levels of arousal experienced by those individuals undertaking such dangerous pursuits. This brings us to the topic of "imagistic religions," which—in contrast to "doctrinal religions"—are characterized by low-frequency but highly arousing dysphoric rituals evincing strong emotions with an enduring impact (Atkinson and Whitehouse 2011). This impact is a prerequisite for the successful absorption of these acts into episodic human memory. Accordingly, rituals are frequently of a traumatic nature, including, for example, initiation cults and vision quests, typically involving extreme forms of deprivation, bodily mutilation and flagellation, and psychological trauma based around participation in shocking acts. These rituals promote not only the long-term rumination but also the mystical significance of the acts and artifacts involved, thus encouraging an intense cohesion among participating individuals or groups. Whether bull baiting can be included among these low-frequency and highly arousing dysphoric rituals at Göbekli Tepe remains a matter for discussion; however, if this were the case, it would imply the existence of socially distinct, "knowledgeable," and perhaps even "elite" groups for whom the erection, modification, and burial of monumental buildings were of greatest concern.

3. Based on insights from ethnographic studies of recent megalithic tomb-building communities in West Sumba, it is posited that the erection/modification/burial of a monumental structure at Göbekli Tepe would have been a time-consuming and expensive undertaking for the "handling community." The works were therefore likely used as a public demonstration of wealth and power. In the case of monument burial, one wonders whether this could have been some extremely elaborate form of "potlatch." On the other hand, we also know that any single PPNA settlement group (~150 people) would have been too small to physically erect some of the tallest and heaviest pillars. The provision of essential labor and victuals by neighboring settlements in the Göbekli Tepe catchment would have bound all participants in extensive networks of obligation and (delayed) reciprocity. The accomplished works would have served to commemorate these commitments. In this same context,

we should be aware that Göbekli Tepe people and groups from adjacent regions belonged to what has been termed a common "cultic community," with shared beliefs and traditions (Dietrich et al. 2012, 684). This is especially evident if we compare animal depictions discovered at Göbekli Tepe with those found applied to a whole range of objects from other Upper Mesopotamian PPNA sites, including shaft straighteners, stone plaquettes, the flat bones of large herbivores, as well as stone cups and bowls.

4. The incorporation of "curated" or "recycled" pillars into newly erected and modified buildings could reflect an entirely pragmatic decision by the "handling community" to save time, energy, and expense. More idealistic reasons are also conceivable. "Handling communities" or indeed "incipient elites" (see point 2) might have considered the monoliths "symbolically charged," their incorporation in new or modified buildings a historical link to the accomplishments of earlier generations. As such, these components could have been used to legitimize claims of ancestral descent, thus vindicating social status and local (perhaps even regional) political influence.

ACKNOWLEDGMENTS

The death of Klaus Schmidt in July 2014 was totally unexpected, not least for those who had been working with him so closely for so many years. In addition to the personal loss of a friend and colleague, there was the vacuum left behind by the passing of such a prominent figure in the academic field of early Neolithic research in Turkey. Not only this, there was also the future of research at Göbekli Tepe to be considered, a site that for twenty years had been the hub of Klaus's professional career. In spite of this loss—in fact, because of this loss—the Göbekli Tepe project continues its important work but now strengthened by new collaborations and the help of esteemed colleagues. Our unrelenting commitment to the site and to Klaus's legacy would not be possible without the valued support of the General Directorate of Cultural Assets and Museums, Ministry of Culture and Tourism of Turkey (Ankara), the Şanlıurfa Museum, and the German Research Foundation (DFG, Bonn). This chapter is a contribution of the John Templeton Foundation project Our Place: Our Place in the World.

REFERENCES

Adams, Ron L. 2005. "Ethnoarchaeology in Indonesia Illuminating the Ancient Past at Çatalhöyük." *American Antiquity* 70 (1): 181–88. https://doi.org/10.2307/40035277.

Adams, Ron L. 2010. "Megalithic Tombs, Power, and Social Relations in West Sumba, Indonesia." In *Monumental Questions: Prehistoric Megaliths, Mounds, and Enclosures*, ed. David Calado, Maxiliam Baldia, and Matthew Boulanger, 2123: 279–84. BAR International Series. Oxford: Archaeopress.

Adams, Ron L., and Ayu Kusumawati. 2010. "The Social Life of Tombs in West Sumba, Indonesia." *Archaeological Papers of the American Anthropological Association* 20 (1): 17–32. https://doi.org/10.1111/j.1551-8248.2011.01025.x.

Atkinson, Quentin D., and Harvey Whitehouse. 2011. "The Cultural Morphospace of Ritual Form: Examining Modes of Religiosity Cross-Culturally." *Evolution and Human Behavior* 32 (1): 50–62. https://doi.org/10.1016/j.evolhumbehav .2010.09.002.

Beuger, Claudia. n.d. "The Tools of the Stone Age Masons of Göbekli Tepe—an Experimental Approach." Unpublished manuscript.

Clare, Lee, Oliver Dietrich, Julia Gresky, Jens Notroff, Joris Peters, and Nadja Pöllath. n.d. "Early Neolithic Ritual and Conflict Mitigation at Körtik Tepe and Göbekli Tepe, Upper Mesopotamia: A Mimetic Theoretical Approach." In *Violence and the Sacred in the Neolithic: Girardian Conversations*, ed. Ian Hodder. Cambridge: Cambridge University Press.

Dietrich, Oliver. 2011. "Radiocarbon Dating the First Temples of Mankind." *Zeitschrift Für Orient-Archäologie* 4: 12–25.

Dietrich, Oliver, Manfred Heun, Jens Notroff, Klaus Schmidt, and Martin Zarnkow. 2012. "The Role of Cult and Feasting in the Emergence of Neolithic Communities: New Evidence from Göbekli Tepe, South-Eastern Turkey." *Antiquity* 86 (333): 674–95. https://doi.org/10.1017/S0003598X00047840.

Dietrich, Oliver, Çiğdem Köksal-Schmidt, Cihat Kürkçüoğlu, Jens Notroff, and Klaus Schmidt. 2013a. "Göbekli Tepe: A Stairway to the Circle of the Boars." *Acta Archaeologica* 5: 30–31.

Dietrich, Oliver, Çiğdem Köksal-Schmidt, Cihat Kürkçüoğlu, Jens Notroff, and Klaus Schmidt. 2014. "Göbekli Tepe: Preliminary Report on the 2012 and 2013 Excavation Seasons." *Neo-Lithics* 1 (14): 11–17.

Dietrich, Oliver, Çiğdem Köksal-Schmidt, Jens Notroff, and Klaus Schmidt. 2013b. "Establishing a Radiocarbon Sequence for Göbekli Tepe: State of Research and New Data." *Neo-Lithics* 1(13): 36–41.

Dietrich, Oliver, Jens Notroff, Lee Clare, Christian Hübner, Çiğdem Köksal-Schmidt, and Klaus Schmidt. 2016. "Göbekli Tepe, Anlage H: Ein Vorbericht Beim Ausgrabungsstand von 2014." *Der Anschnitt: Anatolian Metal VII* 31: 53–69.

Dietrich, Oliver, and Klaus Schmidt. 2010. "A Radiocarbon Date from the Wall Plaster of Enclosure D of Göbekli Tepe." *Neo-Lithics* 2 (10): 82–83.

Düring, Bleda. 2011. *The Prehistory of Asia Minor: From Complex Hunter-Gatherers to Early Urban Societies*. Cambridge: Cambridge University Press.

Herrmann, Richard A., and Klaus Schmidt. 2012. "Göbekli Tepe—Untersuchungen Zur Gewinnung Und Nutzung von Wasser Im Bereich Des Steinzeitlichen Bergheiligtums." In *Wasserwirtschaftliche Innovationen Im Archäologischen Kontext: Von Den Prähistorischen Anfängen Bis Zu Den Metropolen Der Antike*, ed. Florian Klimscha, 57–67. Rahden, Germany: Verlag Marie Leidorf.

Hodder, Ian, and Peter Pels. 2010. "History Houses: A New Interpretation of Architectural Elaboration at Çatalhöyük." In *Religion in the Emergence of Civilization: Çatalhöyük as a Case Study*, ed. Ian Hodder, 163–86. Cambridge: Cambridge University Press. https://doi.org/10.1017/CBO9780511761416.007.

Kromer, Bernd, and Klaus Schmidt. 1998. "Two Radiocarbon Dates from Göbekli Tepe, South Eastern Turkey." *Neo-Lithics* 3 (98): 8–9.

Kurapkat, Dietmar. 2015. *Frühneolithische Sondergebäude Auf Dem Göbekli Tepe in Obermesopotamien Und Vergleichbare Bauten in Vorderasien*. Berlin: Technische Universität Berlin.

Notroff, Jens, Oliver Dietrich, and Klaus Schmidt. 2014. "Building Monuments, Creating Communities: Early Monumental Architecture at Pre-Pottery Neolithic Göbekli Tepe." In *Approaching Monumentality in Archaeology: IEMA Proceedings*, vol. 3, ed. James F. Osborne, 83–105. Albany: State University of New York Press.

Özbaşaran, Mihriban. 2012. "Aşıklı Höyük." In *The Neolithic in Turkey: The Cradle of Civilization, New Discoveries*, ed. Mehmet Özdoğan, Nezih Başgelen, and Peter Kuniholm, 135–58. Istanbul: Arkeoloji ve Sanat Yayınları.

Peters, Joris, Benjamin S. Arbuckle, and Nadja Pöllath. 2014. "Subsistence and Beyond: Animals in Neolithic Anatolia." In *The Neolithic in Turkey*, vol. 6: *10500– 5200 BC: Environment, Settlement, Flora, Fauna, Dating, Symbols of Belief, with Views from North, South, East, and West*, ed. Mehmet Özdoğan, Nezih Başgelen, and Peter Kuniholm, 135–203. Istanbul: Archaeology and Art Publications.

Peters, Joris, and Klaus Schmidt. 2004. "Animals in the Symbolic World of the Pre-Pottery Neolithic Göbekli Tepe, South-Eastern Turkey: A Preliminary Assessment." *Anthropozoologica* 39 (1): 179–218.

Piesker, Katja. 2014. "Göbekli Tepe—Bauforschung in Den Anlagen C Und E in Den Jahren 2010–2012." *Zeitschrift Für Orient-Archäologie* 7: 14–54.

Pustovoytov, Konstantin. 2002. "14C Dating Pedogenic Carbonate Coatings on Wall Stones at Göbekli Tepe (Southeastern Turkey)." *Neo-Lithics* 2 (2): 3–4.

Pustovoytov, Konstantin. 2006. "Soils and Soil Sediments at Göbekli Tepe, Southeastern Turkey: A Preliminary Report." *Geoarchaeology: An International Journal* 21 (7): 699–719. https://doi.org/10.1002/gea.20134.

Pustovoytov, Konstantin, Klaus Schmidt, and Heinrich Taubald. 2007. "Evidence for Holocene Environmental Changes in the Northern Fertile Crescent Provided by Pedogenic Carbonate Coatings." *Quaternary Research* 67 (3): 315–27. https://doi.org/10.1016/j.yqres.2007.01.002.

Rosenberg, Michael, and Richard W. Redding. 2000. "Hallan Çemi and Early Village Organization in Eastern Anatolia." In *Life in Neolithic Farming Communities: Social Organization, Identity, and Differentiation*, ed. Ian Kuijt, 39–61. New York: Kluwer Academic/Plenum.

Russell, Nerissa. 2012. "Hunting Sacrifice at Neolithic Çatalhöyük." In *Sacred Killing: The Archaeology of Sacrifice in the Ancient Near East*, ed. Anne M. Porter and Glenn M. Schwartz, 79–95. Winona Lake, IN: Eisenbrauns.

Schmidt, Klaus. 2000. "Göbekli Tepe, Southeastern Turkey: A Preliminary Report on the 1995–1999 Excavations." *Paléorient* 26 (1): 45–54. https://doi.org/10.3406/paleo.2000.4697.

Schmidt, Klaus. 2006. *Sie Bauten Die Ersten Tempel: Das Rätselhafte Heiligtum Der Steinzeitjäger*. München: C. H. Beck.

Schmidt, Klaus. 2010. "Göbekli Tepe–Der Tell Als Erinnerungsort." In *Leben Auf Dem Tell Als Soziale Praxis: Beiträge Des Internationalen Symposiums in Berlin Vom 26–27, Februar 2007*, ed. Svend Hansen, 13–23. Bonn: Dr. Rudolf Habelt GmbH.

Schmidt, Klaus. 2011. "Göbekli Tepe." In *The Neolithic in Turkey*, vol. 2: *The Euphrates Basin*, ed. Mehmet Özdoğan, Nezih Başgelen, and Peter Kuniholm, 41–83. Istanbul: Archaeology and Art Publications.

Twiss, Kathryn. 2008. "Transformations in an Early Agricultural Society: Feasting in the Southern Levantine Pre-Pottery Neolithic." *Journal of Anthropological Archaeology* 27 (4): 418–42. https://doi.org/10.1016/j.jaa.2008.06.002.

Twiss, Kathryn. 2012. "The Archaeology of Food and Social Diversity." *Journal of Archaeological Research* 20 (4): 357–95. https://doi.org/10.1007/s10814-012-9058-5.

Vinther, Bo Møllesøe, Henrik Brink Clausen, Sigfus Johann Johnsen, Sune Olander Rasmussen, Katrine K. Andersen, Susanne L. Buchardt, Dorthe Dahl-Jensen, Inger K. Seierstad, Marie-Louise Siggaard-Andersen, Jørgen Peder Steffensen, et al. 2006. "A Synchronized Dating of Three Greenland Ice Cores throughout the Holocene." *Journal of Geophysical Research* 111 (D13): D13102. https://doi.org/10.1029/2005JD006921.

Watkins, Trevor, and Klaus Schmidt. 2012. "Our Place: Our Place in the World: Workshop in Urfa Initiates a Three-Year Research Project on Göbekli Tepe and Contemporary Settlements in the Region." *Neo-Lithics* 1 (12): 43–46.

5

Re-presenting the Past

Evidence from Daily Practices and Rituals at Körtik Tepe

Marion Benz, Kurt W. Alt, Yilmaz S. Erdal, Feridun S. Şahin, and Vecihi Özkaya

Near Eastern early Holocene communities were clearly deeply rooted in the Epipaleolithic traditions of their region. Their mode of building and subsistence closely resemble the Natufian of the Levant and the Younger Dryas groups of northern Mesopotamia, respectively (Bar-Yosef and Valla 2013; Belfer-Cohen and Goring-Morris 2013; Benz et al. 2015). However, there were also fundamental changes, which can be observed above all in social and ritual praxis (Hodder 1990; Cauvin 1997; Özdoğan and Özdoğan 1998; Stordeur 2010; Schmidt 2011; Özkaya and Coşkun 2011; Watkins 2012; Benz and Bauer 2013; Yartah 2013; Morenz 2014). These changes mirror new modes of approaching the environment, as well as new modes of communality and probably also of belief systems. Although these processes differ from northern Mesopotamia to the central and southern Levant in their material expressions, the structural processes seem rather similar (Benz 2012; see also Gebel 2017).

With the discovery of the monumental cult buildings at Göbekli Tepe, Barbara Bender's thesis, further developed by Brian Hayden, has gained interest. They suggest that it was power-hungry "aggrandizers" who initiated the intensification and, finally, the production of food to host people at feastings and to impress their fellows (Hayden 2014). As we will show, at Körtik Tepe there is no evidence for male dominance or sex-specific status. It was rather the need for new ethics to avoid conflict and mistrust in these larger permanent

DOI: 10.5876/9781607327370.c005

settlements, which required material expressions of collective memories, group identities, and social commitment (Sütterlin 2000, 2006; Bauer and Benz 2013). The production of resources thus seems to be an epiphenomenon of large, permanent, and territorially bound groups (Benz 2000).

Here, we will not tackle the much-debated question of whether the term *religion* is appropriate to describe the ritual behavior and belief systems of early Holocene hunter-gatherers. This debate strongly depends on either a definition based on cognitive capacities, including all homo sapiens, at least from the Upper Paleolithic (e.g., Boyer 1993; Barrett 2011; Anati 2013), or a more restrictive one, based on historical and material evidence for authority and dogma (e.g., Bloch 2008; Christensen 2010).

We consider the objects and remains of ritual activities from Körtik Tepe, which we present here, "as parts of processes of social interaction" (Morley 2007, xxii). The focus of our investigation includes the use and social impact of imagery and ritual behavior for the creation of—as Maurice Bloch (2008) defined it—the "transcendental social," that is, the imagined communal identity of a social entity.[1] The specific characteristics of symbolic action and depictions and the relationship of humans with these objects will allow us to surmise some theses on the social processes at the transition to sedentism and the formation of "essentialized roles and groups" (Bloch 2008, 2056). Whereas Maurice Godelier (2007) emphasized the importance of a territory for the creation of a society, Maurice Halbwachs highlighted the "mémoire collective" as a decisive means for the construction of community more than sixty-five years ago (Wetzel 2009). This "social memory," itself a constantly changing construction of the present, lends social entities the ideological basis that forges their past, present, and future in one coherent concept. We will outline that the archaeological and anthropological data from Körtik Tepe give evidence to the importance of both concepts: the commitment to a place and the "social memory." Evidence from burial rituals and engagement with preserved objects might indicate that keeping relationships alive beyond death and re-presenting past experiences gained in importance for personal or group identities.

We will argue that these people were on the threshold of new forms of communality. Face-to-face relationships were replaced by symbolic objects, thereby creating enhanced group identities and supra-regional communication networks as well as collective memories that transcended personal affiliations, time, and space. Since the excavations are ongoing and many final investigations still have to be completed, our results should be considered preliminary.

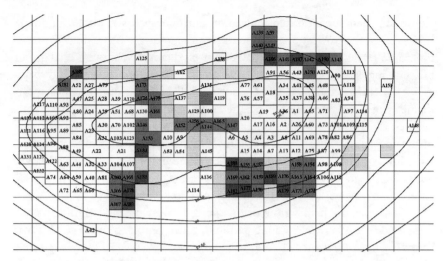

FIGURE 5.1. *Trenches of Körtik Tepe, southeastern Anatolia, 2015. Dark gray: ongoing excavation; midle gray: excavation finished; light gray: planned excavation. Source: Körtik Tepe Excavation Archive.*

THE SETTING

Körtik Tepe is a small (1.5 ha), low mound at the confluence of the Batman Çayı and the Tigris River in southeastern Anatolia (37°48'51.90" N, 40°59'02.02" E) (figure 5.1). It was identified as a prehistoric site during the rescue program preceding construction of the Ilısu Dam and has been under excavation since 2000 (Özkaya and Coşkun 2011, 90; Özkaya, Coşkun, and Soyukaya 2013).[2] Rich water reservoirs and streams from the Taurus Mountains join the Tigris in a wide and fertile floodplain from Diyarbakır to Batman Çayı. They provide, especially in early spring but also during the hot summer months, enough water for a dense vegetation cover. According to satellite images and our stratigraphic studies, both rivers, the Batman and the Tigris, were much closer to the site in earlier times than today. Fishhooks and net sinkers point to the importance of fish in the peoples' diet. The nearby mountain ranges further contributed to the wide spectrum of animal and plant species. This diversified, rich environment was probably one reason for the establishment of this permanent settlement at the latest from 10200 cal BCE to 9200 cal BCE (Benz et al. 2012, 2015; Rössner et al. 2017).

THE CREATION OF "OUR PLACE"

Evidence for permanent living at Körtik Tepe comes from several levels.

Above all, there are multiple renovation and reoccupation phases of single houses since the earliest occupation during the Younger Dryas into the early Holocene (Benz et al. 2012, 2015; Coşkun et al. 2012). Although changes in building materials can be observed over the course of time, the size and shape of the dwellings remained quite similar in all building units, with a mean diameter of about 2.50 m ± 50 cm. The floors of some houses were paved with pebbles. Only three larger buildings with a diameter of up to 3.80 m have been uncovered so far (Özkaya and Coşkun 2011). However, there is no special construction type, building materials, or special activity zones inside these larger buildings. Therefore, they cannot be identified as of a special type. All houses are round to oval and at least partly dug into the soil.

Moreover, many heavy groundstone objects, mortars, and grinding stones were used intensively over a long period until they broke or were worn down to a hole in the bottom. Unusable grinding stones and mortars were often reused as building materials (Schreiber et al. 2014).

Open-area activities, such as fireplaces and dumping areas, were also highly structured. As observed in Trench A91, fireplaces made of selected flat river pebbles were shifted slightly over time but were rebuilt repeatedly in the same area. Finally, a hearth with clay walls (oven?) was built in the same place (figure 5.2). Extensive, thick layers of animal bones in certain areas point to an organized dumping of food remains. For the moment, it must remain an open question whether these remains constitute the animal bones of large communal feastings or of continual organized dumping of food remains. Skulls, antler, and horn deposits in some graves (Özkaya and Coşkun 2011, fig. 6) and in open spaces, predominantly in the center of the settlement, suggest a ritual deposition of these animal body parts. They consist of a mixed assemblage of different species, mostly sheep, goat, aurochs, and deer.

The strongest evidence for the marking of a place that contributes to the ritual and communal identity of a group/household are the human burials placed beneath the living floors of the habitations. Since the earliest occupation at Körtik Tepe, burials were positioned beneath house floors. Because little surface has been uncovered for this period, it is not yet clear whether this tradition had been regularly practiced during the Younger Dryas. But beneath one house with several occupational phases, clearly dating to the second part of the Younger Dryas, a baby had been buried beneath the first floor of the third occupation phase. The multi-layered living floor above this burial proves that life continued above the deceased baby (Benz et al. 2015). Further excavations will show whether adults were also buried beneath house floors at this early time. At the latest, during the early Holocene this practice was definitely

FIGURE 5.2. *Built hearth of the early Holocene settlement in an open area where a sequence of shifting fireplaces, all lined with large, flat river pebbles, was previously located. Source: A. Willmy, Körtik Tepe Photo Archive.*

routine (Özkaya and Coşkun 2011). It is important to note that not all houses had burials beneath their floors, indicating social differentiation. Generally, the burials were oriented along the course of the house walls, but there was a clear dominance of the N-NE to S-SW orientation. Adult burials were more strictly oriented that way than were sub-adults (figure 5.3).[3] Most individuals were buried in a hocker position. Further evidence for local commitment will be given when we discuss the isotope analyses.

The demarcation of the place and households thus functioned on various levels (Sütterlin 2006): from daily subsistence activities of preparing food and cooking to the burial of the dead. The past was structuring the present, and the site was shaped on the one hand by communal action and on the other by smaller entities, probably households, by placing "ancestors" beneath house floors. But in contrast to the visible exposition of, for example, plastered skulls or the small earth mounds marking the place of burials at the Pre-Pottery B site of

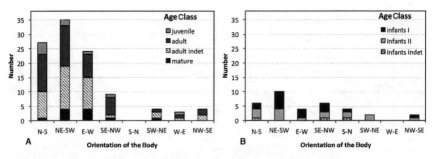

FIGURE 5.3. *Orientation of the individuals according to age class: (a) adults, (b) sub-adults.*

Aswad in Syria (Stordeur and Khawam 2007; Khawam 2014), at Körtik Tepe no traces of such a visible marking were uncovered. Given the very few overlapping or truncations of burials by later burials, we can surmise that the burial places were either marked by archaeologically invisible means or remembered by oral traditions. The houses at Körtik Tepe were thus places that related the past with the present for those who knew the invisible stories about them. Multi-layered floors and several renovation phases indicate a continuous building tradition. Micromorphological analyses will be necessary to specify the time depth and durations of occupations. Households were above all mental constructions and imagined unities of those who were aware of the past or who had been told about it. Further evidence for the mental creation of households was perhaps the small decorated chlorite sherds (see below), which were fabricated from the pieces of smashed stone vessels as souvenirs during burial rituals.

LOCAL COMMITMENTS AND DIETS

The strong commitment to place is reflected in the strontium and oxygen values, which range for strontium ($87Sr/86Sr$) between 0,70798‰ and 0,70848‰ ± 2SD and for oxygen between 14,79‰ and 17,12‰ ± 2SD (VSMOW) (strontium: n = 89; oxygen: n = 87; Benz et al. 2016). This corresponds to the local signal given by water and plant reference samples. Despite the fairly homogeneous geological situation around Körtik Tepe, reference samples from Hasankeyf, Bismil, and the Diyarbakir region show different signals. The people buried at Körtik Tepe thus probably also grew up in that area and were mobile only in a rather restricted territory with similar geological conditions.

The results of the isotope analyses on carbon and nitrogen point to a mixed diet of mostly C_3 plants and meat. Clear signals for freshwater resources are not reflected in the data, but further analyses might complement our results

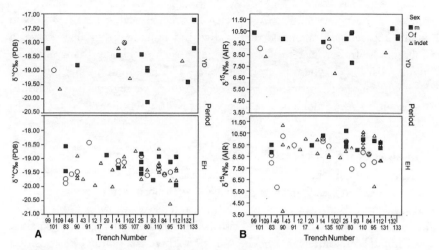

FIGURE 5.4. *Results of the carbon (a) and nitrogen (b) isotope analyses show no clusters of segregating groups. Variabilities within trenches can be as high as those between trenches. With a few exceptions, females have lower δ15N values than the contemporary male individuals. Trenches were sorted so that neighboring trenches were close to each other (from left to right on the x-axis): SE-NE-center-NW-SW. Period: EH = early Holocene, YD = Younger Dryas; sex: f = female, m = male, indet. = undetermined.*

(Dufour et al. 2007; Nehlich et al. 2010; Lillie, Budd, and Potekhina 2011). In light of the tools, such as hooks and net sinkers, as well as thousands of fish remains on the site, fish were definitely a significant contribution to the diet (Coşkun et al. 2010).

Interestingly, $δ^{13}C$ values of males of the Younger Dryas show a higher variance than those during the early Holocene, indicating the consumption of different plants including some C_4 plants. This might indicate a wider range of foraging area during the Younger Dryas or a reduction in species diversity during the early Holocene, as a result of either settlement activities or climatic changes.[4] The $δ^{13}C$ values of females and sub-adults also show a very low variance for the early Holocene occupation (figure 5.4). Whether this reduction in variance was a result of conscious specialization in selected resources, pointing to dietary traditions, or of ecological changes, depletion of plant variety near the settlement, or changing foraging territories remains an open question.

To conclude this section, there is no cluster of individuals with a similar dietary pattern in one household or area, except for a young woman and a child of undetermined age in Trench 84. The diet of these two individuals must have been very similar (figure 5.4). This means that, generally, access to food resources was open and diets varied individually rather than between households. In contrast to later settlements such as Halula (Saña, Tornero, and Molist 2014), where households were—at least in animal herding—distinctive economic entities, no such specializations or segregations of households within the settlement was observed at Körtik Tepe.

DIVERSITY OF BURIAL CUSTOMS

At this stage of the ongoing excavations at Körtik Tepe, it is not possible to present a statistically improved analysis of all burials (for an overview, see Özkaya and Coşkun 2011; Özkaya, Coşkun, and Soyukaya 2013). Generally, an increasing differentiation in burial rituals can be observed, even though most individuals were buried in a hocker position. There is a tendency to position the head of the dead body in the N or NE. This rule is most strongly visible for adults, whereas children can be buried in varying orientations. The orientation of the face and the side on which the dead body was laid vary strongly.

To compare the results of the isotope analyses with the burial customs, we clustered the 151 analyzed individuals for isotope studies according to their grave goods. Irrespective of different cluster methods, two clusters were clearly separated: burials without any grave goods (Type 1) and those with many grave goods (Type 4: at least one vessel, one other object, and more than fifty beads). Two further, not so clearly identified clusters were differentiated: those with only meat parts (animal bones) as grave goods (Type 2) and burials with some functional objects or fewer than fifty beads (Type 3). During the Younger Dryas there were almost no Type 4 burials, but their amount increased to 30 percent during the early Holocene. When comparing these four clusters with the results of our isotope study, no correlation is visible.

However, it is of interest that sub-adults (until age twelve) correlate with Type 4 burials, meaning that the number of "richly" buried children is significantly higher than that for adult burials, whereas mature individuals (> forty years) had almost no or only a few grave goods.

A special burial ritual that should be mentioned briefly existed at Körtik Tepe from the Younger Dryas onward into the early Holocene: some selected individuals were buried with a tortoise shell on or near their heads. Again, these special burials do not correlate with diet or with any of the four types of grave goods assemblages, that is, they can have many or very few burial goods (Coşkun et al. 2010). Individuals of all age classes were buried with tortoise shells, but the number of mature individuals within this burial type was significantly enhanced.

USE AND STANDARDIZATION OF SYMBOLS

As demonstrated above, there are clear social differentiations within the Körtik Tepe community and enhanced ritual traditions within some households. It is tempting to correlate these social differentiations with the possession and accumulation of prestigious material commodities, in particular with the artfully decorated stone vessels and ritual objects such as long chlorite pestles (*bâton poli*), horn-shaped stone objects, and mace heads (Özkaya and Coşkun 2011). However, we will show that modern concepts of wealth accumulation for prestige or status do not seem appropriate. Though the detailed analysis of these items awaits final contextual publication, some aspects of the use and style of the stone vessels and decorated chlorite items give important clues for ritual and symbolic behavior.[5]

The stone vessels were not only produced as grave goods but were highly estimated objects, as repairs and use traces show (figure 5.5). However, for the burial of some individuals, the vessels were deliberately destroyed by a strong stroke on their bottoms (Lichter 2007). The sherds were scattered over the dead body or placed beside it (figure 5.6). Though the content of the burials was sieved, many chlorite vessels were not complete, with one or two pieces missing. It might be suggested that these pieces were kept as souvenirs, possibly reworked, decorated, and circulated (s. Point 3).

The stone vessels comprise several different shapes, from flat convex bowls to slender high vessels, but at least two forms and decorations seem to be standardized: the convex high bowls (figure 5.7a) and slender bucket-shaped vessels (figure 5.7b).

The convex bowls are decorated with symbols of concentric circles/spirals from which at least four bundles of rays forming an X depart. Horizontal zigzag lines, sometimes in the form of snakes, connect the circles/spirals at the height of their center. On some of these vessels a pair of antithetic goats is also depicted (Coşkun et al. 2010, fig. 2a). This combination of motifs, with one exception, is restricted to this vessel type and has not been depicted on the bucket-shaped vessels or on very flat bowls. This specific type of decoration and vessel form has not yet been found at any other early Holocene site, although local adaptations were uncovered (Çelik 2017; Miyake 2013; Rosenberg 2011; Schmidt 2011; Yartah 2013). At Tell Qaramel and Tell 'Abr 3, both in northwestern Syria about 400 km from Körtik Tepe, isolated sherds with an identical decoration were uncovered, one of which has several holes on its border—indicating that it had either been repaired or that the shard was fixed by a string (figure 5.7c; Mazurowski and Kanjou 2012, plates 84.7–8, 85.11, figure 5.7 d–e; Yartah 2013, figs. 173, 34, 134; see also Schmidt 2011, fig. 4). Moreover, the isolated

FIGURE 5.5. *Repaired chlorite vessel. Source: Körtik Tepe Photo Archive.*

sun-like symbol with concentric circles and four bundles of rays was depicted on a bone object at Gusir Tepe (Karul 2013, 28, fig. 3), on two stone slabs from 'Abr 3 (Yartah 2013, figs. 124.1–2), and on a probably contemporary small stone object from Tell Qaramel (Mazurowski and Kanjou 2012, plate 68.4).

The bucket-shaped, more slender vessels with rather straight walls often bear decorations of an abstract bird and vertical bundles of zigzag lines, representing snakes or water, respectively. Sometimes they also have depictions of humans and scorpions, whereby the size of the humans never exceeds the sizes of the animals (Özkaya, Coşkun, and Soyukaya 2013, 61) (figure 5.7b). However, there is no strict standardization for this type of vessel. A similar bucket-shaped chlorite vessel was discovered at Hasankeyf Höyük that is almost plain, except for some geometric decoration on the rim and a very rough graffiti of a human figure on the vessel's wall (Miyake 2013, 45, fig. 1). Depictions of similar abstract birds are known from 'Abr 3 (Yartah 2013).

We can conclude that there was a high standardization of form and motif combinations, akin to what has been observed at other early Holocene sites

FIGURE 5.6. *In the burial of this female juvenile individual, nearly 3,000 beads, remains of two broken chlorite vessels, a broken stone ax, and two further stone tools were discovered. The body was covered with a layer of gypsum. Source: Körtik Tepe Photo Archive.*

(e.g., Köksal-Schmidt and Schmidt 2007; Mazurowski and Kanjou 2012, 76; Benz and Bauer 2013). The figurative style of animal depictions is also highly standardized. Despite this standardized combination of shape and motifs, there is no concentration of these objects in one household or one area at Körtik Tepe. This means that the use of these objects and symbols was not restricted to one special group or a specific context.[6] Symbolic traditions thus comprised the whole community. Although Körtik Tepe was included in a wide regional network of exchange (evidenced by obsidian and other exotic raw materials) and communication (evidenced by the use of a similar symbolic repertoire), it also had very specific symbols, restricted to Körtik Tepe.

The combination of motifs was known to other groups as far as the middle Euphrates, where they were imitated in a local style with other motifs added to the composition. Yet the lack of complete vessels of Type a at other northern Mesopotamian sites might suggest that only the fragments were imported as such, with high symbolic meaning recording the memory of their original provenance.

Further evidence for increasing conscious relations with distant places or past experiences is supported by the existence of small, decorated stone

FIGURE 5.7. *(a) Chlorite bowl with standardized "sun motif," snakes, and wild goats. Source: Körtik Tepe Photo Archive; (b) bucket-shaped vessel with bird, snake, and human depictions as well as concentric circles; (c) chlorite vessel with the goat motif, probably a local production, Tell 'Abr 3, Syria. Source: Yartah 2013, figs. 173.4–5, (d-e); at Tell 'Abr 3 (d) and Tell Qaramel (e), only shards of high-quality vessels with the sun-like motif have been found. Sources: Yartah 2013, figs. 34, 134, Tell Qaramel Photo Archive with photography by R. Mazurowski.*

platelets. For many years, all over the Fertile Crescent, from Shkârat Msaied in southern Jordan to Çayönü on the Upper Tigris (e.g., Harpelund 2011; Cauvin 1985), small, decorated stones with geometric designs have been found.

FIGURE 5.8. *Decorated chlorite platelets from Körtik Tepe. After Özkaya and Coşkun 2011, figs. 31–32. Source: Körtik Tepe Photo Archive.*

Figuratively decorated stone platelets and so-called shaft straighteners were first reported from the excavations at Jerf el Ahmar (Stordeur et al. 1996). Each of these stone platelets is unique concerning the combination of motifs, but the figures that are recombined are quite standardized. The same holds true for their form. Though none is identical to the others, most have an ovoid to rectangular shape and their size is rather homogeneous, measuring 4–6 cm × 3–6 cm (Benz and Bauer 2013, figs. 4–5; Köksal-Schmidt and Schmidt 2007; Mazurowski and Kanjou 2012, 68–83; Yartah 2013, 145, 150, 153, 165–70; for a comparison of similar items in central Turkey, see Baird et al. 2012, fig. 15; Özbaşaran 2012, fig. 18). Many of them have been found broken or damaged.

The decorated chlorite platelets from Körtik Tepe show the same concept but are different in style (figure 5.8). Their size varies considerably between 12 cm × 3 cm and 3.5 cm × 4 cm, and their decorations were most often made by high reliefs, not incisions. Except for three surface findings, they were discovered exclusively in graves (Özkaya, Coşkun, and Soyukaya 2013, 42–43). The motifs comprise horned animals, snakes, and a highly standardized insect, possibly a wasp.[7] Concentric circles were depicted either isolated or on the central part of the animals' bodies (Özkaya and Coşkun 2011, figs. 31–32; Özkaya and San 2004). Given the concave shape of some items, these objects were probably reworked fragments of chlorite vessels (Stordeur 2015, fig. 3.2), deliberately broken during burial rituals. Interestingly, the eleven items with the "insect" depiction were uncovered in two graves: six in Trench A19, Grave M7, and five in Trench A3, Grave M7, in the central area of the site (Özkaya and Coşkun 2011, fig. 31; Özkaya 2004, fig. 9; Özkaya and San 2004, fig. 4). An elongated chlorite object (10 cm × 7.7 cm) decorated with three aligned concentric circles each separated by a pair of round carved dots was found very close by (Özkaya and Coşkun 2011, fig. 21). This object, though smaller, is very similar to an object from Hasankeyf Höyük, about 30 km downstream on the Tigris from Körtik Tepe (Miyake 2013, 42). Further decorated stone objects, one with a four-legged horned animal (probably a deer, cf. Walter 2014, 67) and the other with two concentric circles around a dot, came from Burial M9, Trench A37. Another item with an unidentified quadruped was discovered in the same trench, in Burial M12. A further piece, found on the surface of the tell, shows a scorpion. This item links Körtik Tepe to Gusir Tepe, where a fragment of a nearly identical representation on a stone object was discovered (Karul 2013, 29). All decorated platelets of Körtik Tepe are from the most recent layers, dating to about 9300/9200 cal BCE (Benz et al. 2012).

So far, only one item has been found in the western area[8] and none in the eastern part. The decorated chlorite platelets thus seem to coincide with a rather restricted area in the center of the tell. An anthropological investigation comparing the buried individuals with stone platelets would be of major interest for future research. For the moment, it is not possible to say whether the individuals buried with the platelets had close familial ties or whether the "possession" of such an item indicated the membership of a group beyond familial or household affiliations. However, the reuse of chlorite sherds to produce highly symbolic items relates their owners to the past. In addition, the isolated sherds of Type a vessels from other faraway sites indicate that these sherds were highly appreciated and probably transmitted specific cultural messages over a wide region.

NEW MEDIALITY AND THE CREATION OF LOCAL AND SUPRA-REGIONAL RELATIONS—DISCUSSION

Evidence from Körtik Tepe clearly demonstrates that the inhabitants were involved in a wide exchange and communication network from northwestern Syria and Iraq to the Upper Tigris region (Özkaya and Coşkun 2011). As outlined above, in contrast to former times, symbols were increasingly carved on stone—either on small, easily transportable objects like the chlorite vessels and stone platelets or in a monumental style, as at Göbekli Tepe and Nevalı

Çori (Clare et al., chapter 4, this volume; Hauptmann 2011). This means that the transmission of encoded information became less bound to human presence and thus independent of time and—in the case of transportable objects—also of space (for a discussion of this change, see Renfrew and Scarre 2005). The archaeological data show that the strong standardization of decorations was emulated to the west of Körtik Tepe.

The precise content and meanings of these motifs remain speculative, since it is the strength of symbols that their interpretation allows variation according to the context and individuals involved (cf. Walter 2014; Morenz 2014). But it can be argued that the standardization and fixing of symbols in stone fostered the creation of symbolic traditions through their durable materiality—either through the built environment or the powerful imagery (Sütterlin 2000). Fixing symbols in stone meant an "objectification of memory" (Gell 1968, 255). Although we are not sure about the precise impact the symbols had in practice, they certainly influenced socialization and life to some degree, for example, by seeing a sherd with the depicted animal, not only the deceased was remembered but also the communally practiced ritual with its noisy clashing of vessels.

The emotional impact and the influence of material environments—especially of stone—on humans have been shown in many psychological and anthropological studies (e.g., Tilley 2004; Boivin 2010; Bauer 2013). With the transition to sedentary ways of life, the material environment to construct communality and social identities gained importance. However, the deliberate fragmentation of objects and their ritual deposition at Körtik Tepe show that material possessions clearly did not have the prime role they play today in the construction of identities. It was rather the commitment to the place and the ancestral affiliation within a household that were of importance for the enhancement of group identities.

This is not to say that mobile hunter-gatherer groups did not have strong commitments to certain places. But as far as we know, these were exceptional natural locations, occasionally turned into sacred places by mythology and rituals. In contrast, the people of Körtik Tepe and other northern Mesopotamian sites chose a certain natural place but turned it into a human-made environment and set the conventions for depictions of those environments.

With the transition to larger, permanent communities, face-to-face relationships could no longer be kept alive as intensively as in small-scale communities of about thirty–fifty people (Dunbar 2013). The potential for aggression and fear was thus increased and the sharing ideology of small-scale foraging communities, based on trust and intimate knowledge, was endangered (Benz and

Bauer 2013). Therefore, we can suggest that the enhanced use of symbols in architecture, rituals, and art became necessary to demonstrate traditional ethics, communal commitment, and group identities.

Although the burial customs suggest an increasing social differentiation of the sedentary hunter-fisher-gatherer community at Körtik Tepe, this is not mirrored in daily practices, architecture, or diets (Benz et al. 2016). Moreover, almost all grave goods except for the beads were deliberately destroyed. Even if sedentism afforded the accumulation of objects and even if objects might have become an index of prestige and wealth, the deliberate destruction of these objects annihilated concepts of inheritance and prestige through the accumulation of goods. It could also be argued that instead of the accumulation and inheritance of material wealth, it was the emotionally arousing rituals with the unmistakable noise of the clash of heavy tools on stone vessels, axes, and "mace heads" that created impressive moments of communal experiences, enhancing collective memories and relating survivors to the past and the community.

It thus seems premature to postulate a hierarchization of the Körtik Tepe community by applying our concepts of wealth and commodities to these early sedentary communities (Gebel 2010). Perhaps the long-term commitment to a place included a high number of "ancestors" beneath the house, contributing to the prestige of its inhabitants (Brami 2014, 219–20). As a hypothesis for further investigations into the socioeconomic behavior of the people of Körtik Tepe, we can suggest that social differentiation was not based on restricted access to resources or special objects but rather on mental concepts and "social memories" long before the invention of writing (Assmann 2002).

CONCLUSION

Humans are born to learn. Their cognitive capacities to imitate and store knowledge of past experiences are extraordinary (Watkins 2012; Bauer and Benz 2013). These two capacities are of high relevance for the interpretation of history making and tradition building in prehistoric contexts. Mimetic processes, that is, the incorporation and imitation of behaviors of others, are never blind copying but rather are creative acts of recombining imitation with personal abilities and ideas (Wulf 2005). The storage of knowledge—even if unconscious—determines future decisions and behaviors.

Both capacities lead to different forms of tradition building. Mimetic processes are based on observation, but tradition building based on stored knowledge depends on recalling past experiences. Mimetic processes can lead to more or less strong local traditions according to the degree of exchange with others. The strong commitment to place at Körtik Tepe has been shown in several domains. Whereas the repeated restoration of floors might be interpreted as some kind of mimetic process far from conscious history making, burying the dead beneath the floor and continuing life above the "ancestors" was a conscious choice. The people of Körtik Tepe thus created strong relationships with former group members beyond death, actively incorporating the past into the present. Although the physical presence of ancestors was less intense than with the circulated skulls typical of the early Holocene cultures of the Levant (e.g., Stordeur and Khawam 2007; Benz 2010; Khawam 2014; Kodas 2015), knowledge about sub-floor burials was probably passed down orally for generations.

Commemoration of the dead was possibly enhanced by keeping memorial items from burial rituals, such as chlorite sherds from deliberately destroyed vessels, which were reworked as symbolic objects. Although the possibility cannot be excluded that these items were found by chance during later activities, the fact that many vessels discovered in graves were lacking only a few sherds and that burials were generally untouched before their excavation is inconsistent with this argument. The isolated sherds with decorations typical of Körtik Tepe, which were discovered about 400 km to the southwest, indicate that these sherds still maintained a high symbolic value, possibly preserving some information of distant places and experiences.

To conclude, the built environment and the enhanced use of symbols in stone during the early Holocene contributed to an objectification and permanence of memories and concepts. The material presence had an impact on mimetic processes and on the socialization of future generations. In addition, personal relationships were consciously prolonged beyond death by integrating the dead into the house. The highly arousing burial rituals probably had a strong impact on collective memory (Whitehouse 2000); moreover, knowledge about these past events was recalled by the physical presence of symbolic items.

Thinking the absent present was not only a cognitive prerequisite for the "transcendental social" in general and for religion in particular (Bloch 2008), but it was probably one of the decisive necessities for avoiding conflict in larger communities (Dunbar 2013; Benz and Bauer 2013). It can be suggested that the stone objects and the sub-floor burials, symbolizing affiliations with the past, turned the absence of face-to-face relationships into a mental and material presence, thereby enhancing social commitment and networks.

At Körtik Tepe, ritual and daily practices were deeply interwoven, but later, with the monumental buildings at other sites in the Upper Tigris and Euphrates region, the segregation of communal buildings from domestic houses reached its climax during the late pre-Pottery Neolithic A (PPNA) and PPNB (see Moetz 2014 and references therein).

We suggest that the people of Körtik Tepe lived on the threshold of new ethics of communality and engagement with their place and the past. However, it seems that these new developments were in contrast with traditional ideas of being in the world of hunter-gatherer communities (Willerslev 2007; Guenther 2010; Theweleit 2013; Widlok 2013). These contradictions led to highly ambivalent practices between strong affection for material objects and their deliberate destruction, between accumulation and deposition, between emerging differentiation and equality. The massive destruction of valued objects reminds one of potluck types of feasts in highly competitive trans-egalitarian communities (Hayden 2014, 174), but it also shows that the relationship with objects should be finite, which is in strong contrast to concepts of heirlooms. Mimetic processes as well as the creation and symbolic re-presentation of strong collective memories thus enhanced relations with the past and advanced—despite the integration into a wide exchange and communication network—increasing local group identities.

ACKNOWLEDGMENTS

We are thankful to Ian Hodder for the invitation to contribute to this volume. The German team is grateful to the German Research Foundation for substantial financial support (BE 4218/2-1, 2-2; AL 287/9-1, 9-2). Additional funding was graciously contributed by the Rutzen-Stiftung and the Templeton Foundation. Our sincere thanks to Vecihi Özkaya for his permission to participate in his excavations. We are also grateful to Mirjam Scheeres, Marc Fecher, and Corina Knipper for their cooperation in the isotope analyses and to Demet Delibaş for her support in the excavation of human remains and age and sex determinations. We cordially thank Pauline H. King for her thorough language editing. The Wissenschaftliche Gesellschaft Freiburg im Breisgau granted financial support for the participation in the SAA Conference, for which MB is very much indebted.

NOTES

1. It is out of the scope of this chapter to give any interpretation of the content of the figurative art (cf. Morenz 2014; Walter 2014).

2. Although Körtik Tepe is listed by Algaze et al. (1991, fig. 2b, 63), it was not identified as an Aceramic site.

3. This trend is visible in the 151 investigated individuals for isotope studies. Only the final analyses of all burials will be able to verify this preliminary result.

4. A detailed discussion on these effects will be given in the final interpretation of stable isotopes from Körtik Tepe.

5. The horn-shaped stone objects and pestles can be considered typical items of early Holocene communities in southeastern Anatolia and northern Iraq. But despite their possible symbolic meaning, it seems premature to give any conclusive interpretation here without an in-depth analysis of their contexts and without analyses of stone provenance, fabrication, and use traces.

6. A more in-depth study of the context of these vessels has to await the final publication.

7. Formerly, this figure was identified as a "bee-like" insect (Özkaya, Coşkun, and Soyukaya 2013, 42), but recently Walter (2014, 70–74) argued that it might be a wasp.

8. Two further pieces were found on the surface of the western part and cannot be considered in situ.

REFERENCES

Algaze, Guillermo, Ray Breuniger, Chris H. Lightfoot, and Michael Rosenberg. 1991. "The Tigris-Euphrates Archaeological Reconnaissance Project: A Preliminary Report of the 1989–1990 Seasons." *Anatolica* 17: 173–211.

Anati, Emanuel. 2013. "On Palaeolithic Religion." In *The Handbook of Religions in Ancient Europe*, ed. Lisbeth B. Christensen, Olav Hammer, and David A. Warburton, 36–44. Durham, NC: Acumen.

Assmann, Jan. 2002. *Das kulturelle Gedächtnis: Schrift, Erinnerung und politische Identität in frühen Hochkulturen*, 4th ed. Munich: Beck.

Baird, Douglas, Andrew Fairbairn, Luise Martin, and Caroline Middleton. 2012. "The Boncuklu Project." In *The Neolithic in Turkey: New Excavations and New Research 3: Central Turkey*, ed. Mehmet Özdoğan, Nezih Başgelen, and Peter Kuniholm, 219–44. Istanbul: Archaeology and Art Publications.

Barrett, Justin L. 2011. "Metarepresentation, *Homo religiosus*, and *Homo symbolicus*." In *Homo Symbolicus: The Dawn of Language, Imagination, and Spirituality*, ed. Christopher S. Henshilwood and Francesco d'Errico, 205–24. Amsterdam: John Benjamins. https://doi.org/10.1075/z.168.11bar.

Bar-Yosef, Ofer, and François R. Valla. 2013. *Natufian Foragers in the Levant: Terminal Pleistocene Social Changes in Western Asia*. Ann Arbor, MI: International Monographs in Prehistory.

Bauer, Joachim. 2013. *Das Gedächtnis des Körpers: Wie Beziehungen und Lebensstile unsere Gene steuern*. Munich: Piper.

Bauer, Joachim, and Marion Benz. 2013. "Epilogue: Archaeology Meets Neurobiology; the Social Challenges of the Neolithic Process." *Neo-Lithics* 13 (2): 65–69.

Belfer-Cohen, Anna, and Adrian Nigel Goring-Morris. 2013. "Breaking the Mold: Phases and Facies in the Natufian of the Mediterranean Zone." In *Natufian Foragers in the Levant: Terminal Pleistocene Social Changes in Western Asia*, ed. Ofer Bar-Yosef and François R. Valla, 543–61. Ann Arbor, MI: Monographs in Prehistory.

Bender, Barbara. 1978. "Gatherer-Hunter to Farmer: A Social Perspective." *World Archaeology* 10 (2): 204–22.

Benz, Marion. 2000. *Die Neolithisierung im Vorderen Orient: Theorien, archäologische Daten und ein ethnologisches Modell*. Berlin: Ex oriente.

Benz, Marion. 2010. "Beyond Death—the Construction of Social Identities at the Transition from Foraging to Farming." In *The Principle of Sharing—Segregation and Construction of Social Identities at the Transition from Foraging to Farming*, ed. Marion Benz, 249–75. Berlin: Ex oriente.

Benz, Marion. 2012. "'Little Poor Babies'—Creation of History through Death at the Transition from Foraging to Farming." In *Beyond Elites: Alternatives to Hierarchical Systems in Modelling Social Formations*, ed. Tobias L. Kienlin and Andreas Zimmermann, 169–82. Bonn: Rudolf Habelt.

Benz, Marion, and Joachim Bauer. 2013. "Symbols of Power—Symbols of Crisis? A Psycho-Social Approach to Early Neolithic Symbol Systems." *Neo-Lithics* 13 (2): 11–24.

Benz, Marion, Aytaç Coşkun, Irka Hajdas, Katleen Deckers, Simone Riehl, Kurt W. Alt, Bernhard Weninger, and Vecihi Özkaya. 2012. "Methodological Implications of New Radiocarbon Dates from the Early Holocene Site of Körtik Tepe, Southeast Anatolia." *Radiocarbon* 54 (3–4): 291–304. https://doi.org/10.1017/S00338 22200047081.

Benz, Marion, Katleen Deckers, Corinna Rössner, Alexander Alexandrovskiy, Konstantin Pustovoytov, Mirjam Scheeres, Marc Fecher, Aytaç Coşkun, Simone Riehl, Kurt W. Alt et al. 2015. "Prelude to Village Life: Environmental Data and Building Traditions of the Epipalaeolithic Settlement at Körtik Tepe, Southeastern Turkey." *Paléorient* 41 (2): 9–30.

Benz, Marion, Yilmaz S. Erdal, Feridun S. Şahin, Vecihi Özkaya, and Kurt W. Alt. 2016. "The Equality of Inequality: Social Differentiation among the Hunter-Fisher-Gatherer Community of Körtik Tepe, Southeastern Turkey." In *Rich and Poor—Competing for Resources in Prehistory*, ed. Harald Meller, Hans Peter Hahn,

Reinhard Jung, and Roberto Rich, 147–64. Halle: Landesamt für Denkmalpflege und Archäologie Sachsen Anhalt—Landesmuseum für Vorgeschichte Halle (Saale).

Bloch, Maurice. 2008. "Why Religion Is Nothing Special but Is Central." *Philosophical Transactions of the Royal Society of London: Series B, Biological Sciences* 363 (1499): 2055–61. https://doi.org/10.1098/rstb.2008.0007.

Boivin, Nicole. 2010. *Material Cultures, Material Minds: The Impact of Things on Human Thought, Society, and Evolution*. Cambridge: Cambridge University Press.

Boyer, Pascal. 1993. "Cognitive Aspects of Religious Symbolism." In *Cognitive Aspects of Religious Symbolism*, ed. Pascal Boyer, 4–47. Cambridge: Cambridge University Press.

Brami, Maxime. 2014. "The Diffusion of Neolithic Practices from Anatolia to Europe: A Contextual Study of Residential and Construction Practices 8500–5500." PhD dissertation, University of Liverpool.

Cauvin, Jacques. 1985. "Le Néolithique de Cafer Höyük (Turquie): Bilan Provisoire Après Quatre Campagnes (1979–1983)." *Cahier de l'Euphrate* 4: 123–34.

Cauvin, Jacques. 1997. *Naissance des Divinités: Naissance de l'Agriculture*. Paris: CNRS Éditions.

Çelik, Bahattin. 2017. "A New Pre-Pottery Neolithic Site in Southeastern Turkey: Ayanlar Höyük (Gre Hut)." *Documenta Praehistorica* XLIV: 360–67.

Christensen, Lisbeth B. 2010. "From 'Spirituality' to 'Religion'—Ways of Sharing Knowledge of the 'Other World.'" In *The Principle of Sharing: Segregation and Construction of Social Identities at the Transition from Foraging to Farming*, ed. Marion Benz, 81–90. Berlin: Ex oriente.

Coşkun, Aytaç, Marion Benz, Yilmaz S. Erdal, Melis M. Koruyucu, Katleen Deckers, Simone Riehl, Angelina Siebert, Kurt W. Alt, and Vecihi Özkaya. 2010. "Living by the Water—Boon and Bane for the People of Körtik Tepe." *Neo-Lithics* 10 (2): 59–71.

Coşkun, Aytaç, Marion Benz, Corinna Rössner, Katleen Deckers, Simone Riehl, Kurt W. Alt, and Vecihi Özkaya. 2012. "New Results on the Younger Dryas Occupation at Körtik Tepe." *Neo-Lithics* 12 (1): 25–32.

Dufour, Elise, Chris Holmden, Wim Van Neer, Antoine Zazzo, William P. Patterson, Patrick Degryse, and Eddy Keppens. 2007. "Oxygen and Strontium Isotopes as Provenance Indicators of Fish at Archaeological Sites: The Case Study of Sagalassos, SW Turkey." *Journal of Archaeological Science* 34 (8): 1226–39. https://doi.org/10.1016/j.jas.2006.10.014.

Dunbar, Robin I.M. 2013. "What Makes the Neolithic So Special?" *Neo-Lithics* 13 (2): 25–29.

Gebel, Hans Georg K. 2010. "Commodification and the Formation of Early Neolithic Social Identity: The Issues Seen from the Southern Jordanian

Highlands." In *The Principle of Sharing: Segregation and Construction of Social Identities at the Transition from Foraging to Farming*, ed. Marion Benz, 31–80. Berlin: ex oriente.

Gebel, Hans Georg K. 2017. "Neolithic Corporate Identities in the Near East." In *Neolithic Corporate Identities*, ed. Marion Benz, Hans Georg K. Gebel, and Trevor Watkins, 57–80. Berlin: ex oriente.

Gell, Anthony. 1968. *Art and Agency: An Anthropological Theory*. Oxford: Clarendon.

Godelier, Maurice. 2007. *Au Fondement des Sociétés Humaines: Ce Que Nous Apprend L'anthropologie*. Paris: Albin Michel.

Guenther, Matthias. 2010. "Sharing among the San, Today, Yesterday, and in the Past." In *The Principle of Sharing: Segregation and Construction of Social Identities at the Transition from Foraging to Farming*, ed. Marion Benz, 105–36. Berlin: Ex oriente.

Harpelund, Anne M. 2011. "An Analysis of the Ground Stone Assemblage from the Middle Pre-Pottery Neolithic B Site Shkarat Msaied in Southern Jordan." MA thesis, University of Copenhagen.

Hauptmann, Harald. 2011. "The Urfa Region." In *The Neolithic in Turkey: New Excavations and New Research 2: The Euphrates Basin*, ed. Mehmet Özdoğan, Nezih Başgelen, and Peter Kuniholm, 85–138. Istanbul: Archaeology and Art Publications.

Hayden, Brian. 2014. *The Power of Feasts: From Prehistory to the Present*. Cambridge: Cambridge University Press. https://doi.org/10.1017/CBO9781107337688.

Hodder, Ian. 1990. *The Domestication of Europe*. Oxford: Basil Blackwell.

Karul, Necmi. 2013. "Gusir Höyük/Siirt: Yerleşik Avcılar." *Arkeo Atlas* 1: 22–29.

Khawam, Rhima. 2014. "L'homme et la Mort au Néolithique Préceramique B: L'example de Tell Aswad." PhD dissertation, University of Lyon.

Kodas, Ergul. 2015. "Contexte Architectural des Crânes Surmodelés: Diversité Contextuelle et Funéraire." *Neo-Lithics* 15 (1): 11–23.

Köksal-Schmidt, Çiğdem, and Klaus Schmidt. 2007. "Perlen, Steingefäße, Zeichentäfelchen—handwerkliche Spezialisierung und steinzeitliches Symbolsystem." In *Vor 12,000 Jahren in Anatolien—Die ältesten Monumente der Menschheit*, ed. Badisches Landesmuseum Karlsruhe, 97–109. Stuttgart: Theiss.

Lichter, Clemens. 2007. "Steingefäß." In *Vor 12.000 Jahren in Anatolien: Die ältesten Monumente der Menschheit*, ed. Badisches Landesmuseum Karlsruhe, 304. Stuttgart: Theiss.

Lillie, Malcom, Chelsea Budd, and Inna Potekhina. 2011. "Stable Isotope Analysis of Prehistoric Populations from the Cemeteries of the Middle and Lower Dnieper Basin, Ukraine." *Journal of Archaeological Science* 38 (1): 57–68. https://doi.org/10.1016/j.jas.2010.08.010.

Mazurowski, Ryszard F., and Youssef Kanjou, eds. 2012. *Tell Qaramel 1999–2007: Protoneolithic and Early Pre-Pottery Neolithic Settlement in Northern Syria*. Warsaw: University of Warsaw.

Miyake, Yutaka. 2013. "Hasankeyf Höyük/Batman: Dicle'nin ilk köyü." *Arkeo Atlas* 1: 40–47.

Moetz, Fevzi K. 2014. *Sesshaftwerdung: Aspekte der Niederlassung im Neolithikum in Obermesopotamien*. Bonn: Rudolf Habelt.

Morenz, Ludwig D. 2014. *Medienrevolution und die Gewinnung neuer Denkräume: Das Frühneolithische Zeichensystem (10./9. Jht. v. Chr.) und seine Folgen*. Berlin: EB-Verlag.

Morley, Iain. 2007. "Material Beginnings: An Introduction to Image and Imagination." In *Image and Imagination—a Global Prehistory of Figurative Representation*, ed. Colin Renfrew and Iain Morley, xvii–xxii. Cambridge: Oxbow Books.

Nehlich, Olaf, Dušan Borić, Sofija Stefanović, and Michael P. Richards. 2010. "Sulphur Isotope Evidence for Freshwater Fish Consumption: A Case Study from the Danube Gorges, SE Europe." *Journal of Archaeological Science* 37 (5): 1131–39. https://doi.org/10.1016/j.jas.2009.12.013.

Özbaşaran, Mihriban. 2012. "Aşıklı." In *The Neolithic in Turkey: New Excavations and New Research 3: Central Turkey*, ed. Mehmet Özdoğan, Nezih Başgelen, and Peter Kuniholm, 135–58. Istanbul: Archaeology and Art Publications.

Özdoğan, Mehmet, and Aşlı Özdoğan. 1998. "Buildings of Cult and the Cult of Buildings." In *Light on Top of the Black Hill: Studies Presented to Halet Çambel*, ed. Guven Arsebük, Machteld J. Mellink, and Wulf Schirmer, 581–93. Istanbul: Ege Yayınları.

Özkaya, Vecihi. 2004. "Körtik Tepe." In *Anadolu'da Doğdu: 60 Yaşında Fahri Işık'a Armağan*, ed. Taner Korkut, 585–600. Istanbul: Ege Yayınları.

Özkaya, Vecihi, and Aytaç Coşkun. 2011. "Körtik Tepe." In *The Neolithic in Turkey: New Excavations and New Research 1: The Tigris Basin*, ed. Mehmet Özdoğan, Nezih Başgelen, and Peter Kuniholm, 89–127. Istanbul: Archaeology and Art Publications.

Özkaya, Vecihi, Aytaç Coşkun, and Nevin Soyukaya. 2013. *Körtik Tepe: The First Traces of Civilization in Diyarbakır (Istanbul)*. Accessed September 2015. http://www.diyarbakirkulturturizm.org/yayinlar/19/.

Özkaya, Vecihi, and Oya San. 2004. "Excavations at Körtik Tepe 2001." In *Salvage Project of the Archaeological Heritage of the Ilısu and Carchemish Dam Reservoirs Activities in 2001*, ed. Numan Tuna, Jean Öztürk, and Jâle Velibeyoğlu, 669–93. Ankara: Tacdam.

Renfrew, Colin, and Chris Scarre, eds. 2005. *Cognition and Material Culture: The Archaeology of Symbolic Storage*. Oxford: Oxbow Books.

Rosenberg, Michael. 2011. "Hallan Çemi." In *The Neolithic in Turkey: New Excavations and New Research 1: The Tigris Basin*, ed. Mehmet Özdoğan, Nezih Başgelen, and Peter Kuniholm, 61–78. Istanbul: Archaeology and Art Publications.

Rössner, Corinna, Katleen Deckers, Marion Benz, Vecihi Özkaya, and Simone Riehl. 2017. "Subsistence Strategies and Vegetation Development at Aceramic Neolithic Körtik Tepe, Southeastern Anatolia." *Vegetation History and Archaeobotany*: 1–15. doi.org/10.1007/s00334-017-0641-z.

Saña, Maria, Carlos Tornero, and Miquel Molist. 2014. "Property and Social Relationships at Tell Halula during PPNB." Paper presented at the conference Religion, History, and Place in the Origin of Settled Societies, Çatalhöyük, Turkey, August 2–3.

Schmidt, Klaus. 2011. "Göbekli Tepe." In *The Neolithic in Turkey: New Excavations and New Research 2: The Euphrates Basin*, ed. Mehmet Özdoğan, Nezih Başgelen, and Peter Kuniholm, 41–83. Istanbul: Archaeology and Art Publications.

Schreiber, Felix, Aytaç Coşkun, Marion Benz, Kurt Alt, and Vecihi Özkaya, with contributions from Nicole Reifarth and Elisabeth Völling. 2014. "Multilayer Floors in the Early Holocene Houses at Körtik Tepe, Turkey—an Example from House Y98." *Neo-Lithics* 14 (2): 13–22.

Stordeur, Danielle. 2010. "Domestication of Plants and Animals, Domestication of Symbols?" In *The Development of Pre-State Communities in the Ancient Near East: Studies in Honour of Edgar Peltenburg*, ed. Diane Bolger and Louise C. Maguire, 123–30. Oxford: Oxbow Books.

Stordeur, Danielle. 2015. *Le Village de Jerf el Ahmar (Syrie, 9500–8700 av. J.-C.): L'Architecture, Miroir d'une Société Néolithique Complexe*. Paris: CNRS Éditions.

Stordeur, Danielle, Bassam Jammous, Daniel Helmer, and George Willcox. 1996. "Jerf el-Ahmar: A New Mureybetian Site (PPNA) on the Middle Euphrates." *Neo-Lithics* 96 (2): 1–2.

Stordeur, Danielle, and Rhima Khawam. 2007. "Les Crânes Surmodelés de Tell Aswad (PPNB, Syrie): Premier Regard sur L'ensemble, Premières Réflexions." *Syria* 84: 5–32.

Sütterlin, Christa. 2000. "Symbole und Rituale im Dienste der Herstellung und Erhaltung von Gruppenidentität." In *Symbole im Dienste der Darstellung von Identität*, ed. Paul Michel, 1–15. Bern: Peter Lang.

Sütterlin, Christa. 2006. "Denkmäler als Orte kultureller Erinnerung im öffentlichen Raum." In *Raum—Heimat—fremde und vertraute Welt*, ed. Hartmut Heller, 80–103. Wien: Lit Verlag.

Theweleit, Klaus. 2013. "An Entirely New Interaction with the Animal World?" *Neo-Lithics* 13 (2): 57–60.

Tilley, Christopher. 2004. *The Materiality of Stone: Explorations in Landscape Phenomenology*. Oxford: Berg.

Walter, Sebastian. 2014. "Ungewöhnliche Tiere in der Kunst des frühen Neolithikums (PPN A): Zu Anthropoden-Darstellungen aus Südostanatolien (Göbekli Tepe, Körtik Tepe) und Nordsyrien (Jerf el Ahmar, Tell Qaramel)." *Zeitschrift für Orient-Archäologie* 7: 56–88.

Watkins, Trevor. 2012. "Household, Community, and Social Landscape: Maintaining Social Memory in the Early Neolithic of Southwest Asia." In *As Time Goes By?—Monumentality, Landscapes, and the Temporal Perspective*, ed. Martin Furholt, Martin Hinz, and Doris Mischka, 23–44. Bonn: Rudolf Habelt.

Wetzel, Dietmar J. 2009. *Maurice Halbwachs*. Konstanz, Germany: Universitätsverlag Konstanz.

Whitehouse, Harvey. 2000. *Arguments and Icons: Divergent Modes of Religiosity*. Oxford: Oxford University Press.

Widlok, Thomas. 2013. "Sharing—Allowing Others to Take What Is Valued." *Journal of Ethnographic Theory* 3 (2): 11–31. https://doi.org/10.14318/hau3.2.003.

Willerslev, Rane. 2007. *Soul Hunters: Hunting, Animism, and Personhood among the Siberian Yukaghirs*. Berkeley: University of California Press. https://doi.org/10.1525/california/9780520252165.001.0001.

Wulf, Christoph. 2005. *Zur Genese des Sozialen: Mimesis, Performativität, Ritual*. Bielefeld, Germany: Transcript. https://doi.org/10.14361/9783839404157.

Yartah, Thaer. 2013. "Vie Quotidienne, Vie Communautaire et Symbolique à Tell 'Abr 3—Syrie du Nord: Données Nouvelles et Nouvelles Réflexions sur l'Horizon PPNA au Nord du Levant 10 000–9 000 BP." PhD dissertation, University of Lyon.

6

Sedentism and Solitude

Exploring the Impact of Private Space on Social Cohesion in the Neolithic

Güneş Duru

The question of why humans chose sedentary life and what chain of events and choices led to sedentism is ongoing. Although there are many varying ideas and hypotheses, this, one of humanity's most important turning points, is still a major topic of discussion. Numerous examples of present-day communities leading a nomadic life are well-known all over the world. Since the subject has in general been studied using progressive and developmental approaches, the question of why some groups have not chosen to "progress" to sedentary life and continued living as nomads for tens of thousands of years has drawn little attention among scholars. The Sarıkeçili community, who live at present in modern Turkey, are a nomadic group that refuses to settle even though their numbers are fast dwindling. Their felt shelters as well as their tools and implements have been replaced in time by plastic objects and canvas tents, yet they carry on with their traditional lifestyle. Defying the state's pressure, they refuse to send their children to school but use modern technological devices such as smart phones for communication and cars and lorries to migrate along the Taurus Mountains.

About 22,000 years ago, during the Upper Paleolithic–Epipaleolithic transition, marked changes occurred in hunter-gatherer groups: they began to increase in population and limit their seasonal migration practices to two main camps. These changes were not only the consequence of their experience but were also by necessity in cognitive transformation (Watkins 2015; Dunbar 2013; Renfrew 2008; Donald 2001). These communities would have begun to perceive a "sense of belonging" to certain places, and the many changes we observe in their material culture show that, alongside this emergence of a sense of belonging, they developed the idea of an identity for themselves and a sense of history being made in a specific locale.

Abandonment of the hunter-gatherer, mobile way of life and the transition to a settled lifestyle meant continually encountering many new problems that had to be overcome. Parallel to the attempt to find solutions to these new issues, communal spaces slowly gave way to individualized/private spaces, and co-residents living under the same roof gave way to family groups and extended families. Building groups emerged along with neighborhoods of separate social groups, probably based on ties of kinship (see, however, Pilloud and Larsen 2011). Buildings specifically intended as houses were built, with major investments in planning, inner architectural features, and floor and wall plasters. Settlement layouts were well defined. All of these distinctive associations of sedentism imply a greater sense of historical time depth in terms of both habituated practices and commemorative practices.

Clearly, this transition process was not nearly as straightforward or punctuated as described above. The question remains how the earliest sedentary communities that probably consisted of a minimum of 150–500 people overcame each of these stages of change. In the light of existing data, the most rational explanation for the need for symbolism would be that it regulated the life of the first settlers. Hunter-gatherers collectively produced symbolism by regularly coming together at special ritual sites, which then continued to exist not only through oral history but in material culture and architecture as well. During the late Natufian and pre-Pottery Neolithic A (PPNA), communal storage areas, work areas, and spaces that represent the community's mutual (communal) identity through time triggered the emergence of public spaces that represent power and mechanisms of social organization, as seen in the Skull Building at Çayönü, the Kult Bau at Nevalı Çori, and the Special Purpose Buildings Area at Aşıklı (Özdoğan and Özdoğan 1998; Özdoğan 1999; Hauptmann 1999; Özbaşaran 2011a). In crowded communities, public spaces (providing unity and cohesion while at the same time generating privilege and power) played an active part in the organization of social roles and the increase in specializations or craftsmanship. Yet one has to consider that these crowded public spaces must have started to alienate individuals.

Hypotheses on the shifting layouts of human settlements that emerged from material culture–oriented approaches to cause-and-effect relationships, based on economic and environmental factors, have been replaced today by interpretations acknowledging multiple causes. Such dehumanized, one-dimensional environmental, economic, or technological models focused essentially on how change happened rather than why. In contrast, recent hypotheses, such as Niche Construction Theory, Theoretic Culture, and the Social Brain Hypothesis (Kendal 2011; Dunbar 1998; Donald 2006), which highlight human behavior as a determining factor, discuss interpretations of cultural evolution and cognitive development in relation to the dynamics leading to the transition to sedentary life. These theoretical developments allow us to expand our understanding of social change beyond reductive material culture descriptions of social change and move toward understanding human action in the past.

This chapter aims to discuss not only why humankind chose to settle; beyond this, it is an attempt to discuss how humankind was changed reciprocally by what it built and by the many new elements involved in the organization of the new way of life. Could the spaces in which "public" histories were built, which were regulatory elements in sedentary life, have had an impact on the emergence of individual buildings, with their specific house histories? Could the "public" buildings have been a factor in starting the individualization of spaces and the birth of the concept of family and extended families, differentiated from their neighbors? To put it in a different way, could it be possible that a new concept of "family" emerged as a reaction to public spaces and social stress? Moreover, could the "collapse" during the end of PPNB/PPNC be the result of a search for a new way of life and an exercise of power by these new family/extended family units?

UNDERSTANDING NEOLITHIZATION

Following initial research in Southwest Asia, the first settled groups were defined as the first farming communities (Braidwood 1973; Braidwood and Willey 1962). Although this process contains certain behavioral and cognitive elements, scholars tended to concentrate on aspects of production and technology. This mind-set has long been criticized, yet it still dominates classification, definition, and interpretation of the period in concern, the Neolithic. The chronology, terminology, and typology we use at present are the direct results of this approach (Kenyon 1957, 1981; Braidwood and Howe 1960).

The most important reason for the interpretation of the earliest sedentary communities as full-time farmers, fulfilling their nutritional needs basically through agriculture, stems from various socio-political developments in 1900s. Theoretical approaches to understanding the productive dynamics of the Industrial Revolution led to a similar understanding of the basic economic changes in the Neolithic, defining it, too, as a revolution (Childe 1941). In addition to ideological and practical innovations, advances in philosophy, technology, and science, new discoveries, the emergence of a defined working class, diversification in production, World Wars I and II, financial crises, the rise of nationalism, and the emergence of national states were concepts that shaped further interpretations and identified parameters of the emergence and way of life of early sedentary communities. Archaeologists have always reflected the present era in their research on the past. Their interpretations are based mainly on the economic norms and moral values of their own time, which is probably what we are doing at the moment.

The Terrazzo Building exposed during the first excavation seasons at Çayönü did not fit the model Robert Braidwood had in mind (Özdoğan 2004, 48). Braidwood was convinced that the community at Çayönü was concentrated on food production and that its world of beliefs had not yet materialized. It is for this reason that the upright stones in the cobble-paved building did not raise suspicion or attract attention. Though it is dated later than Çayönü, the discovery of Çatalhöyük and its outstanding finds were the motivating factors to finally consider the relationship of beliefs and early village communities.

Since the end of the 1960s, Çatalhöyük has quickly become a center of attraction not only for archaeology but also for other disciplines and even for New Age religions. In Turkey neither politicians and the public nor archaeologists had cared much about the site and what it signified. During the 1960s, when Çatalhöyük was identified, contrary to the new trends elsewhere in the world, ideologies and movements such as feminism and spiritualism did not exist in Turkey, and the country was politically unstable. Even though Çatalhöyük was included in the Turkish school curricula, the reason was the acknowledgment of its role globally rather than as a national appreciation. The foremost interpretation was that the "world's oldest village" was located on the Konya plain, and the community was "artistically" well developed.

From one of Turkey's then least-developed cities, Konya, James Mellaart wrote articles on Çatalhöyük using new terms for the Neolithic to describe art, matriarchal societies, obsidian trade, death, shrines, faith, and rituals (Mellaart 1967). He emphasized that Çatalhöyük was the oldest known village in the world. Turkish archaeologists preferred to concentrate on the outstanding artistic finds exposed, since most adhered to the positivist and Marxist schools of thinking suited to progressive interpretations of this artifact class. Discussing belief systems and female deities was not considered relevant until the 1990s by prehistorians in Turkey.

Neo-liberal politics of the 1980s brought new approaches and generated new research that concentrated on individuals and diverse identities beyond Marxist concepts of social classes and consciousness (Eagleton 1983). Studies and research focusing solely on class and economy lost prominence and were replaced by studies and investigations of individuals, daily life, gender, and different identities. Efforts to understand symbolism and belief systems were part of this shift.

Things have been quite different in Göbekli Tepe, the site identified approximately thirty years after Mellaart discovered Çatalhöyük. Göbekli Tepe is one of the most interesting sites in Turkey and has become fairly well-known among the public in a very short time. Although many may not know exactly why Göbekli Tepe is so important, they are aware that it has something to do with religion. The German magazine *Bild* put Göbekli Tepe on its cover and quoted the biblical term "Garden of Eden," while in Turkey the media attention gained some significance during the days of rising tension between Republican secularists and Islamic conservatives. Excavations at Göbekli Tepe had started in 1993, but the site only began to be known among the public around 2003 when the conservative Justice and Development Party (Adalet ve Kalkınma Partisi, AKP) came to power. Even though Göbekli Tepe features many pagan symbols and figures, which are forbidden by Islam, the site is located near Urfa, the birthplace of the prophet Abraham; therefore it caught the attention of the government, local authorities, and sponsors. The significance of Göbekli Tepe lies in the notion that "religion triggered the sedentary lifestyle." As conservatism took over in Turkey and the AKP gained power, the public began to take more and more interest in the supernatural. Religious and cultural tourism drew increasing interest. Archaeologists in Turkey could not ignore this. On the one hand, awareness about the Neolithic had changed quickly; on the other, Göbekli Tepe was well publicized. Despite the opinion of some scholars who think there may not just be temples but also residential buildings at Göbekli Tepe (Banning 2011; Lee Clare, personal communication, 2015), the site and its belief system have come to have enormous importance in discussions about the first sedentary communities.

During the past twenty-five years, scholars have concluded that the factors triggering sedentary life were not just food production (Rowley Conwy 2004, 97; Bellwood 2005, 26–27) but that environmental and climatic factors (Willcox, Buxo, and Herveux 2009; Rosen and Rivera Collazo 2012), psycho-cultural changes (Cauvin 1994), and proximity to sacred places (Notroff, Dietrich, and Schmidt 2014) also played important roles. There are a number of sites and communities who gather and harvest wild plants, fish and hunt, and carry out agriculture on a limited scale (Smith 2001). It is obvious that archaeology did not collect the data on these results suddenly but rather that new surveys, excavations, and fieldwork in the region have helped us understand the Neolithic on a different scale.

The idea, spearheaded by Jacques Cauvin at the end of 1970s (Cauvin 1978), that intense interaction along with technological and cultural exchange happened to a high degree in a certain geographic region was developed by Ofer Bar-Yosef and Anna Belfer-Cohen and defined as a "PPN Interaction Sphere" (Bar-Yosef and Belfer-Cohen 1989a). This model suggests that this region encompassing the Sinai Peninsula and the Negev in the south, the eastern shores of the Mediterranean including Cyprus, the northern Taurus Mountains, the Zagros Mountains in the east, central Anatolia in the west, and possibly Anatolia's Aegean coast was a dense interaction region for exchange of technologies and raw materials. Seashells from the Mediterranean and Red Seas, obsidian from central Anatolia, silex, bitumen used to stick arrowheads onto wooden shafts, and arrowhead traditions of the Helwan, Byblos, Amuq, and Ugarit were found at sites throughout this region and were accepted as strong indications for this aforementioned exchange (Bar-Yosef and Belfer-Cohen 1989a). It has been almost twenty years since Bar-Yosef and Belfer-Cohen came up with the PPN Interaction Sphere, and since then many new sites and finds have been exposed and investigated that open this term up to debate, even if many scholars prefer to use the term and follow their approach.

Cauvin believed the core area was the Middle Euphrates, and he stated that changes in the region corresponded with the emergence of the bull and female symbols (Cauvin 2000). According to him, it was not ecological factors but psycho-cultural changes and belief systems that triggered the transition to agriculture (Cauvin 1994, 92). The ever-changing socioeconomic situation brought with it a change in frame of mind and ways of thinking, and it led to some kind of belief system that spread across a wide area. The rise of agriculture was the direct consequence of this transformation (Cauvin 2000). In summary, Cauvin proposed that the developments in the Middle Euphrates spread to other regions through ideology. In the beginning of the 1990s new discoveries at Jerf el Ahmar, then at Göbekli Tepe and similar sites, prompted Olivier Aurenche and Stefan Karol Kozlowski to propose a new model with Cauvin's initial core region at the center. They concluded that the Middle Euphrates, Zagros, and southeast Anatolia formed a "Golden Triangle" that had intense interaction and communication between sites/communities, which caused a more rapid spread of ideas and innovations than in other regions (Aurenche and Kozlowski 1999). The distinct similarities among the various lithic tool industries, architectural forms, subsistence means, and symbolic elements found throughout the different regions of Southwest Asia imply a vast area of exchange (Aurenche and Kozlowski 1999; Bar-Yosef and Belfer-Cohen 1989b; Braidwood and Howe 1960; Cauvin 2000; Gebel 2004; Kozlowski and Aurenche 2005).

This interaction surely did not appear suddenly with the start of the Neolithic. In the Epipaleolithic mobile hunter-gatherers were already in touch with one another. Obsidian from central Anatolia, seashells, various tool industries, and common symbolic elements all point to an interregional exchange of knowledge in this era (Bar-Yosef and Belfer-Cohen 1989a; Bar-Yosef and Belfer-Cohen 1989b; Belfer-Cohen and Goring-Morris 2014). These areas of exchange gained autonomous characteristics through time, and in the middle PPNB they became defined and even gained definable borders. The North and South Levant were also part of this, while the earliest sedentary communities in central Anatolia stayed away from this dense interaction and kept contact with others limited (Özbaşaran 2011b); in other words, contrary to the unity in the Levant, central Anatolian communities likely chose not to participate in this "unity" but instead to keep protecting their own identity and historical traditions (Duru 2013).

Evidence for the early and middle Epipaleolithic comes from Kharaneh IV and Jilat 6 and for the late Natufian from Rosh Horesha, where mobile hunter-gatherers came together and exchanged knowledge, ideas, and technology (Goring-Morris and Belfer-Cohen 2010). This is even more evident at Göbekli Tepe, where various groups gathered at regular intervals. There are six other centers within a 10 km radius around Göbekli Tepe (Çelik 2007; Güler, Çelik, and Güler 2013), which indicates that aggregation sites were not limited to Göbekli Tepe. The existence of these sites also implies the introduction of competition. Moreover, such symbolic centers must have addressed most of the newly emerged issues of the transitional period—from foraging to sedentary life—by generating some sort of "order" through the use of symbols and rituals. These symbols and rituals were repeated through time, creating a temporal and ancestral structure that ordered daily life. Competition between the sites, such as between Göbekli Tepe and other nearby sites, to produce "the most perfect" symbolic structure contributed to the emergence of common cosmologies and histories. It is likely that the newly emerged meanings organized the dynamics of living together. Indeed, the transmission of these common expressions to the early sedentary "villages" indicates that the concepts of a public and a public history—which organize, regulate, and keep the communities together—continued within the separate settlements.

EMERGENCE OF PRIVACY

As mentioned, aggregation sites that started to be generated in the Epipaleolithic directed the communities toward sedentism. Sedentary life is one of the results of these long-lasting experiences and of many other different causes. More than just an economic revolution, it is one of the hardest trials in human history. Transition to a sustainable sedentary life means the emergence of new social interrelations, organization of these relations, changes in culture, realization of the difficulties of adaptation to a new way of life, and changes in the social structure of the community—all of which necessitate that archaeologists assess and examine the cognitive and behavioral aspects involved and move beyond description of technological and economic traits. As Trevor Watkins underlines in "Neolithization Needs Evolution, as Evolution Needs Neolithization," where he brings together Braidwood's approach highlighting the question of "why then? why not earlier?" with that of Cauvin's (2000) interpretation of "psycho-cultural transformation in the way that people imagined their world," the period in concern needs to be assessed within the context of the cognitive changes of Homo sapiens (Watkins 2013, 3).

The first centuries of sedentary life would have had big social, economic, and ecological challenges. Establishment of the new way of life and of consciousness in sustaining this new life might not have been fully managed (Bandy and Fox 2010, 2). Provisional, flexible, innovative, and productive moments may have been generated where unique social histories were created. Robin Dunbar (2013, 25) argues that humans' natural psychological mechanisms were not yet equipped to deal with the higher levels of tension and stress. Everyday life in a settlement with thousands of inhabitants would have brought about problems that had to be resolved daily. Nevertheless, the first villagers would have mastered the challenges of constructing "permanent" shelters, generating an autonomous order and a common history, and tying together the individuals of the community to manage living together. In addition to domesticating plants and animals, they would have domesticated themselves, feeling more involved in and entangled with the settlement and the community.

The limited surplus of produce in settlements was kept in communal pits in external areas; when the surplus grew, these storage areas turned into bigger public units as documented in many sites, with Jerf el Ahmar (Stordeur 2000) the most significant example. What was stored here was not just produce but also experiences and symbolisms stretched out over time. As interaction grew, communities generated communal "symbolic storages" (Donald 2006; Watkins 2015).

Through time, storage units, communal production, and consumption activities moved inside buildings. Aşıklı Höyük in central Anatolia is one of the best examples that evidences the process of transfer of outdoor activities to indoors. One can clearly follow in sequence from the mid-ninth millennium cal BCE to the mid-eighth millennium cal BCE how daily life and daily activities were transferred from outdoors to indoors; how residential units, the houses, expanded from one-room to two-room buildings; and how building groups and neighborhoods emerged (Özbaşaran 2011a). Such a development is also known in numerous other sites in Southwest Asia. Communal spaces started to turn into more individualized spaces; internal architectural features and interior organization in buildings led to changes in building size and the emergence of various plan types. More important, apart from common symbolic storage areas, individual storage areas appeared, simultaneously storing the social experiences of the co-residents/families. At Aşıklı and Çatalhöyük, architectural features such as hearths did not change their location within buildings for centuries. They stayed in practically the same spot even when they or the buildings around them were renewed. Burials placed under the floor were also part of this continuity. One could say that this continuity phenomenon is the "symbolic memory stick" of co-residents and families. The continuity of internal arrangements is not determined by physical constraints but is a clear example of habituated practices.

Despite this continuity in central Anatolia, the PPN world lived through rapid changes. The "external symbolic storage" system mentioned in Merlin Donald (2001) acted on a regional scale. In various settlements, specifically during PPNB, building types underwent changes at the same time, as did their technology. This uniformity suggests that groups of mobile craft specialists traveled all across the region, producing similar tools, transferring knowledge, and constructing similar buildings (Duru 2005)—a similar case that we see in pottery movements in later periods. The PPNB likely became a geopolitical structure that controlled the environment and resources while organizing the socio-cultural landscape on a regional scale. Meanwhile, the individual became a part of this public world. He/she began to see him/herself as part of a bigger system than his or her own family or community.

The middle PPNB saw the advent of agriculture and animal domestication in the northern Levant; sheep, goat, cattle, and pig were domesticated in the region. Agriculture and animal management became essential to the new way of life. This increase in material "prosperity" attracted communities with collective technology, symbols, and ideas to this new way of life. This was how mobile forager groups came to accept becoming part of the new system, establishing their own villages or joining existing settlements (Duru 2015).

The number of settlements reached into the hundreds. Catchment areas and raw material sources overlapped and access routes became more complex, all of which likely contributed to a new social organization among communities. Whether this was a central authority or centralized management is hard to say, but the increasingly complex social configuration was obvious from the presence and products of craftsmen and experts. The specialist symbology and technology of the mobile traders, craftsmen, or experts wandering across the region and communicating within the "interaction sphere" were the means by which these new trends spread, as well as perpetuating the concept of outsourcing specialist activity to an itinerant group. Similar examples of such traders, *çerçi*, were well-known in the past (and still today, although few) in Anatolia (Tozlu 2014), where they exchanged not only goods but also information.

The close interaction and exchange of ideas, knowledge, and raw material had likely led to an interdependence between communities. Through time they would have lost their local individuality/originality and sustained much of their lives within the social circle of the large regional organization, the prime mover of which was communication and interaction.

Common spaces—such as agricultural lands, pastures, and areas of foraging, water, and similar resources—as well as holy landscapes would have necessarily expanded just as the population did. Conflicts of interest would inevitably have arisen between communities. However, there is no firm evidence that such tensions resulted in any recognizable discord or organized violence (Erdal and Erdal 2012). In this regard one can say that a consciously built system, or one that was formed by itself, prevented regional problems.

The fundamentals of this regional system were formed at aggregation sites such as Göbekli Tepe, where the common symbolic rituals were collectively created. The material culture woven with rituals and stories should have sped up the formation of a regional cosmology. According to Dunbar (2004), primitive belief systems gained doctrinal traits through time, and these doctrines penetrated the population through anthropomorphic supernatural expressions, represented by ritual and practice.

The Natufian period and the early Neolithic (PPNA) signify the transition between Dunbar's two phases. After the first thousand years of sedentism, the symbolic elements that kept communities together likely transformed into a doctrinal belief system, forming an almost "public" administration/organization. Anthropomorphic and supernatural symbols, which peaked in monumental architecture at Göbekli Tepe, began to show themselves on transportable artifacts at Jerf el Ahmar, Tell 'Abr 3, Hallan

Çemi, Tell Qaramel, and Körtik Tepe, as well as at sites like Nevalı Çori and Çayönü, though in the latter two in a more solid and concrete way, where special purpose buildings were differentiated from the houses (Aurenche and Kozlowski 1999; Hauptmann 1999; Mazurowski 2007; Özdoğan 1999; Özkaya and Coşkun 2009; Yartah 2005).

Ian Hodder used the terms *domus/agrios* in *The Domestication of Europe*, explaining conceptually the terms *indoors* and *outdoors* (Hodder 1990). Although he used these terms in a different context than is the case here, the increasing contrast of "inside" and "outside" is part of a similar process of transformation. In the PPNB an increase in indoor activities could have caused a decrease in willingness to take part in communal, external activities. However, the underlying doctrine that organizes social rules in a settlement strengthens over time and even penetrates indoors. Therefore, the conflict/interaction between public and private, inner and outer, and between two types of history, public and domestic, becomes more apparent in the PPNB.

CASE STUDY: AŞIKLI HÖYÜK

The tension between public and private at Aşıklı on the central Anatolian Plateau was quite different than that seen from other sites within the PPN interaction sphere. The dichotomy between public and private space emerged in a more gradual, balanced, and unique way, without sharing common symbolic meanings with the rest of the region. As in the late Natufian and early PPNA, activities such as food processing, tool making, and so on took place in open areas surrounded by enclosed structures. The primary role of the enclosed spaces was to provide shelter. As mentioned, the dead were buried underneath the floors of these buildings, providing a link between social identities and this use of space. This pattern gradually gave way to more individual, private use of space; through time, formerly public collective activities such as cooking and tool making entered into the one-room buildings.

Buildings were always renewed on the same spot, reproduced exactly, including the internal architectural features (figure 6.1). This continuity of structure, which lasted about 600 years, stands as evidence of the transmission of the structural and also social biographies of these houses and their residents to the following generations. Continuity in under-floor burials and in architectural features, such as hearths within the houses, indicates this pattern of social transformation in the community. The theme of continuity was so powerful that it makes one think that the interior of the buildings was not truly separate from the exterior world. In other words, the buildings constructed

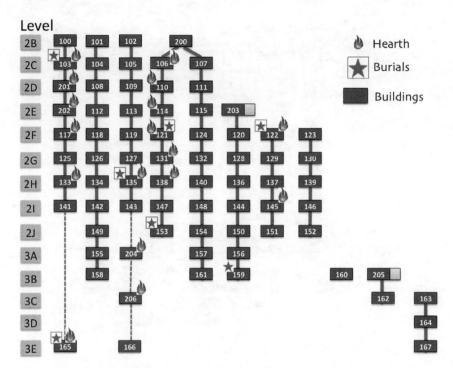

FIGURE 6.1. *Buildings at Aşıklı Höyük are rectangular, always built on top of each other using the old walls as the foundation of new ones. A remarkable continuity of buildings can be seen in this figure; the fire icon represents the buildings with hearths, and the stars represent buildings with burials.*

individualized histories but life was still collective and embedded in site-wide habituated practices, as opposed to the pattern observed in the rest of the PPN world. Social and economic roles inside and outside the houses were managed in a balanced manner, without significant distinctions.

Toward the end of this 600-year phase, buildings started to cluster. Clusters were separated by narrow spaces or passages, and personal ornaments such as necklaces and bracelets were placed in burials alongside the dead—all of which indicates that the process of individualization had gained traction. Building clusters generated neighborhoods. Almost every neighborhood had its own midden area. The organic growth in the settlement was replaced by a more systematized social organization. Clusters of building groups were designed radially in respect to the "gravel street" that separated the residential area from the Special Purpose Buildings Area (figure 6.2). This orientation

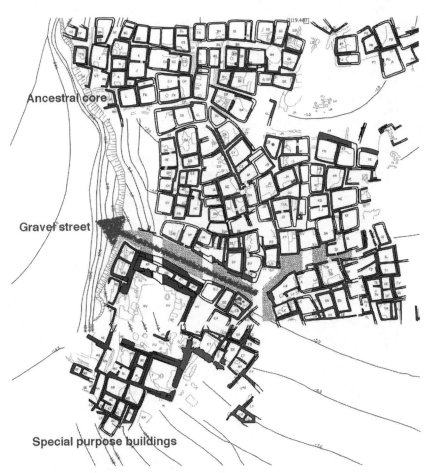

FIGURE 6.2. *Settlement consists of two distinct areas, the residential area in the north and the public area in the south. The areas are separated from each other by a monumental gravel street.*

was likely generated as a result of the choices of the households and not by any kind of authority. The "public" buildings showed similar continuity of meaning and role just like the domestic buildings. The five-times-renewed Building T (figure 6.3), with its red and yellow painted floor and wall plasters, is an example. Evidence of communal feasting activities observed in the same area, beyond being public, indicated the continuity of the collective way of living. The entire process of spatial reorganization at Aşıklı seems to be the consequence of a compromised social structure caught between the

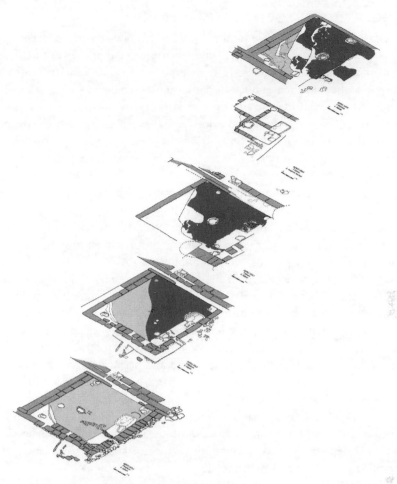

FIGURE 6.3. *Five phases of renewal of Building T.*

public and the private. Although a strong individualization process seems to have been initiated within the houses, buildings have not been uniformly transformed into isolated houses as happened in the PPN world. Food production, distribution, and consumption and tool making were still collective tasks. Public life had significance as a common value to be maintained. As opposed to the situation of the PPN interaction sphere, the evidence from Aşıklı indicates a collective structure that was designed by the preferences of individual households. The tension between private and public history making was balanced.

REVOLUTION OF SOCIAL INTERACTIONS

If the Epipaleolithic communities were not interested in the technological and social innovations of the other groups, they might have not shared them and instead chosen to clash. These groups were self-sufficient; the main motivation behind their regular gathering must have been a more intangible social and personal interest. In time the curiosity about diversity and innovations created the common symbolic ideological world, collectively constructed as was the case in Göbekli Tepe. The behaviors and values of the hunter-gatherers who came from long distances to arrive at such centers began to be transformed through dense interaction with each other. They simultaneously created symbolic meanings and lived their own self-transformations. The fact that six other similar sites are known within a 10 km radius around Göbekli (Schmidt 2006) implies not only the existence of other social gathering sites but probably also the emergence of competition between groups. Symbolic communication is a part of every community's social behavior, and it usually expresses the central ideal or need of the community. Following Donald (2001), some would argue that sedentism and its new symbolic communication method sped cognitive transformation at a rate never experienced before.

Homo sapiens not only created new meanings through material culture but also aimed to share them (Sterelyn and Watkins 2015). The meanings of the symbols and markers on objects spread and permeated others in various ways. These meanings functioned, on one hand, as means of communication between communities and, on the other, with the increase in population, as a source of stress. Scholars largely agree that a cognitive transformation happened in Southwest Asia simultaneous with the changes in the Epipaleolithic and Neolithic and gained momentum with the transition to sedentary life (Watkins 2013; Benz and Bauer 2013).

Increasing interaction between communities had various consequences, of which some—such as the fast spread of culture, language, beliefs, information, and technology—were accepted in the past as positive developments. Assimilation and the loss of past identities could also have occurred. One also has to consider the negative consequences, such as the spread of contagious diseases that increased mortality or the increase in the instability between the powerful and the weak—dynamics that began to penetrate social life in this period.

In settled life, individuals, while navigating and understanding their various social relationships, also begin to resemble each other in material culture, architecture, and even burial customs. Through the regional communication network, various communities engaged in a fast-paced cultural sharing. Within these similar systems, social roles and rules and new ethical norms and behaviors were reorganized rapidly. Social Identity theory identifies a series of self-identities, based on perceptions of individual, interpersonal, and social group relationships and similarities (Brewer and Gardner 1996). In the PPNB the individual's social role and status became ever more prominent. This interaction obviously did not start with the transition to sedentism, yet relationships began to individualize.

In communities consisting of hundreds or even thousands, the individual feels the need to find those to whom he or she feels close. Rules people had to obey regulate their interactions with others to avoid conflict. The individual fulfills various roles within the community, and this may result in a position the individual imagines or does not feel comfortable with. The individual is not alone; many others are aware of the same. This competition incites the concept of power within a group of people. Specialties and hierarchies become more evident and pronounced. The individual stays in constant interaction with certain groups. Even though everyone lives in the same community, some individuals or groups only come together with others for specific occasions and rituals. When the choices the individual has the power to make are generated in private space, the house becomes the center of micro-power. To give an example from an archetypal twentieth-century nuclear family: when a workman, who works under his boss and foreman, goes home, he finds his meal ready on the table. He eats together with his wife and children. Almost everything goes according to his wishes and plans. In the morning when he leaves home and his neighborhood to go to work, his social status and ego begin to dwindle. The moment he steps into the factory, he becomes the underling once again. This cycle continues over and over. His role in the world of "indoors" is totally different than that of his outer world.

This is not necessarily limited to negotiating power relationships in a family situation. An example comes from present-day Iran in a different context. The dominant social and political ideology in Iran has generated two different ways of living, in which individuals live almost completely differently in public and private. Contrary to the strict rules that apply in public spaces, Iranians live a more relaxed life in private. Although it is highly illegal, some Iranians go to house parties where alcohol and drugs are consumed, or they communicate with the outside world through satellite television. By doing so, they live and construct their identities totally differently in private than in public.

The first sedentary communities do not present such distinct differences, however, and it is only after the first thousand years of sedentary life that one can observe the basis of differentiation between the private and public space. The mutuality between the public and the private starts to shape each one by the other. This follows the concepts described by Jürgen Habermas: "The line between private and public sphere[s] extended right through the home. The privatized individuals stepped out of the intimacy of their living rooms into the public sphere of the salon, but the one was strictly complementary to the other" (Habermas 1989, 45). The earliest example of such a situation would have been known first in the Neolithic. Sedentism, through time, brought isolation to humans, and they differentiated themselves from the collective group by building walls. Meanwhile, ethical rules for constraining behavior and private intimacy started to build up.

Collectivity caused by the converging and uniting of individuals, communities, and things ends up with the integration of different groups. Integration causes each community to resemble every other community. As the system takes over, these similarities and collectivities expand and spread through regional exchange or interaction, gaining power over the existing economic, ethical, and ideological norms of individual communities and becoming a common regime. The dynamics that bring both individuals and communities together may not only be constructive but may give rise to social restrictions, oppression, and conflict as well.

CONCLUSION

As many scholars have expressed, beyond the significant economic and technological innovations, the transition to sedentism was an era when hitherto unforeseen changes rapidly affected social life and new personal relationships within a community evolved. As many foraging nomadic groups with diverse social structures came together, they not only shared innovations but also generated a common history, language, and identity. Surely this was the result of a lengthy process of compromise.

The symbols these groups created and shared formed a sort of "upper identity." The conceptualization of identities into upper and lower here draws from Erik Erikson (1972) in separating a personal individual identity from the social or collective identity. The "upper" in this usage correlates with the external, social identity and the "lower" with the personal. Every hunter-gatherer group gained a new sub-identity in addition to the previous individual qualities that defined each group. By the advent of sedentary life, these "upper and lower identities" intertwined, grew, and became more visible and palpable. While an individual feels that he or she belongs to a certain village, to a region, or to a common past, the sub-identity of the community gets reinforced in the scale of co-residents or extended families. Kent Flannery was largely right when he suggested that the transition from round to rectangular houses was a turning point in family structures and the household (Flannery 2002, 421). The sub-identities develop in these spaces and start to be isolated from the public. This proposed process can be seen in F. Brian Byrd's idea of the change in house sizes and the relation between the increasing number of rooms to larger lineage groups (Byrd 2005). Definition of a space, as the materialized form of social behaviors (Giddens 1984) and as the basic element in the reproduction of habituated practices (Bourdieu 1977, 87–94), suggests concepts similar to but articulated in a different way from what Hodder suggests for *domus/ agrios* (Hodder 1990). In particular, there is the increasing individualization evidenced by the socioeconomic and structural changes seen within the space (Kuijt 2004). Whether it is the families or the groups of people living under the same roof, houses and the lifeways within the houses began to be differentiated from those of the public space beyond.

With the development of sedentary life, symbolic meanings that brought the community together found themselves under a strong institutionalized structure: the upper identity. Their meanings were expressed in collective activities, in public spaces, and in shared burial customs linking back to ancestors. Public group interaction coalesced around public history making can be seen as the means by which the community held together and eased the stress and conflict of cohabitation by carrying out ceremonies, feasting, and rituals.

The upper identity in the northern Levant is not restricted to one settlement; it encompasses many other sites in the area. The uniformity of changes that occur in the region concerning burial customs, plan and size of buildings, and similar factors is the consequence of this shared identity. Individuals who see and feel themselves as part of a village or area within this sphere then become part of a wider cultural geography. So, when inside the home, what is left outside is not only the immediate community and workload but also an entire cultural system. The symbolism at the center of this system puts pressure on individuals in the form of increasing regulation of public behavior, and individuals isolate this stress in their own private spaces: the houses. The tension between the public and the private, or between upper and lower identities, generates the need for privacy and pushes the community toward the development of the private histories of family units (a similar process to the development of "house societies" discussed by Joyce in chapter 8, this volume).

Could these newly constituted sub-identities have led to a quest for a new way of life? As the upper identity becomes more powerful and more demanding, the more the conflict between the private and the public becomes significant. The symbolic meanings and rituals that were the end results of the common cultural interaction/exchange may have triggered sedentism, but at the same time the reactions of individual entities to the same meanings prompted a self-transformation in their own living spaces and created sub-groups tied to each other by history-making processes (whether at family, kinship, or fictive kinship levels). Therefore, I ask: Could it be possible that the upper identity stemming from the Epipaleolithic and the lower identity growing in private spaces clashed and brought the deformation and collapse that we see in many settlements at the end of PPN?

The rising number of settlements producing more conflicts on account of resources and the increasing mortality rate brought about by the domestication of animals and agriculture could have contributed to disillusionment with the upper identity. During the seventh millennium BCE, settlements lost much of their old splendor and were abandoned. This could be the consequence of the conditions stated above. Pottery, stamp seals, amulets, portable artifacts bearing various decorative elements, and the diverse collective meanings they indicate—which we observe in the subsequent Pottery Neolithic period—must be the visible indications of differentiated communities (Atakuman 2014, 36).

The very early sedentary groups may not have had internal social or interactional systems that were prepared for the experience of living in crowded groups in the same place for years. They also may have had difficulties in transforming symbolic meanings into a doctrinal belief system. It seems likely that humanity experienced not only its first rise in the Neolithic but also its first downfall.

It is not that easy to answer clearly why communities became sedentary. It is obvious that the answer is not one-dimensional. The question poses one of the most difficult challenges in understanding an important turning point in history. Although archaeologists have concentrated on different answers in different times, the sustainability of the sedentary way of living seems to be a more important question. The nature of the psychosocial-cultural transformations humans experienced in sedentary life, the consequences of this transformation, and last but not least, understanding the impact of these transformations on our present-day norms perhaps have more significance than the question of "how" in conceptualizing the Neolithic transition.

ACKNOWLEDGMENTS

I would like to thank Mihriban Özbaşaran for her constructive arguments and critical assessments. I would also like to thank Brenna Hassett for facilitating the transmission of this article to the English-speaking world.

REFERENCES

Atakuman, Çiğdem. 2014. "Architectural Discourse and Social Transformation during the Early Neolithic of Southeast Anatolia." *Journal of World Prehistory* 27 (1): 1–42. https://doi.org/10.1007/s10963-014-9070-4.

Aurenche, Olivier, and Stefan Karol Kozlowski. 1999. *La Naissance du Néolithique au Proche Orient*. Paris: Errance.

Bandy, Matthew, and Jake R. Fox. 2010. *Becoming Villagers: Comparing Early Village Societies*. Tucson: University of Arizona Press.

Banning, Edward B. 2011. "So Fair a House." *Current Anthropology* 52 (5): 619–60. https://doi.org/10.1086/661207.

Bar-Yosef, Ofer, and Anna Belfer-Cohen. 1989a. "The Levantine 'PPNB' Interaction Sphere: People and Culture in Change." British Archaeological Reports. *BAR International Series* 508: 59–72.

Bar-Yosef, Ofer, and Anna Belfer-Cohen. 1989b. "The Origins of Sedentism and Farming Communities in the Levant." *Journal of World Prehistory* 3 (4): 447–98. https://doi.org/10.1007/BF00975111.

Belfer-Cohen, Anna, and Nigel Goring-Morris. 2014. "On the Rebound—a Levantine View of Upper Palaeolithic Dynamics." In *Modes de Contacts et de Déplacements au Paléolithique Eurasiatique*, ed. Marcel Otte and Foni Le Brun-Ricalens, 27–36. Liège: Université de Liège.

Bellwood, Peter. 2005. *First Farmers: The Origins of Agricultural Societies*. Malden, MA: Blackwell.

Benz, Marion, and Joachim Bauer. 2013. "Neurobiology Meets Archaeology: The Social Challenges of the Neolithic Processes." *Neo-Lithics* 2: 11–24.

Bourdieu, Pierre. 1977. *Outline of a Theory of Practice*. Cambridge: Cambridge University Press. https://doi.org/10.1017/CBO9780511812507.

Braidwood, Robert. 1973. "The Early Village in Southwestern Asia." *Journal of Near Eastern Studies* 32 (1–2): 34–39. https://doi.org/10.1086/372218.

Braidwood, Robert, and Bruce Howe. 1960. *Prehistoric Investigations in Iraqi Kurdistan*. Chicago: University of Chicago Oriental Institute.

Braidwood, Robert, and Gordon R. Willey, eds. 1962. *Courses toward Urban Life: Archaeological Considerations of Some Cultural Alternates*. Chicago: Aldine.

Brewer, Marilynn, and Wendy Gardner. 1996. "Who Is This 'We'? Levels of Collective Identity and Self-Representations." *Journal of Personality and Social Psychology* 71 (1): 83–93. https://doi.org/10.1037/0022-3514.71.1.83.

Byrd, F. Brian. 2005. *Early Village Life at Beidha, Jordan: Neolithic Spatial Organization and Vernacular Architecture, the Excavations of Mrs. Diana Kirkbride-Helbaek*. Oxford: Oxford University Press.

Cauvin, Jacques. 1978. *Les Premiers Villages de Syrie-Palestine du IXéme au VIIéme Millénaire Avant J.C.* Lyon: Maison de l'Orient.

Cauvin, Jacques. 1994. *Naissance des Divinités, Naissance de L'agriculture: La Révolution des Symboles au Néolithique.* Paris: Flammarion.

Cauvin, Jacques. 2000. *The Birth of the Gods and the Origins of Agriculture.* Cambridge: Cambridge University Press.

Çelik, Bahattin. 2007. "Şanlıurfa Yeni Mahalle -Balıklıgöl Höyüğü." In *Türkiye'de Neolitik Dönem*, ed. Mehmet Özdoğan, Nezih Başgelen, and Peter Kuniholm, 165–78. Istanbul: Arkeoloji ve Sanat Yayınları.

Childe, V. Gordon. 1941. *Man Makes Himself.* London: Watt and Co.

Donald, Merlin. 2001. *A Mind So Rare: The Evolution of Human Consciousness.* New York: Norton.

Donald, Merlin. 2006. "Art and Cognitive Evolution." In *The Artful Mind: Cognitive Science and the Riddle of Human Creativity*, ed. Mark Turner, 3–20. Oxford: Oxford University Press. https://doi.org/10.1093/acprof:oso/9780195306361.003.0001.

Dunbar, Robin. 1998. "The Social Brain Hypothesis." *Evolutionary Anthropology* 6 (5): 178–90. https://doi.org/10.1002/(SICI)1520-6505(1998)6:5<178::AID-EVAN5 >3.0.CO;2-8.

Dunbar, Robin. 2004. *The Human Story: A New History of Mankind's Evolution.* London: Faber.

Dunbar, Robin. 2013. "What Makes the Neolithic So Special." *Neo-lithics* 2 (13): 25–29.

Duru, Güneş. 2005. "Neolithic Architecture of Central Anatolia." MA dissertation, Istanbul Technical University.

Duru, Güneş. 2013. "Human-Space, Community-Settlement Interactions during the End of the 9th and the Beginning of the 7th Mill cal BC: Aşıklı ve Akarçay Tepe." PhD dissertation, Istanbul University Prehistory Department.

Duru, Güneş. 2015. "Kamusal İnsan'ın (ilk) Çöküşü." In *İletişim Ağları ve Sosyal Organizasyon*, ed. Adnan Baysal, 195–210. Istanbul: Ege Yayınları.

Eagleton, Terry. 1983. *Literary Theory.* Oxford: Blackwell.

Erdal, Yılmaz Selim, and Dilek Erdal. 2012. "Organized Violence in Anatolia: A Retrospective Research on the Injuries from the Neolithic to Early Bronze Age." *International Journal of Paleopathology* 2 (2–3): 78–92. https://doi.org/10.1016/j.ijpp.2012.09.014.

Erikson, Erik. 1972. *Childhood and Society.* Middlesex: Penguin Books.

Flannery, Kent. 2002. "The Origins of the Village Revisited: From Nuclear to Extended Households." *American Antiquity* 67 (3): 417–33. https://doi.org/10.2307/1593820.

Gebel, Hans Georg. 2004. "Central to What? The Centrality Issue of the LPPNB Mega-Site Phenomenon in Jordan." In *Central Settlements in Neolithic Jordan*, ed. Hans-Dieter Bienert, Hans Georg K. Gebel, and Reinder Neef, 1–20. Berlin: Ex-oriente.

Giddens, Anthony. 1984. *The Constitution of Society*. Cambridge: Polity.

Goring-Morris, Nigel, and Anna Belfer-Cohen. 2010. "Great Expectations or the Inevitable Collapse of the Early Neolithic in the Near East." In *Becoming Villagers: Comparing Early Village Societies*, ed. Matthew S. Bandy and Jake R. Fox, 62–77. Tucson: University of Arizona Press.

Güler, Gül, Bahattin Çelik, and Mustafa Güler. 2013. "New Pre-Pottery Neolithic Sites and Cult Centres in the Urfa Region." *Documenta Prehistorica* 40: 291–303. https://doi.org/10.4312/dp.40.23.

Habermas, Jürgen. 1989. *The Structural Transformation of the Public Sphere: An Inquiry into a Category of Bourgeois Society*. Cambridge: Polity.

Hauptmann, Harald. 1999. "The Urfa Region." In *Türkiye'de Neolitik Dönem*, ed. Mehmet Özdoğan, Nezih Başgelen, and Peter Kuniholm, 65–87. Istanbul: Arkeoloji ve Sanat Yayınları.

Hodder, Ian. 1990. *The Domestication of Europe*. Oxford: Blackwell.

Kendal, Jeremy. 2011. "Cultural Niche Construction and Human Learning Environments: Investigating Sociocultural Perspectives." *Biological Theory* 6 (3): 241–50. https://doi.org/10.1007/s13752-012-0038-2.

Kenyon, Kathleen. 1957. *Digging Jericho*. London: E. Benn.

Kenyon, Kathleen. 1981. *The Architecture and Stratigraphy of the Tell: Excavations at Jericho*. London: British School of Archaeology in Jerusalem.

Kozlowski, Stefan, and Olivier Aurenche. 2005. *Territories, Boundaries, and Cultures in the Neolithic Near East*. Oxford: Archaeopress.

Kuijt, Ian. 2004. "When the Walls Came Down: Social Organization, Ideology, and the 'Collapse' of the Pre-Pottery Neolithic." In *Central Settlements in Neolithic Jordan*, ed. Hans-Dieter Bienert, Hans Georg K. Gebel, and Reinder Neef, 183–200. Berlin: Ex-oriente.

Mazurowski, Ryszard. 2007. "Tell Qaramel: Excavations 2005." *Polish Archaeology in the Mediterranean* 17: 483–99.

Mellaart, James. 1967. *Çatalhöyük: A Neolithic Town in Anatolia*. London: Thames and Hudson.

Notroff, Jens, Oliver Dietrich, and Klaus Schmidt. 2014. "Building Monuments, Creating Communities: Early Monumental Architecture at Pre-Pottery Neolithic Göbekli Tepe." In *Approaching Monumentality in Archaeology: IEMA Proceedings*, vol. 3, ed. James F. Osborne, 83–105. Albany: State University of New York Press.

Özbaşaran, Mihriban. 2011a. "Aşıklı 2010." *Anatolia Antiqua* 19: 27–37.

Özbaşaran, Mihriban. 2011b. "The Neolithic on the Plateau." In *The Oxford Handbook of Ancient Anatolia 10,000–323 BCE*, ed. Sharon Steadman and John G. McMahon, 99–124. Oxford: Oxford University Press.

Özdoğan, Asli. 1999. "Çayönü." In *Türkiye'de Neolitik Dönem*, ed. Mehmet Özdoğan, Nezih Başgelen, and Peter Kuniholm, 36–63. Istanbul: Arkeoloji ve Sanat Yayınları.

Özdoğan, Mehmet. 2004. "Neolitik Çağ-Neolitik Devrim-İlk Üretim Toplulukları Kavramının Değişimi ve 'Braidwoodlar.'" *Tuba-Ar* 7: 43–51.

Özdoğan, Mehmet, and Aşli Özdoğan. 1998. "Buildings of Cult and the Cult of Buildings." In *Light on Top of the Black Hill: Studies Presented to Halet Çambel*, ed. Guven Arsebük, Machteld J. Mellink, and Wulf Schirmer, 581–93. Istanbul: Ege Yayınları.

Özkaya, Vecihi, and Aytaç Coşkun. 2009. "Körtik Tepe, a New Pre-Pottery Neolithic A Site in South-Eastern Anatolia." *Antiquity*. Accessed April 2, 2017. http://www.antiquity.ac.uk/projgall/ozkaya/.

Pilloud, Marin A., and Clark Spencer Larsen. 2011. "'Official' and 'Practical' Kin: Inferring Social and Community Structure from Dental Phenotype at Neolithic Çatahöyük, Turkey." *American Journal of Physical Anthropology* 145 (4): 519–30.

Renfrew, Colin. 2008. "Neuroscience, Evolution, and the Sapient Paradox: The Factuality of Value and of the Sacred." *Philosophical Transactions of the Royal Society Biological Sciences* 363 (1499): 2041–47. https://doi.org/10.1098/rstb.2008.0010.

Rosen, Arlene, and Isabel Rivera-Collazo. 2012. "Climate Change, Adaptive Cycles, and the Persistence of Foraging Economies during the Late Pleistocene/Holocene Transition in the Levant." *Proceedings of the National Academy of Sciences of the United States of America* 109 (10): 3640–45. https://doi.org/10.1073/pnas.1113931109.

Rowley-Conwy, Peter. 2004. *Debates in World Archaeology*. Abingdon: Routledge.

Schmidt, Klaus. 2006. *Sie Bauten die Ersten Tempel: Das Rätselhafte Heiligtum der Steinzeitjäger: Die Archäologische Entdeckung am Göbekli Tepe*. Munich: Beck.

Smith, Bruce. 2001. "Low-Level Food Production." *Journal of Archaeological Research* 9 (1): 1–43. https://doi.org/10.1023/A:1009436110049.

Sterelyn, Kim, and Trevor Watkins. 2015. "Neolithization in Southwest Asia in a Context of Niche Construction Theory." *Cambridge Archaeological Journal* 25 (3): 673–91. https://doi.org/10.1017/S0959774314000675.

Stordeur, Danielle. 2000. "New Discoveries in Architecture and Symbolism at J'erf el Ahmar (Syria), 1997–1999." *Neo-Lithics* 1: 1–4.

Tozlu, Necdet. 2014. "Kültürümüzde Çerçilik ve Çerçi Esnaflığının Bayburt Gümüşhane Erzincan Yöresi Örneği." *Erzincan Üniversitesi Sosyal Bilimler Enstitüsü Dergisi* 8 (2): 23–26.

Watkins, Trevor. 2013. "Neolithization Needs Evolution, as Evolution Needs Neolithization." *Neo-Lithics* 2 (13): 5–10.

Watkins, Trevor. 2015. "The Cultural Dimension of Cognition." *Quaternary International* 30: 1–7.

Willcox, George, Ramon Buxo, and Linda Herveux. 2009. "Late Pleistocene and Early Holocene Climate and the Beginnings of Cultivation in Northern Syria." *Holocene* 19 (1): 151–58. https://doi.org/10.1177/0959683608098961.

Yartah, Taher. 2005. "Les Bâtiments Communautaires de Tell'Abr 3 (PPNA, Syrie)." *Neo-Lithics* 1 (5): 3–9.

7

"Every Man's House Was His Temple"

Mimetic Dynamics in the Transition from Aşıklı Höyük to Çatalhöyük

MARK R. ANSPACH

DOI: 10.5876/9781607327370.c007

The major temples ... invited imitation.
RAMSAY MACMULLEN

Religion resided not in the temples but in the house.

FUSTEL DE COULANGES, *La Cité Antique*

This chapter presents two separate but interlocking arguments. The first concerns a specific architectural feature: the hearth. I will insist on the vital role hearths have played in ritual and history making. There is an abundance of ethnographic and historical evidence for the importance of the perpetually burning hearth fire as a sacred marker of ancestral attachment to place. This may shed light on the repetitive building of hearths that archaeologists have encountered in the Near East and Mediterranean as early as the Neolithic.

Thus, excavations under the Central Court at Knossos suggest that activity was formerly centered on large stone-bordered hearths whose locations did not change between the late final Neolithic and early Minoan I, "indicating that they could have served as fixed points within the Court to which people continued to return over several generations," while hearths in final Neolithic III and IV strata below the Palace at Phaistos "exhibit similarly deep continuities in location and significance" (Tomkins 2012, 43). At Abu Hureyra 2, "Hearths were often set in the same place in successive houses" (Moore, Hillman, and Legge 2000, 265; quoted by Hodder, this volume). At Körtik Tepe, outdoor fireplaces made of flat pebbles, though shifting slightly over time, "were rebuilt repeatedly in the same area" (Benz et al., this volume). And at Boncuklu, hearths seem to have been located "in exactly the same spot" in a sequence of five buildings in Area K (Baird et al. 2012, 225–26). I focus here on Aşıklı Höyük, where I will suggest that the repetitive building of hearths in an even longer sequence can be interpreted as a case of commemorative history making.

More broadly, I will question the idea that hearths are necessarily domestic. For us, hearth and home are synonymous, but at Aşıklı Höyük not every house has a hearth. Indeed, we cannot even be sure that the hearth buildings *are* houses. In this regard, Aşıklı Höyük differs from Çatalhöyük. Another difference between the two sites is the presence of special ritual buildings at Aşıklı Höyük. We can sum up these two differences as follows:

1. At Aşıklı Höyük, certain buildings have hearths; at Çatalhöyük, hearths are a part of every house.
2. At Aşıklı Höyük, certain buildings may be identified as ritual buildings; at Çatalhöyük, there are no such separate buildings, but ritual features are a part of every house.

I contend that these are *two aspects of the same phenomenon*. If we recognize that hearths, whether domestic or not, originally possessed religious value, then the integration of hearths into all the houses at Çatalhöyük is a corollary of the more general integration of religious symbolism into the houses there. But how can we account for this more general phenomenon? Here we come to my second argument, which concerns the role of *imitation* as a possible explanatory factor in the transition from specialized ceremonial centers to more elaborate individual houses.

Humans are mimetic creatures (Garrels 2011). We learn what to do through imitation. Children imitate their parents and, in traditional societies, adults imitate their ancestors. At its simplest, imitation may produce the habituated repetition of existing practices. But it can also be a dynamic force, leading to change when prestigious models give rise to widespread emulation. Imitation in this sense cannot be assimilated to either habituated or commemorative behavior; it is a related but distinct phenomenon.

The kind of imitation I am talking about has been the subject of much discussion among archaeologists studying Bronze Age Crete. Jan Driessen traced how the Minoan Hall became "a fashionable element in domestic architecture" on Crete, spreading from palaces to upper-class houses (Driessen 1982, 58).

Examining data from Ayia Irini on Keos, Manolis Melas noted that Minoan goods and architectural refinements, originally confined to a single house, achieved broad distribution at a later stage (Melas 1991, 184). Malcolm Wiener (1984) coined the term *Versailles effect* to describe the diffusion of Minoan cultural traits, while Jeffrey Soles speaks of "Knossos effects," ascribing the influence of the culture's birthplace to its religious status as a "cosmological center" (Soles 1995). Monumental buildings modeled on the Palace there "were constructed in the outlying countryside in imitation of Knossos to provide a setting for ritual performances that imitated those performed at Knossos itself" (Soles 2002, 131). While these Minoan examples do not precisely correspond to what we find in Neolithic Anatolia, taken together they are very suggestive of how elements originally belonging to special ceremonial buildings might end up distributed widely across individual houses.

SHRINE OR HOUSE (OR BOTH)?

When James Mellaart first excavated Çatalhöyük in the early 1960s, he singled out certain buildings as "shrines" based on the presence of such features as horn installations, leopard reliefs, or vulture paintings. Mellaart's approach would dovetail with the later tendency to view the Near Eastern Neolithic through the prism of the Levant, where clearly identifiable special purpose cult buildings are common. Mehmet and Aslı Özdogan emphasize the distinctiveness of the central Anatolian plateau in this regard. They contrast an Anatolian cultural area, including parts of western Turkey, with a "traditional" Near Eastern zone comprising the Levant, Syro-Mesopotamia, and southeastern Turkey. In the latter zone, cult buildings differ in size, plan, and details of construction. They are "ostentatious when compared to domestic houses" and may be seen as "forerunners of the later monumental temples of Syro-Mesopotamia" (Özdogan and Özdogan 1998, 584–86).

In the much more egalitarian social setting of Çatalhöyük, there are no such monumental structures. Ian Hodder found Mellaart's distinction between "shrine" and "house" impossible to maintain and refers to all buildings on the site as houses. The basic floor plan varies little from one building to the next, and the artwork and ritual installations are not confined to a few shrines or temples but are widely distributed and associated with burial of the dead beneath the floor. Despite the lack of functional differentiation, Hodder suggests that certain buildings may be deemed in some sense "dominant," if only because they have a larger concentration of burials and display a greater concern for maintaining a well-defined internal floor scheme, while some architecturally non-elaborate buildings have no burials at all (Hodder 2006, 151–52). However, the observable differences between buildings do not correspond to differences in indicators of material wealth. Obsidian point densities cannot be correlated with either the degree of architectural elaboration or the number of burials, and bigger or more elaborate buildings do not have a greater number of burial goods. "While some differences between house types were found in some markers of disease and workload, the overall impression," Hodder concludes, "is of a fierce egalitarianism" (Hodder 2014, 5). A fierce egalitarianism that may also be expressed in the notable absence of any ritual buildings other than the houses themselves.

There seems to be no doubt that these buildings *are* houses. After all, there is abundant evidence that people lived in them. Yet each house at Çatalhöyük was also a ritual center in its own right—a veritable small-scale temple. What are we to make of this?

FROM TEMPLE TO HOUSE: HISTORY MAKING, EMULATION, AND THE SPREAD OF SACREDNESS

Perhaps we need to question some commonsense ideas about the nature of houses (Banning 2011). "For us a house is a purely secular building," Lord Raglan observes, "and we never think of it as having anything in common with a temple or church. For many peoples, however, a house is anything but a purely secular building" (Raglan 1964, 9). The example of the Ainu of Japan is instructive:

> The house of the Ainu was a place for worship of the *kamui* almost as much as a house. The Ainu seem never to have had temples. The nearest thing to such was the house of the village chief. This was much larger than ordinary houses and was often resorted to on great occasions such as the Bear Festival . . . In times of emergency or communal anxiety special services of supplication might be held there, but as a rule every man's house was his temple, enshrining the hearth-fire sacred to Kamui Fuchi. (In Neil Gordon Munro's *Ainu Creed and Cult* [1962], quoted in Raglan 1964, 10)

On important occasions, the chief's house served as the village temple, while on an everyday basis, an individual's own house served as the family temple. This illustrates for Raglan "the way in which sacredness spreads from the originally sacred building to ordinary houses" (Raglan 1964, 10).

This process is especially transparent among the Herero of Southwest Africa, where the sacred hearth fire in the chief's compound was the source from which villagers fetched a light in the evening to kindle their own hearths. The hearth was also sacred to the ancient Greeks and Romans. The altar of Hestia at Delphi or Vesta at Rome was regarded as the center of the earth and had its counterpart in each city and each home (Raglan 1964, 79–81). Because of its preeminent ritual significance, the hearth fire was integral to history making at both the collective and household levels.

When the Herero moved their village to a different site, a firebrand carried from the old sacred hearth was used to light the new one (Raglan 1964, 81). Similarly, when the Greeks set out to found a colony, they established a bond of continuity and kinship with their place of origin by carrying with them "seeds of flame" in earthen pots. "Sacred fire from the public hearth of the mother-city would be transferred to the new colony, thus linking the two in a ritual rich with symbolic values," writes Irad Malkin, adding: "In the subsequent history of each colony the memory of its foundation continued to play a central role through the heroic cult accorded to the deceased founder" (Malkin 1987, 2, 133–34).

In Roman legend, the fire of Vesta was identified with the flame Aeneas brought to Latium from Troy. The first Roman emperor, Augustus Caesar, claimed descent from Aeneas through the Julians. When Augustus added to his titles that of pontifex maximus, he highlighted this glorious ancestry. "As the priest responsible for the cult of the state hearth," comments Beth Severy, he became "even more the descendant and reincarnation of Aeneas" (Severy 2003, 107). Rather than moving to the chief priest's residence, or *domus publica*, near the temple of Vesta in the Forum, Augustus dedicated a new shrine to Vesta in a "public" part of his own house on Palatine Hill. The anniversary of this historic event became an annual holiday (Severy 2003, 123), and Ovid celebrated the coming of "a priest risen from Aeneas" with verses invoking the goddess of the hearth: "Vesta, protect your kin's head! The fires, whose sanctities he nurses with his hands, live well" (*Fasti* 3.419–28, quoted by Severy 2003, 102).

The hearth fire was equally important to history-making processes at the household level. Just as Greeks carried fire from a mother city to a new colony, a Greek bride's mother would accompany her to her new house with a torch; the marriage ceremony began at the bride's childhood hearth and ended at the hearth in her husband's home (Malkin 1987, 122–23). Walter Burkert describes a ritual performed in Argos that connects the collective and household levels: "The hearth of a house in which someone has died is extinguished, and after the prescribed period of mourning, new fire is fetched from the state hearth, and the domestic hearth is kindled anew with a sacrifice" (Burkert 1985, 61).

The houses at Çatalhöyük have a hearth on one side and a tomb on the other. Fustel de Coulanges (1984, 30) argues that the domestic hearth originated with the cult of the dead; he cites Servius, a Latin commentator on Virgil, who traces the domestic cult of the Lares and Penates—household gods associated with the hearth—back to what he says was the very ancient usage of burying the dead in the house. Raglan notes that "it has been a common practice to build temples over tombs" and that "in former times the central feature of a church was an altar containing the bones of a saint" (Raglan 1964, 172, 174). If we recognize that houses were originally modeled on temples and indeed built to serve as small-scale temples, then the practice of burial beneath the floor will seem less strange.

In Egypt, the pyramids were originally the tombs of divine kings, but they did not remain the prerogative of the latter forever. Subsequently, remarks A. M. Hocart, "great nobles indulged in small ones," while in modern times, the ancient fashion survived among the peasantry—stepped pyramids, the oldest known type, could still be found in village cemeteries in Hocart's (1952, 132) day.[1] Raglan, Hocart's friend and disciple, points to a similar process of imitation down the social scale in the evolution of English architecture over the previous 600 years: "What we learn from careful dating is that the features of all houses are copied from palaces [i.e., palatial mansions] of a somewhat earlier date" (Raglan 1964, 5). Just as every Ainu's house was his temple, every Englishman's house is his castle—and what is a castle if not the local equivalent of the chief's house?

In Rome, the chief's house stood on the Palatine, the origin of the word *palace*. When Augustus brought the cult of Vesta into his own house, he launched a new fashion. Unlike the Greek Hestia, who figured equally in both civic and domestic worship, the Roman goddess—tended in splendid seclusion by her Vestals—was primarily an embodiment of the *res publica*. In private homes, the cult of the hearth centered mainly on the family's household gods, the Lares and Penates (Bailey 1932, 49, 159; Carandini 2015, 26). But Augustus merged his family cult with the cult of the state. The new shrine to Vesta in his Palatine residence, while not replacing the one in the Forum, restored to the cult the domestic character it may have possessed in an earlier age. Vesta was now, as Francesca Caprioli writes, "also the goddess of the hearth in the house that sheltered her" (Caprioli 2007, 83). In a number of dwellings at Pompeii, depictions of Vesta flanked by household gods attest to her newfound domestic prominence. Andrea Carandini asks whether "the duplication of the cult of Vesta in the public part of Augustus's house is not at the origin of this return of the older domestic Vesta" (Carandini 2015, 76). As Ramsay MacMullen notes, people spontaneously imitated the emperor even in the way they "decorated the walls of their homes" (MacMullen 2000, 113). The example of Vesta demonstrates that such imitation could extend to cult elements. The presence of the hearth goddess in the ruler's Palatine home served as a model for those below. What began as an effort at history making on the part of Augustus set off a chain reaction of emulation in less august houses.

"The desire to emulate one's betters has been a most potent, perhaps the most potent, force in the diffusion of customs," proclaims Hocart (1952, 129). The operation of this force in the case of modern mortuary ritual can be observed in the spread of cremation through the emulation of upper-class preferences in twentieth-century Britain (Parker Pearson 1982, 105). In an influential article, Daniel Miller (1982) showed how the emulation of higher castes in India structured the material expression of rules of purity and pollution, as seen in the evolution of eating vessels from the early Iron Age to the present day. Following Miller's lead, Olivier Nieuwenhuyse (2009) uses emulation to shed light on the "painted pottery revolution" in the late Neolithic Near East: the sudden widespread adoption of technologically advanced and stylistically elaborated Fine Ware across Upper Mesopotamia at the beginning of the Halaf period (6200–5300 cal BCE).

In the present chapter, which compares two Neolithic sites in central Anatolia, I propose that an analogous phenomenon may be at work in the regional evolution from specialized ceremonial centers to more elaborate individual houses. The "desire to emulate one's betters"—or what cultural theorist René Girard calls "mimetic desire," a desire imitated from a model perceived as superior (see in Younés 2012)—might help account for the distinctive configuration of the settlement at Çatalhöyük, where every man's (or woman's) house is a shrine. As we shall see, Çatalhöyük differs in this important respect from the prior settlement at Aşıklı Höyük. But the two sites have something essential in common. They both display a degree of building continuity that is unparalleled outside the central Anatolian Neolithic (Düring 2006, 93).

For centuries, the residents of each community built and rebuilt their dwellings on the same spot and in the same alignment as earlier buildings. In other words, they took as a model the building whose walls they used as a foundation and imitated that model as closely as possible. Against this backdrop of continuity, the break in continuity between Aşıklı and Çatalhöyük is all the more mysterious. Imitation explains repetition, but can it also explain change?

The "painted pottery revolution" offers an example of imitation producing swift and durable change following many centuries of uniformly repetitive behavior. "Curiously enough," writes Nieuwenhuyse, "for over 800 years after the initial adoption of fired ceramics in the Near East ... people had more or less completely abstained from decorating their vessels." They went on making plain pottery just as their predecessors had. But once fancier vessels were introduced as prestige items, they became the new models to imitate and soon replaced plain pottery. At the northern Syrian site of Tell Sabi Abyad, the proportion of decorated ceramics rose from less than 20 percent at around 6300 BCE to more than 80 percent within a century or two (Nieuwenhuyse 2009, 82).

Hocart and Raglan show that ritual institutions spread as they are transmitted down the social scale. People prefer to imitate high-status models—and in societies like that of Çatalhöyük, status is associated with ritual. Manolis Melas observes that cults and rituals "are usually subject to wide imitation or adoption" (Melas 1991, 179). The migration of symbolically charged features from communal ritual centers to individual houses may spring from the same impulse that induces Raglan's African villagers to appropriate a spark of the sacred fire from the chief's hearth and bring it into their own homes. It is understandable that ritual buildings would come to serve as architectural models; they are, in Mehmet Özdogan's phrase, "prestige buildings." Writing of monumental non-domestic structures in the Neolithic, Özdogan notes how "the construction of these prestige buildings, such as special cult buildings or temples," has played an essential role in developing "architectural designs, which in time were adopted to the domestic buildings" (Özdogan 2010, 29).

The adoption in domestic buildings of features borrowed from temples or cult buildings is, I argue, a mimetic phenomenon: it is the result of emulation. Özdogan and Özdogan write that most of the cult buildings in Neolithic sites of southeastern Turkey are "ostentatiously different from all domestic structures" (Özdogan and Özdogan 1998, 585). Turning this formula around, one could add that most of the domestic structures at Çatalhöyük, with its fierce egalitarianism, are *ostentatiously the same* as cult buildings elsewhere. In the following pages, I put this notion to the test by taking a new look at Aşıklı Höyük and asking whether cult buildings there—including some structures previously unrecognized as such—do not foreshadow certain features of houses at Çatalhöyük.

AŞIKLI HÖYÜK AS A PREDECESSOR OF ÇATALHÖYÜK

Although relatively little known, Aşıklı Höyük is one of the most extensively explored Neolithic sites in Western Asia, with 12 percent of its 4.5 hectares having been excavated in the course of numerous digs undertaken since 1989 by archaeologists working under the aegis of Istanbul University (Özbaşaran 2011). First documented by Ian Todd in 1964, the site is located alongside the Melendiz River in an obsidian-rich area on the southwestern outskirts of the volcanic plateau of Cappadocia. In a region where the majority of the settlements use volcanic rock for construction, Aşıklı stands out as an exception. The main building material at Aşıklı throughout most of its 500 years of habitation was *kerpiç*, the mudbrick commonly used in the Konya plain. For this reason, Güneş Duru (2002) argues persuasively that the founders of Aşıklı, the earliest known settlement in West Cappadocia, may originally have come from the Konya plain, bringing their traditional construction technique with them.

Mudbricks were also used in southeast Anatolia, but the building plans and settlement patterns there were very different. From the latter standpoint, as we shall see in a moment, Aşıklı bears a much greater resemblance to Çatalhöyük. The initial samples for C-14 dating taken by Todd "provided results earlier than those obtained at Çatalhöyük, hence giving rise to the idea that Aşıklı would clarify many questions pertaining to Çatalhöyük" (Duru and Özbaşaran 2005, 15). The levels of occupation are now known to extend from approximately 9000 to 7400/7000 cal BCE (Stiner et al. 2014; Özbaşaran 2011), so that, rather tantalizingly, the end of the settlement at Aşıklı would appear to coincide very roughly with the beginning of habitation at Çatalhöyük.

Of course, the chronology alone cannot establish that Çatalhöyük was founded by migrants from Aşıklı, 90 miles to the east. Another clue might be the relatively sudden appearance of domestic sheep on the Konya plain around the time Çatalhöyük was first settled. Douglas Baird suggests that this "may mark the introduction of domesticates from Cappadocia" (Baird 2012, 440), where there is evidence for selective manipulation of sheep at Aşıklı by 8200 cal BCE (Stiner et al. 2014). But there are several other communities in the region from which Çatalhöyük's settlers could have come. Most recently, excavations at the nearby ninth millennium BCE site of Boncuklu, discovered by Baird during the Konya Plain Survey, have turned up a number of the symbolic traits that characterize Çatalhöyük. In Hodder's enumeration, these include "burial in houses, removal of heads, installations on walls, paintings on walls and floors, and separations between 'clean' and oven-related 'dirty' parts of floors." As Hodder notes, some of these elements are missing from the much larger contemporary site of Aşıklı, where "elaborate symbolism is not found within domestic houses." It therefore seems reasonable to see Çatalhöyük as combining elements from more than one source: "Presumably several of these local settlements came together in the foundation of Çatalhöyük" (Hodder 2014, 10).

Although Aşıklı lacks the wall art and installations found at Boncuklu, it is much closer to Çatalhöyük from an architectural standpoint. The buildings at

Boncuklu are "small oval houses set within extensive midden areas" (Hodder 2014, 9–10). At Aşıklı, older-style oval buildings have been found only at the lowest levels (Özbaşaran and Duru 2015). Otherwise, the buildings are rectangular or trapezoidal loam brick structures, densely packed together but without shared walls, making Aşıklı "the first example of the clustered neighbourhood settlements of the Central Anatolian Neolithic." As at Çatalhöyük, there are no doors in external walls of buildings; the only means of entrance would have been from above (Düring 2006, 72, 77–78).

Outwardly then, Aşıklı Höyük looks a lot like Çatalhöyük. Unlike the latter, however, the settlement at Aşıklı includes monumental complexes of apparent public or ceremonial buildings. The main such complex, on the southwest, is separated from the "residential" neighborhoods to the northeast by a wide pebbled street. As described by Ufuk Esin and Savaş Harmankaya, one of the most noteworthy structures in this complex is Building T. "Inside T were post-holes for large wooden posts and a large hearth situated against the east wall," which stood on stone foundations. The floor and internal walls were red. "Along the north, west and south walls ran a low bench also covered with red plaster." Drawing a comparison with the "Terrazzo" Building at Çayönü and the "Temple" at Nevalı Çori, Esin and Harmankaya conclude that Building T at Aşıklı "may also have been a shrine used for religious ceremonies" (Esin and Harmankaya 1999, 124).

Apart from Aşıklı's satellite at Musular, monumental buildings of this kind have not been found at other central Anatolian Neolithic sites. At the same time, while Aşıklı's monumental building complexes surely stand out within the context of central Anatolia, they are decidedly modest by the standards of southeastern Anatolia and the Near East. In public buildings at Nevalı Çori and Çayönü, Bleda Düring writes, "a wealth of stone sculptures, sacrificial slabs, and human remains were found, pointing to a series of ritual activities taking place in these buildings. By contrast, at Aşıklı Höyük these types of features are completely absent in the building complexes" (Düring 2006, 106–7).

Except for the intramural burials of human remains, the same apparent lack of ritual features characterizes the non-monumental buildings at Aşıklı. The features documented in the literature are essentially limited to hearths and "braziers," pits, postholes, bins, and a few grindstones embedded in floors, while ovens seem to be found only in the main building complex. In addition, Düring notes, "isolated references are made to benches and platforms, but it is far from clear how common they are, and where they are located" (Düring 2006, 83–84). In any case, the benches are not hemmed in by aurochs horns; there are no faunal installations of any kind, no animal reliefs or paintings on the walls. Compared to Çatalhöyük, students of ritual may find the pickings slim. Is there no religious symbolism at all in the buildings at Aşıklı?

RETHINKING THE HEARTHS OF AŞIKLI

I would argue that the principal ritual features found at Aşıklı are the *hearths*. As we saw, a large hearth is the most striking feature present in ceremonial building T. This might tip us off to the religious significance a hearth could potentially possess for the inhabitants of Aşıklı. A number of lesser hearths have also been found in small individual buildings. Unlike the one in the large ceremonial building, however, these hearths are commonly assumed to be purely domestic.

It is true that the two kinds of hearth are not identical. Düring characterizes the one in T as "a large round hearth, of a type not found elsewhere in the settlement." The hearths in the other buildings at Aşıklı are quadrangular (Düring 2006, 84, 103–4). However, the evidence from Aşıklı's non-domestic annex at Musular suggests that the shape is of secondary importance. Duru and Özbaşaran describe the A building at Musular, with its red painted floor, postholes, benches, and "fairly large hearth," as "quite similar to the Aşıklı T building" and thus a likely site for ritual activities. Yet the hearth in the A building is quadrangular (Duru and Özbaşaran 2005, 18), suggesting that the similar shape of Aşıklı's smaller hearths would not disqualify them from being ceremonial hearths.

We saw that the ceremonial hearth in Building T was set against a wall with stone foundations. Now, not only are the floors of Aşıklı's other hearths paved with stones, they are surrounded on all sides by upright flat stones forming a wall up to 30 cm high (Bıçakçı and Özbaşaran 1991, 139). As it happens, stone walls are among the defining traits of the ceremonial building complexes at Aşıklı (Düring 2006, 102, 105). On a smaller scale, the stone walls around each hearth fire could also have had ceremonial value.

Moreover, the repetitive construction of hearths in the same locations—an impressive manifestation of the growing concern for temporal depth associated with the emergence of sedentism in the region (Hodder 2007)—strongly suggests that these features were endowed with symbolic importance. In one room of the deep sounding, hearths have been found in every phase of rebuilding. Not only that, but they tend to be located in the same position, occupying first the southeast corner of the room through three successive phases, then shifting to the northwest corner and occupying it over the next three phases (Düring 2006, 95). In another room of the same trench, hearths were present over four successive phases, generally in the southwest corner; then, after a gap of one phase in which the southern part of the room was "blanketed with a fill of loose yellow earth and put out of use," a new hearth was found in the southwest corner, built directly over the locations of earlier hearths (Esin and Harmankaya 1999, 121–23).[2] These sequences demonstrate a remarkable degree of continuity despite the fact that deposits of up to 40 cm of soil may separate successive hearths (Düring 2006, 94–95).

How can we explain the continuity observed in the placement of hearths within individual buildings? Düring remarks elsewhere that marked building continuity has also been found in Mesopotamian temple sequences, but in those cases, he adds, the continuity "was religiously sanctioned and differed from the foundation practices of domestic buildings" (Düring 2006, 94). Could not certain foundation practices at Aşıklı have been religiously sanctioned? Like the use of stone perimeters, continuity in the placement of hearths may be a clue that the buildings containing them were not (or not simply) "domestic" in the modern sense.

In many cultures, the hearth fire must be kept perpetually burning because it is a spiritual presence embodying the continuity of a place. Among the Ashanti, Raglan tells us, it is a crime to change the location of the fire without good cause. Often, the hearth fire has been an object of worship. Raglan cites the example of the fire god Agni in India. Every morning the household gathered around the fire and pronounced these words: "We approach thee, O fire, daily with reverential adoration" (Raglan 1964, 77–78). In the same vein, Fustel quotes these verses from the Rig-Veda: "Before all the other gods Agni must be invoked. We will pronounce his venerable name before that of all the other immortals. O Agni, whatever god we honor with our sacrifice, it is always to you that the holocaust is addressed." The ancient Greeks likewise accorded the hearth fire pride of place. When they assembled at Olympia, the first sacrifice offered was for the hearth fire; Zeus had to make do with the second (Fustel de Coulanges 1984, 27).

Raglan gives many more examples from far-flung times and places, but there is no need to list them all here. I am not making the claim (nor does Raglan) that worship of the hearth fire is universal, only that it is sufficiently widespread to warrant consideration as a possibility when we evaluate hearths in an archaeological context. In the specific case of the hearths at Aşıklı, my thesis is that they are indeed of ritual significance. Naturally, there is no way to prove this, since we know nothing about the positive content of the religion practiced by the inhabitants. Nevertheless, it is possible to identify several more pieces of direct or indirect evidence that point to the symbolic or ceremonial importance of the hearths. Let us first approach the question indirectly by asking whether the hearths are "domestic" or "non-domestic."

ARE AŞIKLI'S HEARTHS "DOMESTIC"?

Marion Cutting associates the presence of hearths with "everyday living" (Cutting 2005, 46), and Bleda Düring takes them to be, in and of themselves, a sign of domesticity: "Hearths seem clear evidence for domestic activities, which one would assume to take place in a house" (Düring 2006, 90). At Aşıklı, however, the latter assumption may be unjustified. Düring remarks on the relative scarcity of internal features "related to domestic production, consumption, and storage, such as grinding stones, bins, and 'braziers', which seem to be present in only a few buildings" (Düring 2006, 92). In a report on the initial excavations, Erhan Bıçakçı and Miriban Özbaşaran originally suggested that houses might have served only for such activities as sleeping "and not for productive purposes" (Bıçakçı and Özbaşaran 1991, 140).

In fact, Düring cites indications that, to an unusual degree, the inhabitants of Aşıklı performed their domestic tasks *outside the house*, in open midden areas: "The excavators argue that these spaces were in use for a variety of activities, including the production of bone, antler and obsidian tools, butchering activities, and processing of plant foods, as well as for the disposal of refuse ... Thus, many of the activities that would occur in a domestic context elsewhere took place in such communal open areas at Aşıklı Höyük" (Düring 2006, 76–77).

Regardless of whether domestic tasks were also performed inside houses, the internal space could conceivably have been subdivided into "living" and "sleeping" areas. Some buildings at Aşıklı contain two or three rooms. If, as Düring surmises, the rooms with hearths are the "living rooms" of domestic units, then one might expect to find the hearths located more often in multi-room buildings. By Düring's own calculations, however, the percentage of multi-room units containing a hearth is smaller than that of the single-room units (Düring 2006, 90), while in their report on the initial excavations, Bıçakçı and Özbaşaran (1991, 139) said they found no hearths at all in multi-room buildings.

Düring's figures differ greatly from those of Bıçakçı and Özbaşaran. As Düring explains, he has redefined the categories of buildings present at Aşıklı. Bıçakçı and Özbaşaran distinguished "one-room buildings" from "houses." Their "houses" correspond roughly to Düring's "multiple-room units." However, Düring has grouped together in the same "units" rooms that share an outer wall, even when, as is often the case, no doorways connect them (Düring 2006,

90). In contrast, Bıçakçı and Özbaşaran reserved the term *house* for buildings in which the enclosed rooms are connected by internal doors. Not only is this definition of a house intuitively satisfying, it has the virtue of producing a very neat opposition between one-room buildings, each of which has a hearth, and houses, which generally lack hearths (Bıçakçı and Özbaşaran 1991, 138–39).[3] If "houses" are defined in this way, then we may class the buildings with hearths as non-domestic structures.

Düring terms the distinction Bıçakçı and Özbaşaran established between houses and buildings with hearths "enigmatic" (Düring 2006, 90). He does not consider the possibility that hearths could have anything but a domestic use. Once one recognizes that hearths may originally have possessed a ritual function, the enigma dissolves. The thesis defended here thus helps shed light on what might otherwise look like an anomalous characteristic of the settlement at Aşıklı: the fact that hearths are predominantly located in special one-room buildings to which there is no direct access from neighboring houses.

The construal of the rooms with hearths as domestic living rooms does not seem to be very compatible with the overall tenor of Düring's theoretical argument, which is to challenge the assumption that individual buildings in the clustered neighborhoods belonged to autonomous households.[4] I find the general thrust of Düring's argument very persuasive. A dilemma he faces is that this theoretical position leaves no good way to calculate the population of the site, always a speculative undertaking in the best of circumstances. Without any clear correspondence between buildings and households, there is little basis for guessing how many people used each building. In the end, Düring estimates the population by "taking the number of living rooms as a measure of the number of households," falling back on the conventional assumption that "living rooms may have been used by core families" whose size can be determined from the size of the living room (Düring 2006, 101). Here, the number of rooms with hearths must serve as a proxy for the number of living rooms, a questionable solution given that the distribution of hearths in the settlement "does not allow for the definition of clusters of rooms centred on a room with a hearth" (Düring 2006, 92).

Where Düring defines the rooms with hearths as living rooms, Cutting proposes that they served as kitchens: "The many small one-roomed buildings containing hearths may have served as dedicated cooking areas" (Cutting 2005, 46). Many features associated with the hearths "suggested that they were used to prepare food," she writes, citing in particular the presence of storage facilities (Cutting 2005, 44). However, this criterion does not seem to be conclusive because she also notes that buildings "with and without hearths were similar in terms of storage facilities" (Cutting 2005, 46). Moreover, Cutting observes in passing that the hearths were "kept very clean" (Cutting 2005, 41). This is a striking contrast to what has been found at Çatalhöyük, where the areas around domestic hearths and ovens look "dirty" to the naked eye and are often called "occupation" floors since they are typically dense with charcoal, everyday discard, and faunal remains (Hodder 2006, 53, 119–22). At Aşıklı, the rooms with hearths are apparently free of such telltale traces of everyday living.

So far, I have tried to show that the hearths at Aşıklı are arguably not domestic hearths. Just as a "non-domestic" building may be a ceremonial building, so a non-domestic hearth may be a ceremonial hearth. This is an indirect or negative argument for the possible religious significance of the hearths. But their cleanliness may well signify more than a mere lack of dirt; it could also be a positive marker of ritual purity. This would hold true regardless of whether food was cooked on the hearths. A hearth's sacred character does not necessarily preclude its being used for cooking, as long as contamination is avoided; thus, the Ainu will place a pot over the hearth fire but take great care not to let it boil over (Raglan 1964, 48).

Plastered floors are another possible marker of ceremonial status; while not uncommon at Aşıklı, they are nearly always present in the rooms with hearths (see Cutting 2005, 43–44). At Çatalhöyük, excavators noticed a fairly consistent opposition—one that seems to hold until the upper levels are reached (Hodder 2014, 15)—between the "dirty" occupation floor, ordinarily found in the southern part of the house, and a "clean" area with white plastered floors in the northern part of the house, where most of the wall art and ritual installations are located. The latter part of the house, with its clearly marked ceremonial status, is also where intramural burials are concentrated (Hodder 2006, 50–51).

HEARTHS AND BURIALS AT AŞIKLI AND KHIROKITIA

This last point brings us to what may be the most compelling reason to view the rooms with hearths at Aşıklı as something more than dedicated cooking areas: the fact that many of these rooms feature sub-floor burials (usually of adult women and children; see Cutting 2005, 36). Intramural burial is the only form of human burial so far discovered at Aşıklı, but the evidence indicates that it was an exceptional practice, not the ordinary way of treating the dead. Düring estimates that burials have been found in only about 10 percent of the rooms excavated. He suggests that the number of burials unearthed to date may under-represent the true number, but he also notes that in the deep sounding, which offers unusually detailed data from a large number of settlement phases, burials were found in less than 5 percent of the rooms (Düring 2006, 86–89). Since the great majority of rooms at Aşıklı do not have burials, one may reasonably attribute a special status to those that do.

It seems all the more significant, then, that sub-floor burials are more likely to be located in the minority of rooms containing hearths. Ufuk Esin already pointed to this pattern in her report on the initial 1989 and 1990 excavation seasons: "Human burials were found mostly under the floors of the rooms containing hearths" (Esin 1991, 130). The frequent presence of hearths was the only discernible regularity Düring could find among the rooms containing sub-floor burials (Düring 2006, 89). While recent excavations indicate that the correlation between burials and hearths is not as consistent as originally thought, it remains noteworthy. According to Güneş Duru (personal communication, 2015), of the approximately eighty-five burials now discovered at Aşıklı, two-thirds have been found in buildings with hearths.

An especially interesting example of the association between hearths and burials comes from the deep sounding discussed earlier. We saw that one room there was found to contain a hearth throughout every phase of its existence. As Düring points out, of the five sub-floor burials found in the deep sounding, three are located in this room: one at the lowest phase and two more in the third phase up (Düring 2006, 95). I would add that the two new burials are associated with the addition of a second hearth in the northwest corner of the same room (Esin and Harmankaya 1999, 122), thus implying an even tighter linkage between the building of hearths and the burial of the dead. After that, a hearth continues to occupy the same corner for three phases before moving to the center of the room in the following phase (see Düring 2006, 94, fig. 4.9). Altogether, that makes for a series of at least seven consecutive phases during which hearths were continuously rebuilt in the same room.

Although one cannot deduce too much from a single example, the discovery of human burials in the bottom strata of this extraordinary sequence is very suggestive. The rekindling of a hearth flame in each successive phase may have been a way to keep alive the memory of important individuals buried beneath or of important events marked by the burials. In that case, the construction of the hearths would serve the purpose of commemorative history making. If the hearths were always built in the same place, one might postulate a purely habituated behavior, but the change occurring in the third phase indicates that more is involved. As we just saw, new burials in this phase are accompanied by the addition of a second hearth in a different position, and hearths in the following three phases then shift to the new location. These facts are most compatible with an interpretation in terms of history making.

In any event, the overall association found at Aşıklı between burial of the dead and hearths is, I think, strong evidence for the hearths' religious significance.[5] Intriguingly, a similar association has been documented at the Aceramic Neolithic site of Khirokitia (Cyprus), where many small pit graves were found under floors of the round houses. George R.H. Wright underscored what he termed the "idiosyncratic" positioning of these graves. Whereas at other sites such as Çatalhöyük, he commented, "house burials are largely arranged in and under benches, at Khirokitia, in so far as there is a particular association, it is with the hearth. Burials are dug against or near hearths, or equally hearths are arranged over burials" (Wright 1992, 61). Wright presumes that a burial found near a hearth is a "house burial." But is a structure containing a hearth and a burial necessarily a house?

In a recent reassessment of the evidence, Vasiliki Koutrafouri questions this assumption made by the original excavator, pioneering Cypriot archaeologist Porphyrios Dikaios. "All buildings at Khirokitia were houses to Dikaios," she explains, "and as such they ought to have had a hearth for strictly domestic purposes" (Koutrafouri 2009, 271). Whenever burning had taken place on a platform covering a burial, Dikaios labeled the platform a hearth. Yet a study of the different contexts in which traces of fire have been found, including burial fills,[6] leads Koutrafouri to conclude that burning often had a ritual function (Koutrafouri 2009, 331). Rather than being motivated by the intentions of cooking or heating commonly associated with hearths, the repeated act of lighting a fire and burning some substance on a platform over a burial may well have been "related to what lay beneath" (Koutrafouri 2009, 290–91). In short, "the hearth may not have been simply a hearth." Likewise, "some dwellings prove not to have been simply 'houses,'" while others come "closer to what we would think as 'temples'" (Koutrafouri 2009, 238–39).

Somewhat fewer than a third of the burials at Khirokitia are concentrated in structures that, though generally not larger than other buildings, feature outsized, architecturally redundant pillars whose symbolic importance came to be recognized in the wake of discoveries made on the mainland at the southeastern Anatolian sites of Nevalı Çori and Göbekli Tepe (Koutrafouri 2009, 258, 262–63). Nonetheless, Koutrafouri cautions that the word *temple* would not necessarily be appropriate for the buildings at Khirokitia if the term were taken in the modern sense to imply worship of a superior being. This caveat certainly applies in the case of Aşıklı as well. By the same token, though, we must take care not to let our analysis of prehistoric sites be limited by modern preconceptions of what a temple or cult building should look like.

RETHINKING ÇATALHÖYÜK'S HEARTHS IN THE LIGHT OF AŞIKLI'S

No architectural features comparable to the pillars of Khirokitia, let alone those of Nevalı Çori or Göbekli Tepe, distinguish the buildings I have singled out for attention at Aşıklı. This renders their special status more difficult to recognize. Still, only a minority of buildings at Aşıklı contain hearths, and even fewer include sub-floor burials. Moreover, there seems to be little evidence that the small one-room buildings with hearths are domestic structures. They are clearly not monumental buildings, but that does not make them houses. I would argue that the hearth buildings at Aşıklı, especially the ones with sub-floor burials, are best understood as *non-monumental cult buildings*. At Çatalhöyük, by contrast, there is nothing special about the coupling of hearths and sub-floor burials in the same building. Both hearths and burials are widely integrated into domestic houses as part of a broader shift that Ian Hodder has described in the following terms: "In central Anatolia at Aşıklı Höyük in the late ninth and early eighth millennia BCE, there are ceremonial buildings but houses are much less elaborate than at Çatalhöyük in the ensuing millennia. At Çatalhöyük a wide range of functions from burial, ritual and art to storage, manufacture and production are more clearly drawn into the house" (Hodder 2006, 242).

Within this range of functions, hearths may be the hardest to classify. Do they belong at the more symbolic end of the spectrum, along with burial, ritual, and art, or is their place with more down-to-earth productive activities? In reality, the two choices are not mutually exclusive. A hearth produces heat and light; it may also represent a symbolic or ritual focal point, as the etymology of the word *focus* attests. In the present chapter I have emphasized the ritual function because this aspect has too often been missing from discussions of indoor hearths at Aşıklı and Çatalhöyük. That may have to do with *where* the hearths are situated inside Çatalhöyük houses: near the ovens on the "occupation floor" in the southern part of the main room, not in the symbolically elaborate northern area where adult burials, ritual installations, and art are concentrated. Hearths and sub-floor burials are present together in these houses, but they are also largely separated. To the extent that hearths are located within a "dirtier" domestic area, it will seem natural to ascribe a purely domestic function to them. And if these hearths are taken as a model for understanding the hearths of Aşıklı, then the latter will be seen as purely domestic, too.

On the other hand, if the hearths of Aşıklı are taken as a starting point and their possible ritual function is recognized, then we may begin to look differently at the hearths in Çatalhöyük houses as well. Such a reversal in perspective could contribute to the growing reconsideration of a too sharply drawn opposition between a sacred part of the houses and a wholly mundane and domestic part. Recently, Hodder highlighted the limits of this opposition: "While it is still the case that there are differences between the activities and features in the southern (hearth) and northern (burial) parts of houses, this is not a distinction between domestic and ritual." Children in particular are often interred in southern areas, while hearths and ovens are occasionally found in northern ones. Countless acts "blur the boundaries" between the everyday and the special, including the deliberate placement under "domestic" floors of obsidian caches and foundation or abandonment deposits (Hodder 2010, 16).

Hints to the special status of hearths at Çatalhöyük may be found by examining another category of deposits that "represent a previously unrecognized aspect of ritual at the site": the collections of animal parts and other items placed under floors during occupation that Nerissa Russell, Louise Martin, and Katheryn Twiss dub "commemorative deposits." These deposits participate in a process of commemorative history making. Thought to be made up of objects that had been used in ceremonies, such deposits "primarily occur in the area of the house (south and west) where ovens and hearths are usually found, often near and in three cases directly beneath these fire installations," suggesting that "these physical memories were associated with the symbolism of the hearth" (Russell, Martin, and Twiss 2009, 121).

One hearth-related deposit in particular deserves scrutiny. Built into the lining of the fire pit for hearth F. 646 in Building 3 (ca. Level VI), it is made up of two human ribs: "The ribs stand out from the rest of the bone in the unit, which is typical of construction material. There is no nearby burial, so they appear to have been deliberately placed here" (Russell, Martin, and Twiss 2009, 108). This is the only instance of the use of human remains in a commemorative deposit, a fact that receives no special comment in a study whose emphasis is on animals. The singular appearance of human ribs in such a non-burial context raises questions that cannot be addressed here. However, the placement of this unique deposit in the lining of a fire pit does suggest that the hearth may possess an unsuspected ritual dimension.

Also worth mentioning is a deposit discovered in Building 33 (ca. Level II). This building belongs to a late Neolithic stratum in the area on the crest of the East Mound excavated by Team Poznań (TP). In addition to horn cores from a wild sheep and a probable wild goat and many small stones, the deposit in question includes large pieces of bone from the meaty portions of the bodies of wild animals (cattle and equids), something more usually seen in feasting deposits. This atypical commemorative deposit "lies below a small hearth and directly above an infant burial," from which it is separated by several centimeters (Russell, Martin, and Twiss 2009, 112). The fact that the hearth itself is located above the infant burial seems significant, especially since this burial is the only one in the building (Russell, Martin, and Twiss 2009, 121). Craig Cessford and Julie Near have pointed to "a possible link between fire and the burial of neonates" at Çatalhöyük (Cessford and Near 2005, 179).

Two additional fire installations are located in the central section of the same building, while an oven is relegated to a small niche in the southwest corner. Arkadiusz Marciniak and Lech Czerniak emphasize that the central location of the fire installations in this building marks "a significant departure from the Early Neolithic patterning" (Marciniak and Czerniak 2012). It is a harbinger of things to come. Central hearths are increasingly found not only in the upper late Neolithic strata of the TP Area on the East Mound but also on the early Chalcolithic West Mound as well as at Canhasan, the other main early Chalcolithic site in the Konya plain (Erdogu 2009, 134). As the hearth rises to prominence in the late phases of occupation at Çatalhöyük, ovens find themselves sidelined. They "increasingly seem to occur in marginal areas in houses or in external areas," Hodder comments. "It is as if the central room of the house became more associated with commensalism, display and social exchange" (Hodder 2014, 15).

Once the hearth occupies pride of place in the central room of the house, it becomes synonymous with home. But the hearth was originally something more than a symbol of domesticity. Its ritual beginnings explain its later centrality. During the main phases of occupation at Çatalhöyük, the ritual status of the hearth is easy to overlook amid the spectacular profusion of religious imagery for which the site is famous. Coexisting with ovens in the more workaday part of the house, away from the area where adult burials are concentrated, the hearths hardly appear glamorous to us. But the disparate destinies of oven and hearth are instructive. The ultimate centrality of the hearth is the outcome of a trajectory that commenced in the more austere ritual environment of Aşıklı Höyük, where hearths played a more visible role in the religious life of the inhabitants.[7]

That, at least, is what I hope to have shown here. At Aşıklı, I contend, the ceremonial buildings include not only those hitherto recognized as such—namely, the monumental complex with a large hearth and the parallel "non-domestic" site nearby at Musular—but also the non-monumental yet arguably non-domestic buildings that contain smaller indoor hearths and sub-floor burials. Then, at Çatalhöyük, all non-domestic ceremonial buildings disappear when both hearths and burials are drawn into the domestic living space as part of a process of emulation that turns every house into a temple.

But when every house is a temple, a temple is no longer what it once was. The main reason hearths appear less important to us in their new setting is simply that they have become more commonplace. Once hearths moved from special buildings at Aşıklı Höyük to ordinary houses at Çatalhöyük, a relative drop in their status was inevitable. Something similar happened in the case of the "painted pottery revolution" cited earlier. As Olivier Nieuwenhuyse remarks, the rapid spread of Fine Ware "resulted in a loss of 'exclusivity' of this once rare novelty. Indeed, during the early Halaf almost every vessel was a Fine Ware vessel" (Nieuwenhuyse 2009, 86). This is the paradox of imitation: when a model perceived as special is copied on a large scale, it is special no more. So it is that the hearth may lose some of its glow.

ACKNOWLEDGMENTS

I am very grateful to Ian Hodder for his comments and suggestions and to Imitatio, a project of the Thiel Foundation, for supporting my work. Special thanks to Güneş Duru for reviewing a draft of the chapter in the light of his firsthand knowledge of the site at Aşıklı; he was kind enough to fill me in on some of the as-yet-unpublished data emerging from the latest excavations. I would also like to express appreciation for stimulating discussions with my colleagues at the LIAS (Institut Marcel Mauss, École des Hautes Études en Sciences Sociales, Paris). I am solely responsible for the interpretations presented here and for any errors of fact.

NOTES

1. Hocart knew contemporary Egypt firsthand, having succeeded E. E. Evans-Pritchard in the chair of sociology at the University of Cairo.

2. In this sequence, Düring assigns hearths to three successive phases (2H-MS, 2G-MS, and 2F-ME) and then one more (2C-C) after a gap of two phases. However, Esin and Harmankaya report a hearth "of the same construction" in 2E-JV, situated directly over the southern part of 2F-ME (Esin and Harmankaya 1999, 122), making for an uninterrupted sequence of four phases, with a gap of only one phase before a hearth returns in the same location. The continuity would thus seem to be even stronger than stated by Düring.

3. In a later (1998) report, Özbaşaran assigns hearths to one-room units in all but two cases (Düring 2006, 84).

4. For a good statement of the case against viewing households as discrete, clearly bounded entities within the characteristic clustered neighborhood settlements of the central Anatolian Neolithic, see Düring and Marciniak 2006.

5. A possible association between the dead and fire is also worth noting. Of forty-eight individual skeletons studied, twenty-six showed various degrees of burning, conceivably as a result of the proximity of hearths; there was no sign that the buildings themselves had burned. While noting how hard it is to link traces of burning on bones to a particular mortuary practice, Le Mort et al. classify Aşıklı as one of a few Near Eastern sites (along with Kebara, Jerf el Ahmar, Dja'de el Mughara, and Nahal Hemar) where the burning of human remains might have had ritual significance (Le Mort et al. 2000, 45–46). Recent excavations at the lower levels of the site have turned up an eight- or nine-year-old child lying with its head on top of the cobblestone floor of a hearth in Building 1 (Özbaşaran and Duru 2015, 48; see also 45, fig. 5).

6. Although burning of skeletons was apparently less common than at Aşıklı (see the previous note), "there were a few cases of human bones found charred." More often, "the burial fill was found full of charcoal and evidence of fire action" (Koutrafouri 2009, 330). Charcoal was often found among the bones at Aşıklı (Cutting 2005, 36).

7. Central hearths also existed at the very beginning of the trajectory, as demonstrated by recent excavations at the early levels of the settlement at Aşıklı. Özbaşaran and Duru (2015, 44–45) report finding a round central fireplace, "entirely different" in form and location from the "typical hearths" of the eighth millennium, in a ninth millennium semi-oval building (B.3, Level 4) and a similar central hearth in a stratigraphically slightly later ninth millennium building, B.1 (the latter hearth is the one where the child's skeleton was found, discussed in note 5).

REFERENCES

Bailey, Cyril. 1932. *Phases in the Religion of Ancient Rome*. Berkeley: University of California Press.

Baird, Douglas. 2012. "The Late Epipaleolithic, Neolithic, and Chalcolithic of the Anatolian Plateau, 13,000–4000 BC." In *A Companion to the Archaeology of the Ancient Near East*, ed. Daniel T. Potts, 431–65. Oxford: Blackwell. https://doi.org/10.1002/9781444360790.ch23.

Baird, Douglas, Andrew Fairbairn, Louise Martin, and Caroline Middleton. 2012. "The Boncuklu Project: The Origins of Sedentism, Cultivation, and Herding in Central Anatolia." In *The Neolithic in Turkey: New Excavations and New Research*, vol. 3: *Central Turkey*, ed. Mehmet Özdoğan, Nezih Başgelen, and Peter Kuniholm, 219–44. Istanbul: Archaeology and Art Publications.

Banning, E. B. 2011. "So Fair a House: Göbekli Tepe and the Identification of Temples in the Pre-Pottery Neolithic of the Near East." *Current Anthropology* 52 (5): 619–60. https://doi.org/10.1086/661207.

Bıçakçı, Erhan, and Miriban Özbaşaran. 1991. "Aşıklı Höyük 1989, 1990: Building Activities." *Anatolica* 17: 136–45.

Burkert, Walter. 1985. *Greek Religion: Archaic and Classical*. Trans. John Raffan. Cambridge, MA: Harvard University Press.

Caprioli, Francesca. 2007. *Vesta Aeterna: L'Aedes Vestae e la Sua Decorazione Architettonica*. Rome: "L'Erma" di Bretschneider.

Carandini, Andrea. 2015. *Il Fuoco Sacro di Roma*. Rome: Laterza.

Cessford, Craig, and Julie Near. 2005. "Fire, Burning, and Pyrotechnology at Çatalhöyük." In *Çatalhöyük Perspectives: Themes from the 1995–99 Seasons*, ed. Ian Hodder, 171–82. Cambridge: McDonald Institute for Archaeological Research, British Institute at Ankara.

Cutting, Marion V. 2005. *The Neolithic and Early Chalcolithic Farmers of Central and Southwest Anatolia: Household, Community, and the Changing Use of Space*. Oxford: Archaeopress.

Driessen, Jan. 1982. "The Minoan Hall in Domestic Architecture on Crete: To Be in Vogue in Late Minoan IA?" *Acta Archaeologica Lovaniensia* 21: 27–92.

Düring, Bleda S. 2006. *Constructing Communities: Clustered Neighbourhood Settlements of the Central Anatolian Neolithic ca. 8500–5500 Cal BC*. Leiden: Nederlands Instituut voor het Nabije Oosten.

Düring, Bleda S., and Arkadiusz Marciniak. 2006. "Households and Communities in the Central Anatolian Neolithic." *Archaeological Dialogues* 12 (2): 165–87. https://doi.org/10.1017/S138020380600170X.

Duru, Güneş. 2002. "Architectural Indications for the Origins of Central Anatolia." In *The Neolithic of Central Anatolia: Internal Developments and External Relations during the 9th–6th Millennia Cal BC*, ed. Frédéric Gérard and Laurens Thyssen, 171–80. Istanbul: Ege Yayınları.

Duru, Güneş, and Mihriban Özbaşaran. 2005. "A 'Non-Domestic' Site in Central Anatolia." *Anatolia Antiqua* 13 (1): 15–28. https://doi.org/10.3406/anata.2005.1034.

Erdogu, Burcin. 2009. "Ritual Symbolism in the Early Chalcolithic Period of Central Anatolia." *Journal for Interdisciplinary Research on Religion and Science* 5 (July): 129–51.

Esin, Ufuk. 1991. "Salvage Excavations at the Pre-Pottery Site of Aşıklı Höyük in Central Anatolia." *Anatolica* 17: 124–35.

Esin, Ufuk, and Savaş Harmankaya. 1999. "Aşıklı." In *Neolithic in Turkey: The Cradle of Civilization*, ed. Mehmet Özdogan and Nezih Başgelen, 114–32. Istanbul: Arkeoloji ve Sanat Yayınları.

Fustel de Coulanges, Numa-Denis. 1984 [1864]. *La Cité Antique*. Paris: Flammarion.

Garrels, Scott R., ed. 2011. *Mimesis and Science: Empirical Research on Imitation and the Mimetic Theory of Culture and Religion.* East Lansing: Michigan State University Press.

Hocart, Arthur Maurice. 1952. *The Life-Giving Myth and Other Essays.* New York: Grove.

Hodder, Ian. 2006. *Çatalhöyük: The Leopard's Tale.* London: Thames and Hudson.

Hodder, Ian. 2007. "Çatalhöyük in the Context of the Middle Eastern Neolithic." *Annual Review of Anthropology* 36 (1): 105–20. https://doi.org/10.1146/annurev.anthro.36.081406.094308.

Hodder, Ian. 2010. "Probing Religion at Çatalhöyük: An Interdisciplinary Experiment." In *Religion in the Emergence of Civilization: Çatalhöyük as a Case Study*, ed. Ian Hodder, 1–31. Cambridge: Cambridge University Press. https://doi.org/10.1017/CBO9780511761416.001.

Hodder, Ian. 2014. "Çatalhöyük: The Leopard Changes Its Spots: A Summary of Recent Work." *Anatolian Studies* 64: 1–22. https://doi.org/10.1017/S0066154614000027.

Koutrafouri, Vasiliki G. 2009. "Ritual in Prehistory: Definition and Identification; Religious Insights in Early Prehistoric Cyprus." PhD dissertation, University of Edinburgh.

Le Mort, Françoise, Aslı Erim-Özdogan, Metin Özbek, and Yasemin Yilmaz. 2000. "Feu et Archéoanthropologie au Proche-Orient (Épipaléolithique et Néolithique): Le Lien avec les Pratiques Funéraires: Données Nouvelles de Çayönü (Turquie)." *Paléorient* 26 (2): 37–50. https://doi.org/10.3406/paleo.2000.4708.

MacMullen, Ramsay. 2000. *Romanization in the Time of Augustus.* New Haven, CT: Yale University Press.

Malkin, Irad. 1987. *Religion and Colonization in Ancient Greece.* Leiden: Brill.

Marciniak, Arkadiusz, and Lech Czerniak. 2012. "Çatalhöyük Unknown: The Late Sequence on the East Mound." In *Proceedings of the 7th International Congress on the Archaeology of the Ancient Near East*, vol. 1: *Mega-Cities and Mega-Sites*, ed. Roger Matthews and John Curtis, 3–16. Wiesbaden: Harrassowitz Verlag.

Melas, Manolis. 1991. "Acculturation and Social Mobility in the Minoan World." In *Thalassa: L'Egée Préhistorique et la Mer (Aegaeum 7)*, ed. Robert Laffineur and Lucien Basch, 169–88. Liège: Université de Liège.

Miller, Daniel. 1982. "Structures and Strategies: An Aspect of the Relationship between Social Hierarchy and Cultural Change." In *Symbolic and Structural Archaeology*, ed. Ian Hodder, 89–98. Cambridge: Cambridge University Press. https://doi.org/10.1017/CBO9780511558252.010.

Moore, Andrew M.T., Gordon C. Hillman, and Anthony J. Legge. 2000. *Village on the Euphrates: From Foraging to Farming at Abu Hureyra*. Oxford: Oxford University Press.

Nieuwenhuyse, Olivier. 2009. "The 'Painted Pottery Revolution': Emulation, Ceramic Innovation, and the Early Halaf in Northern Syria." In *Méthodes d'Approche des Premières Productions Céramiques: Étude de Cas dans les Balkans et au Levant*, ed. Laurence Astruc, Alain Gaulon, and Laure Salanova, 81–91. Rahden/Westfalen, Germany: Verlag Marie Leidorf.

Özbaşaran, Mihriban. 2011. "Re-starting at Aşıklı." *Anatolia Antiqua* 19 (1): 27–37. https://doi.org/10.3406/anata.2011.1087.

Özbaşaran, Mihriban, and Güneş Duru. 2015. "The Early Sedentary Community of Cappadocia: Aşıklı Höyük." In *La Cappadoce Méridionale: De la Préhistoire à la Période Byzantine*, ed. Dominique Beyer, Olivier Henry, and Aksel Tibet, 43–51. Istanbul: Institut Français d'Études Anatoliennes.

Özdogan, Mehmet. 2010. "Transition from the Round Plan to Rectangular—Reconsidering the Evidence of Çayönü." In *Neolithic and Chalcolithic Archaeology in Eurasia: Building Techniques and Spatial Organisation*, ed. Dragos Gheorghiu, 29–34. Oxford: Archaeopress.

Özdogan, Mehmet, and Aslı Özdogan. 1998. "Buildings of Cult and the Cult of Buildings." In *Light on Top of the Black Hill: Studies Presented to Halet Çambel*, ed. Güven Arsebük, Machteld J. Mellink, and Wulf Schirmer, 581–93. Istanbul: Ege Yayınları.

Parker Pearson, Michael. 1982. "Mortuary Practices, Society, and Ideology: An Ethnoarchaeological Study." In *Symbolic and Structural Archaeology*, ed. Ian Hodder, 99–114. Cambridge: Cambridge University Press. https://doi.org/10.1017/CBO9780511558252.011.

Raglan, Lord. 1964. *The Temple and the House*. New York: Norton.

Russell, Nerissa, Louise Martin, and Katheryn C. Twiss. 2009. "Building Memories: Commemorative Deposits at Çatalhöyük." *Anthropozoologica* 44 (1): 103–25. https://doi.org/10.5252/az2009n1a5.

Severy, Beth. 2003. *Augustus and the Family at the Birth of the Roman Empire*. New York: Routledge. https://doi.org/10.4324/9780203211434.

Soles, Jeffrey S. 1995. "The Functions of a Cosmological Center: Knossos in Palatial Crete." In *Politeia: Society and State in the Aegean Bronze Age (Aegaeum 12)*, vol. 2, ed. Robert Laffineur and Wolf-Dietrich Niemeier, 405–14. Liège: Université de Liège.

Soles, Jeffrey S. 2002. "A Central Court at Gournia." In *Monuments of Minos: Rethinking the Minoan Palaces (Aegaeum 23)*, ed. Jan Driessen, Ilse Schoep, and Robert Laffineur, 123–42. Liège: Université de Liège.

Stiner, Mary C., Hijlke Buitenhuis, Güneş Duru, Steven L. Kuhn, Susan M. Mentzer, Natalie D. Munro, Nadja Pöllath, Jay Quade, Georgia Tsarstidou, and Mihriban Özbaşaran. 2014. "A Forager-Herder Trade-Off, from Broad-Spectrum Hunting to Sheep Management at Aşıklı Höyük, Turkey." *Proceedings of the National Academy of Sciences of the United States of America* 111 (23): 8404–9. https://doi.org/10.1073/pnas.1322723111.

Tomkins, Peter. 2012. "Behind the Horizon: Reconsidering the Genesis and Function of the 'First Palace' at Knossos (Final Neolithic IV–Middle Minoan IB)." In *Back to the Beginning: Reassessing Social and Political Complexity on Crete during the Early and Middle Bronze Age*, ed. Ilse Schoep, Peter Tomkins, and Jan Driessen, 32–80. Oxford: Oxbow Books.

Wiener, Malcolm H. 1984. "Crete and the Cyclades in LM I: The Tale of the Conical Cups." In *The Minoan Thalassocracy: Myth and Reality*, ed. Robin Hägg and Nanno Marinatos, 17–26. Stockholm: Swedish Institute in Athens.

Wright, George R.H. 1992. *Ancient Building in Cyprus*. Leiden: Brill.

Younés, Samir, ed. 2012. *Architects and Mimetic Rivalry*. Winterbourne, UK: Papadakis.

8

Interrogating "Property" at Neolithic Çatalhöyük

Rosemary A. Joyce

In his introduction to this volume, Ian Hodder calls for an exploration of two different kinds of history making, through which place and history were articulated in the Neolithic. He draws attention to the role of religion and ritual in making histories through which people in delayed-return systems (Woodburn 1982) were held together long enough for those delayed returns to be beneficial. History making, Hodder writes, can take two forms: "the repetition of practices within buildings as the result of habituated behavior" and "commemorative behavior in which people consciously build social memories and historical links into the past." As Hodder notes, the two forms of history making are intertwined and raise questions of ownership, "rights to land or to animals, to buildings or ancestors."

My contribution to thinking through this issue at Çatalhöyük begins with questions of ownership. I attempt to expand our thinking as archaeologists to encompass what has been called "immaterial property": the rights to things themselves, the performances through which they may be asserted (which could include rituals), and knowledges of a variety of kinds that can be owned and leave their traces in things produced through ownership of knowledge. My specific goal was to explore whether proprietary knowledges might be evident in some of the more modest kinds of materials present in Neolithic sites, not only in the more dramatic kinds of materials, such as burials and architectural elaboration, that have allowed recognition of conscious history making at Çatalhöyük.

To undertake this exploration of ownership and its intersection with history making, I return to the literature on the concept of a social house, based on the writing of Claude Lévi-Strauss (1982), which has been critical in the development of locally grounded concepts at Çatalhöyük such as "history houses" (Hodder and Cessford 2004; Hodder and Pels 2010). This literature is my point of departure in considering what kinds of properties might have been held and transmitted mindfully as part of an estate with which persistent multi-generational groups of people could have identified and on behalf of which delaying returns would have been seen as desirable. The concept of a "house society" is a heuristic—real as a model we create, not as a "stage," "type," or even "social formation" (Gillespie 2000, 2007, 30; Hugh-Jones 1993, 116). In a discussion of Amazonian Tukanoan people, Stephen Hugh-Jones (1993, 95) suggested that as heuristic models, house societies allow simultaneous consideration of two different ways of thinking about social relations—one "emphasizing hierarchy and exclusiveness," which he found "more pertinent to mythological and ritual contexts," and the other "more egalitarian, inclusive," which he saw as "more pertinent to daily life." Both understandings of social relations, he emphasized, worked to describe Tukanoan society. The tension between them is also arguably one that underpinned the development of institutionalized social distinctions over the course of the Neolithic.

By focusing my attention on how apparently interchangeable ceramic vessels were made, I risk seeming to explore only the habitual kind of reproduction of continuity, one of the two kinds of history making under discussion in this volume. There is a long tradition of treating the making of things we think of today as everyday necessities as something mechanical. I suggest that at Çatalhöyük, pottery may sometimes have been part of conscious history making of a commemorative kind analogous to the conscious history making involved in burying the dead in house platforms and embedding animal parts in the walls of some houses. Ultimately, I hope to contribute to understanding what Hodder calls "history making in place" by looking at the final placement of pots, a few complete, others fragmented. As Hodder notes, it is up to the archaeologist to demonstrate "specificity of memory construction." At the moment of origin of settled life, the "curation, circulation, and deposition" of even things we have come to think of as simply made to meet demands of survival, like pottery, have to be seen as potentially forming part of a more conscious creation of links to place and history. While I am concerned directly with the knowledge used in making pottery, other knowledges, including those required for ritual commemoration of ancestors, can be thought of as immaterial property whose ownership distinguished between different groups within Çatalhöyük even before we can clearly see those distinctions in other material registers.

IMMATERIAL PROPERTY

To begin, we need to clarify what kinds of property are significant in the identity of social houses like those used as analogues for Çatalhöyük. Susan Gillespie (2007) argues that the English translation of the original French definition of a *société à maisons* introduced misleading emphases, including equating the estate with "wealth." This she seeks to correct with a new translation in which a house is a "moral person, keeper [*détentrice* is literally a detainer] of a domain composed altogether of material and immaterial property, which perpetuates itself by the transmission of its name, of its fortune and of its titles in a real or fictive line held as legitimate on the sole condition that this continuity can express itself in the language of kinship or of alliance, and, most often, of both together" (Gillespie 2007, 33).

No priority is given to a material estate or wealth over "immaterial property," including names and titles, whose transmission gives continuity to a house. The domain "kept" by the house must be sufficiently well recognized to distinguish it from other houses but need not result in marked material distinctions that an archaeologist could see as evidence of unequal access to resources. This core of the estate is about identity and the creation of enduring history, as in the use of names and titles over generations.

To explore a potential immaterial domain of house estates at Çatalhöyük, I needed to identify some material that would result from the use of differentially inherited rights. Because my own expertise is in ceramic analysis, I was drawn to this material. The first question I asked was, can we see variability in the manner of manufacture of morphologically similar vessels made from the two main Neolithic wares defined at the site? As demonstrated below, the answer to this question is yes. The variation present is sufficient to suggest that otherwise similar pots consumed at the site were products of makers employing different knowledge.

The second research question then became, did members of different social houses use different techniques to produce vessels otherwise classified together in ware and form? The response to this question is a bit more complicated. As discussed below, there is a very limited sample of pottery that can be securely associated with the occupation of specific buildings. The analysis of differential use of alternative techniques requires larger samples than are available from such contexts. Sufficiently large samples were available from midden deposits that built up in spaces adjacent to houses. Assuming that midden deposits reflect behaviors of residents of neighboring houses (or even making a more relaxed assumption, that middens were built up from activities by groups of residents who had access to specific spaces through some form of social identification), we can take middens as proxies for actions of members of different social houses.

The midden deposits analyzed showed different frequencies of the technological styles identified, allowing us to continue to entertain the possibility that knowledge of how to make pots was one form of closely held possession at Çatalhöyük. A complication is introduced by current arguments that much of the Dark Line Ware vessels were made outside the site and brought into it. This would be analogous to an ethnographic case discussed below, in which Native Californian Yurok possess as property the right to own and use objects made elsewhere by others.[1] So we can amend this conclusion to say that knowledge of how to make pots, or knowledge of and rights to acquire pots, was a possession differentiating some residents at the site from others.

The final research question that would ideally be addressed is whether the most distinctive buildings—those with multiple burials, those with elaborate features such as wall paintings and installations of animal parts, and especially those identified as "history houses"—had distinctive patterns of consumption of pottery of different technological styles when compared to other buildings. Analyses of sherds from midden deposits are suggestive of a positive answer to this question, but history houses do not stand out absolutely from a category composed of these three kinds of more complex buildings.

The conclusion that pots that appear to us today to be interchangeable tools of everyday life may actually have been products and signs of distinctive social relations is entirely consistent with the general argument that for the Neolithic residents of Çatalhöyük, there was no division of everyday and ritualized life. The finding that pots of different manufacture with similar pragmatic utility were differentially available is consistent with the image of a settlement where, while absolute hierarchies based on material privilege were absent or muted, social distinction was emergent. The placement of selected pots in the floors of some houses, finally, should encourage us to see pots, as much as animal parts and human remains, as materials used to make histories through commemorative acts.

HOUSE SOCIETIES AND THEIR ESTATES

In an overview of ethnographic studies of house societies, I noted the centrality in many cases of the house building, land, and built features on the land as central to the house estate (Joyce 2000). Yet the estate represented physically by the land and its built features was, in all the cases I discussed, subordinate to two other kinds of property: names and histories. Names of buildings and names or titles given to people are central to house estates because it is through the receipt and use of names that affiliation with a house is publicly evident. Some names are common to multiple house members; others belong to specific roles or positions in houses. Names may seem entirely fleeting and thus immune to archaeological investigation; but names sometimes find expression as physical things, like the crest-bearing objects of the peoples of the Northwest Coast of the United States, a possible parallel to the selective use of different animals for installations and ornaments at Çatalhöyük.

In many of the ethnographic examples I discussed, names are attached to locations in the landscape, whether fields, buildings, or "natural" features. Adopting these names links people not just with human persons who held the name previously but also with the places where those histories are located. Histories are the second widespread kind of house property that is not immediately, self-evidently, visible in physical form—except in places like Çatalhöyük. There, "history houses" concentrated the bodies of persons and other things, along with wall paintings, in spaces whose precise rebuilding over time emphasized the kinds of histories central in house societies: those of continuity and persistence.

Gillespie (2007, 36) notes that names, titles, and histories are only some of the kinds of immaterial properties that can be part of a house estate, adding to the list "ritual and technological knowledge, and the right to craft or display certain objects . . . the rights to perform certain ceremonies, songs, and dances." For example, in a discussion of Hopi curation of oral traditions, Wesley Bernardini (2008, 484) suggests that what have been called "clans" among the Hopi are better described as Lévi-Straussian houses that "express identity primarily through performance of a proprietary ceremony" in which oral traditions reiterate the history of "places where the proprietary ceremony has been performed by a succession of custodial house-groups." Bernardini (2008, 490) argues that it is the "control of esoteric knowledge" that gives one household control of the ceremony that defines that group.

Knowledge is the central example of the category of possessions that are not directly visible archaeologically that I consider in this chapter. Knowledge actually makes up the determinative portion of house estates in the indigenous towns of the Northwest Coast of North America (Lévi-Strauss 1982). While Euroamerican scholars celebrate the *forms* knowledge took (such as house posts and screens, masks, boxes, and other regalia), strictly speaking, these objects are vessels for the important property that houses of high rank display on charged social occasions: crests, titles, songs, and stories.

Ritual procedures are one domain of knowledge that is well documented in ethnographic accounts of house societies, including in cases where houses were not distinguished by material differences. The Saribas Iban of Borneo occupy longhouses that impose uniform apartment size and structure, and no occupants of the quarters in these longhouses accrued differential wealth. Yet in each longhouse, the occupant of one set of quarters was identified as the longhouse "source" who uniquely possessed the rules for rituals necessary for the persistence of the longhouse and its residents (Sather 1993, 72). The "longhouse source" is described as the "caretaker of the central 'source post'" that "centres the house both ritually and in terms of the internal orientation of its parts" (Sather 1993, 70–71).

Another longhouse-dwelling group in Borneo, the Dayak studied by Christine Helliwell (1993), provides a second example. The residents of each apartment literally own the architectural elements that make up that space; yet at the same time, those possessions form part of a single sheltering structure. Each apartment's hearth was "ritually linked" to a root or origin apartment. The senior resident of that apartment Helliwell (1993, 55) describes as "morally, and in some cases legally, responsible for what takes place in such affiliated apartments. In the case of seriously unacceptable behavior . . . he will intervene in the affairs of the household and demand a change in behavior."

Possession of moral authority and ritual responsibility, as in the longhouses of these two groups in Borneo, might well leave no materially visible signs; or it might be the case that the dwelling of a moral authority of this kind would become a focus of important marking, as were history houses and elaborated houses at Çatalhöyük.

The interdependence of ritual authority and pragmatic responsibility for the persistence of the shared building structure are reiterated among the Saribas Iban in the custodianship of each apartment's "heritable estate, including ritual sacra that symbolize the continuing life of the family" (Sather 1993, 72). These estates were primarily composed of whetstones and strains of rice, redundant objects that each group had to have, things we might archaeologically see as interchangeable. These everyday objects can also be seen as material parts of a network supporting practices in which ritual and pragmatic knowledge were mobilized together to support persistence.

KNOWLEDGE AS PROPERTY

Some house-based resource differences, while too subtle to support identifying dominant houses in a hierarchical order, are evident at Çatalhöyük (Hodder 2013b, 26, 2013a, 17). Sheep controlled by the residents of certain houses may have been kept separated, as they displayed distinct diet and pathologies from those of other houses. Diets of human beings buried in specific houses were found to be similar. Subtle distinction was noted in technologies employed within some history houses, including "a slight tendency to greater diversification of ground stone types" (Hodder 2013a, 19). Some groundstone objects, stone beads, stone figurines, and possibly clay figurines were possibly produced in specific houses (Hodder 2013a, 19–20). All of this suggests that in a variety of provisioning activities, the people associated with an individual building shared what Lévi-Strauss called a "fortune," even though those fortunes did not translate into marked advantages in consumption on the part of residents of the longer-enduring history houses.

Particularly intriguing are the reported concentrations at Çatalhöyük of stalactites, concretions, and stone figurines made from them in a few elaborate houses (Nakamura and Meskell 2013, 206). The existence of figurines made from different materials implies distinctions in technical knowledge and knowledge of (access to) these materials among producers. The use of materials like this can also involve the possession of rights to obtain, to work, and to use the material. An ethnographic example may clarify this point.

The Yurok of Northern California were one of Lévi-Strauss's original examples of house societies, based in part on what early anthropologists had described as a lack of rules of social organization, that is, a failure to adhere to normative expectations derived from social theory that aligned social difference and economic difference (Lévi-Strauss 1982, 171–73). Yurok house buildings, individually named and with unique histories, were built using the same materials and had similar interior layout and fixtures (Joyce 2000, 194–95). Stored within houses were the movable material parts of the estate: carved food serving implements (spoons and mush paddles), boxes holding regalia used in dances, and containers (baskets) and tools used for everyday tasks (Joyce 2000, 202). All of these were heritable, and named Yurok houses passed this property on intergenerationally.

Yurok social houses also owned as property the rights to perform ceremonies and to use specific materials in ceremonies (Joyce 2000, 200–202). A striking example was the right to use obsidian bifaces from a specific source (Glass Mountain) in performances of the world-renewal White Deer Skin Dance (Dillian 2002, 84–85, 90–91; Joyce 2000; Kroeber 1905). This source was outside the territory occupied by Yurok towns. Carolyn Dillian (2002, 111–15) analyzed historical sources and concluded that the bifaces owned and used by Yurok people were made by indigenous knappers working at Glass Mountain, a sacred site within the territory attributed to the Modoc people, that she argues were obtained by their Yurok owners through social ties to Karuk intermediaries.

Ownership of particular bifaces was one of the parts of house estates that distinguished leading Yurok houses. Yurok owners could use these bifaces in performances or allow their use by others in return for payment; in either case, they received social prestige for their contributions to the ceremonies necessary for world survival. Much of this would be impossible to intuit from the objects themselves; but the existence of large, strikingly worked objects of distinctive materials implies knowledges: knowledge of sources (geological or human), technological knowledge (directly held or indirectly accessed through social ties), and knowledge of ceremonial requirements and esoteric significance.

KNOWLEDGE IN POTTERY

My trial exploration of differences in technological knowledge at Çatalhöyük treats pottery as a material product of knowledge in the same way obsidian bifaces were for the Yurok. Making pots requires a command of a range of technical expertise developed through social learning. The skills employed in making pottery and the approaches to forming pots have been shown to vary even in a small, early Neolithic assemblage, implying that pottery-making knowledge could have been closely held by multiple groups within a single settlement (Vitelli 1998).

Previous investigators at Çatalhöyük have laid the groundwork for a study of technological knowledge variation through detailed morphological work, showing that a restricted repertoire of vessel forms exists, made in a wide range of sizes (Yalman, Tarkan Özbudak, and Gültekin 2013). The execution of vessels from miniatures to very large examples of the same form is one indication already that there is a distinction in technical ability among potters, like that shown by obsidian biface knappers in California who created very large bifaces for ceremonies and by Native Californian basket makers who exhibit their skills by creating tiny miniatures (Abel-Vidor, Brovarney, and Billy 1996).

Pottery was also apparently not used in large frequencies or even numbers in different houses at Çatalhöyük. Using Estimated Vessel Equivalents as their measure, Nurcan Yalman, Duygu Tarkan Özbudak, and Hilal Gültekin state that based on calculations from sherds, only half of the excavated areas had access to a jar and only 34 percent to a bowl (Yalman, Tarkan Özbudak, and

Gültekin 2013, 180–81). While a projection from sherds is always open to concerns about site formation processes, these authors further describe the contexts of a total of twenty-two complete or semi-complete vessels found in situ. These vessels, they note, are more commonly found in history houses: "'History houses' match with the spaces that contain deliberately placed pottery ... always located in the southeast of the main room ... on the east of the main fire installation(s) or in relation to them, and in relation to the ladder base as well ... It is clear that the placement of these pots with these deposits is not related to their primary function. It might instead be a lineage specific behavior transmitted through generations" (Yalman, Tarkan Özbudak, and Gültekin 2013, 181).

By describing the placement of whole vessels as possibly "a lineage specific behavior transmitted through generations," these authors define what we would expect to see as the outward expression of knowledge possessed by a house. The archaeological understanding of pottery has been framed by the assumption that it arose and persisted as a utilitarian adaptation to needs, impeding our recognition of knowledge of pottery production as something potentially valuable, potential property. Yet Karen Vitelli (1998) reminds us that while pottery became utilitarian, we cannot take the end point of a technology as an explanation of its beginnings.

Inspired by the arguments of the ceramic specialists from Çatalhöyük, I explored variation in traces of techniques used in forming apparently uniform vessels. Examination of ceramics from multiple middens shows variation in technical style of manufacture. Invisible at the morphological level (vessel shape) that we associate with function, variability in forming techniques echoes variability in ware composition previously documented for the assemblage (Doherty and Tarkan Özbudak 2013; Yalman, Tarkan Özbudak, and Gültekin 2013, 147–50).

One way of thinking about these ways of making pottery would be to equate them entirely with habitual action, removing them from consideration as possible techniques of history making. Yet there is more to technology than habit. The kind of traces of different ways of making pottery that I recorded are products of what Heather Lechtman (1977, 6) defined as technological style: "Technical modes of operation, attitudes towards materials, some specific organization of labor, ritual observances—elements which are unified nonrandomly in a complex of formal relationships."

These aspects of technological style include activities that cannot be thought of merely as habits. Olivier Gosselain (2000, 191–93) argues that the primary forming of vessels is produced through deeply engrained motor habits expressed in bodily gestures that are unlikely to be consciously altered once acquired. Primary forming may not leave visible traces, as these can be eradicated by actions that come later. The visible variation produced by secondary forming actions, Gosselain (2000, 191) notes, "allows a wide range of people to be aware of potters' behavior and, consequently, to influence potters' choices of techniques." Observations related to secondary forming, the focus of my analysis, lend themselves to examining more conscious forms of history making.

In a situation where pottery is relatively rare and not uniformly distributed, its differential consumption or production may reflect uneven control of knowledge or uneven rights of access, like those that are more obvious when the objects in question match our expectations of "ritual" things. This is particularly well illustrated by analysis of pots installed in building floors and buried during the renewal of buildings. I found that these practices deploy pots with unusual technological style more often than would be expected. Using pottery in installations marks the products of this technology as a material similar in use and significance to bucrania, reemphasizing that at Çatalhöyük, domestic/ritual distinctions cannot be projected onto people's lives.

TECHNOLOGICAL STYLE AT ÇATALHÖYÜK

Identifying traces of variation in technological style began with examination of sherds in a type collection organized by the ceramic specialists and recording of complete or almost complete vessels. Next, sherds from two series of midden deposits were completely recorded and the presence of different technological features was analyzed. Finally, a large sample of sherds in two additional middens and smaller samples from a series of houses were reviewed. In all, data were recorded on 5,665 sherds or vessels, including 75 from Mellaart's collections.[2]

Type collection sherds demonstrated the use of slab building in primary forming (called "the lamination method" in Yalman, Tarkan Özbudak, and Gültekin 2013, 155). In slab building, pieces of clay are overlaid and joined to each other through pressure supplied by the secondary forming paddle and anvil technique, noted as hammering in recent publications. Observations for this study were intended specifically to document these primary and secondary forming techniques and offer a fuller picture of their use than have previous studies. Overlapping slabs were noted on thinner jar forms as well as thicker bowls, where the slab construction is more evident because of the thickness of vessels, especially on and near the base.

The use of slab building was noted on 196 sherds, including bowls (27 of 112 examples identified) and hole mouth jars (16 of 377 examples identified).[3]

Analysis of a sample of 9 sherds with visible slab junctions selected for study using X-rays, consisting of 8 Dark Line sherds and 1 Light Line sherd, confirmed that they were built of overlapping slabs of clay. The lower proportion of hole mouth jars with noted evidence of slab building is most likely a result of more complete finishing of these vessels and of a more complete binding of overlapping slabs possible on thinner-walled vessels. Where slab boundaries were noted on hole mouth jars, the slabs were visible because of incomplete binding leading to separation between slabs, laminar erosion of vessel walls, and the presence of notable voids where slabs came together.

The fact that slab-built pottery was further shaped using the paddle and anvil technique provided a way to begin to assess the existence of different technological styles among the potters provisioning Çatalhöyük. Paddling leaves distinctive marks on both the exterior (where the paddle makes contact) and interior of vessels (where a support is used) (figures 8.1 and 8.2). Guillermo De la Fuente (2011, 93) and Patrick Carmichael (1986, 35–38) describe and illustrate the main signs of use of paddles: facets on the exterior, corresponding concavities on the interior, and in X-rays, flattened voids and cracks radiating from mineral fragments. Use of this technique produces vessels with variable thickness that does not align parallel to the rim or base and may yield an undulating surface. Marks can be eradicated by later steps in vessel construction, such as burnishing.

Adopting a conservative approach, paddle marks were only recorded when a complete facet or facets on the exterior of the vessel could be defined (excluding examples where facets had produced an undulating surface but facet boundaries were softened by later treatment, such as burnishing). An interior mark corresponding to the use of a support was also required. Even with this conservative definition, paddle marks were visible on at least 903 of the sherds reviewed (16%). A large proportion of the sherds recorded were burnished (26%), some very completely (6%). This includes sherds on which traces of the paddle and anvil technique were still visible (46% showing exterior burnishing, 32% showing interior burnishing). Thus, the 16 percent figure for paddle marks is a minimum.

The use of paddles allowed recording of three aspects of secondary forming that reflect different technological style and thus different knowledge. These were the size and shape of the paddle, the spacing of paddle marks, and the placement of paddle marks (whether evenly spaced in rows or placed in overlapping clusters).

Paddle size and shape is inferred from the size and shape of the impact zone on the exterior of sherds. The majority of the impact zones were from 0.5 to 2 cm at their greatest dimension. Most of these were asymmetric (oval). On some,

FIGURE 8.1. *One of the more typical oval marks on the exterior of a sherd. Photograph by Russell Sheptak.*

FIGURE 8.2. *Showing the internal facet that is the trace of the anvil. Photograph by Russell Sheptak.*

pits were noted at the ends of the long axis. A very small number of sherds (less than 2% of the sherds with paddle marks) had well-defined round impact marks ranging from 1 cm to 2 cm in diameter. A slightly larger number of paddle-marked sherds (close to 4%) had significantly larger impact zones, some with well-defined margins suggestive of carefully shaped paddles with rectangular or squared outlines. The use of multiple larger paddles is indicated by variation in dimensions. Two size ranges of roughly square paddles are evident, one approximately 5 cm wide, a second 2.5–3.5 cm wide. Two different rectangular paddles are evident, one 5 cm long by 1.5 cm wide, the other 6 cm long by 3.5 cm wide. The use of different paddle shapes and sizes is an indication that there were multiple ways of making these pots (multiple technological styles).

Placement of paddle marks provided further indications that the makers of these pots employed multiple approaches. Spacing of paddle marks could only be recorded when more than one paddle mark was visible on a sherd, a total of 398 paddle-marked sherds (44%). Three patterns of distribution of paddle marks were identified: regularly spaced rows, in which the distance between paddle marks was consistent (16% of the sample of paddle-marked sherds); rows in which the distance between paddle marks was not consistent (12% of the paddle-marked sherds); and application of paddle marks in clusters, sometimes overlapping (11%). There was no correlation between spacing of paddle marks and vessel form, ware, or area of the vessel that was marked. These were three different gestural repertoires for accomplishing secondary forming, learned by potters within their own communities of practice, deployed by them as the right way to go about the task. They can be seen as the trace of different knowledge.

Based on the detailed recording of paddle use showing that there were multiple ways to complete the secondary forming of what otherwise look like similar vessels, the next step was an analysis of the presence of products of these different approaches in a sequence of middens to see if there was variation in consumption of pots made in these different ways. Initially, this involved recording of two sequences of middens (table 8.1) from superimposed Spaces 259, 260, 261, and 132 and from Spaces 129, 130, 131, 299–305, 314, 319, and 339 (Regan 2014; Sadarangani 2014b), all in the South Area. Subsequent recording added data on midden deposits in Spaces 226 and 279 from the 4040 Area. The goal was to explore a number of potential dimensions of variation: proximity to (and possible association with) a history house sequence, change over time, and variation between houses with multiple burials and other elaboration and those without. Finally, recording of pots installed in house floors was added to the analysis.

While the number of sherds generally precludes statistical testing of significance, there were marked differences in the presence of different approaches to secondary finishing in every area recorded. The conclusion that there were distinct localized groups using pottery made by different practitioners was further reinforced by the results of recording of pots installed in house floors. These unique vessels, whose placement can be compared to other installations

TABLE 8.1. Stratigraphic relationships of buildings (B) and spaces (S) comprising midden deposits from which sherds were recorded

Buildings	Middens	Buildings	Middens
		B 10	Sp 131
		B 44	Sp 129
			Sp 130
			Sp 319
B 42	Sp 259	B 56	Sp 339
B 53	Sp 260	B 65	Sp 299
	Sp 261		Sp 299–305 Sp 314
	Sp 132	B 75	Sp 333 Sp 329

in houses, had traces of the most uncommon and varied techniques found in all the pottery recorded. Masked by the appearance of uniformity in preference for mineral-rich clay bodies firing to dark colors and by the emphasis on production of a limited range of vessel forms, differences in knowledge about how to make pots crosscut Çatalhöyük.

Middens in the South Area

The sequence of Buildings 10, 44, 56, and 65, adjacent to middens deposited in Space 129 and the others listed with it in table 8.1, is identified as a history house (Hodder and Farid 2014, 5). Buildings 53 and 42, which were sequent but not part of a continuous history of reuse, were occupied during the same phases as Buildings 65 and 56. Because they were definitely not rebuilt, these are not part of history houses, although they might (if completely excavated) have been included in the category of houses with multiple burials. Building 53 contained at least eight burials (Sadarangani 2014b, 191). Building 42 contained six burials, one interpreted as a foundation burial, as well as installations of pots in the floor (Sadarangani 2014b, 203–8). Installations of pots in house floors were also found in the history house sequence from Building 65 to Building 10.

A total of 1,816 sherds from the South Area middens were recorded. With one exception (Space 131, discussed below), the proportions of paddle-marked sherds in each midden were within 7 percent of the overall 24 percent proportion in the entire assemblage from these contexts. If all technological styles were equally available or equally appreciated, we would expect the proportions

TABLE 8.2. Results of analysis of technological style represented in sequence of middens associated with the history house occupying Buildings 75 to 10

Space	Paddle-Marked Sherds (N = 434, 24%)	Regular Spacing (N = 117, 27%)	Irregular Spacing (N = 18, 4%)	Clusters, Overlapping (N = 54, 12%)
131 (N = 49)	3 (6%)	1 (33%)	0	0
129/130/319 (N = 659)	164 (25%)	42 (25%)	4 (3%)	18 (11%)
339 (N = 305)	67 (20%)	19 (28%)	3 (4%)	6 (9%)
299, 299–305, 314 (N = 363)	78 (18%)	16 (21%)	3 (4%)	12 (15%)
329, 333 (N = 138)	33 (24%)	7 (21%)	1 (3%)	6 (18%)

of each to closely mirror the overall proportions of the three technological styles in the entire midden assemblage (regular spacing 27%; irregular spacing 4%; clusters 12%). This was not the case. Midden deposits adjacent to the history house sequence that formed during the beginning of the sequence diverged, with a higher proportion of sherds shaped with clusters of paddle marks (table 8.2). During roughly the same period of time, middens adjacent to the earlier of the sequent buildings that did not constitute a history house (table 8.3) initially demonstrated proportions of the three technological styles consistent with the assemblage overall (Space 132) but later demonstrated an even higher frequency of vessels with paddle marks in clusters (Spaces 260/261).

There is no technological advantage to using clusters of paddle marks; it is a minority technique overall. While the sample size for the midden that followed in this sequence (Space 259) is small, the overall proportion of sherds with paddle marking is consistent with the assemblage as a whole (26% in this space, compared with 24% overall; table 8.3). The proportion of paddle-marked sherds that were finished in clusters is the highest of all the contexts recorded in this project. The higher proportion of sherds made this way in specific middens is arguably a product of differences in access to or preferences for pottery made in distinct ways.

In the later phases of the history house sequence, the frequencies of the three different technological styles in middens in Spaces 339, 129, 130, and

TABLE 8.3. Results of analysis of technological style represented in sequence of middens associated with the multiple burial(?) houses Building 53 and Building 42

Space	Paddle-Marked Sherds (N = 434, 24%)	Regular Spacing (N = 117, 27%)	Irregular Spacing (N = 18, 4%)	Clusters, Overlapping (N = 54, 12%)
259 (N = 46)	12 (26%)	2 (17%)	1 (8%)	5 (42%)
260/261 (N = 168)	53 (31%)	9 (17%)	1 (2%)	11 (21%)
132 (N = 88)	26 (29%)	8 (31%)	1 (4%)	4 (15%)

319 mirrored the general proportions in the overall assemblage (table 8.2). In this sequence of middens, it is during the earlier periods of occupation when we see variability in knowledge of pottery production techniques reflected in varying use of products of different communities of practice. This is not a generalized chronological distinction; the proportions of different technological styles present in middens of the same phase in different locations vary. A higher proportion of vessels finished with paddle marks in clusters stands out in the analysis of these middens as a distinctive characteristic, an observation considered more below.

MIDDENS IN 4040 AREA

Space 226 (Sadarangani 2014a) and Space 279 (Yeomans 2014) have been identified as middens with complex histories, locations of a variety of activities assigned to the same stratigraphic level but with middens in Space 226 developing later than those in Space 279. These middens produced very large ceramic assemblages, making them ideal places to explore whether the kinds of patterned differences seen in smaller samples would still be found in larger samples or would disappear, implying that they were effects of sample size.

The midden in Space 279 buried Building 66, and Buildings 70 and 71 were built over it. The space in which this midden formed was bordered by earlier Buildings 55, 57, and 64 and by Building 60, the final stage of history house B.59-60 (Farid 2014, 122–24; Hodder and Farid 2014, 5). Space 226 is a later midden in the same general area, bordered on the north by Building 54 and on the south by the earlier Building 45 (Farid 2014, 124, 126–27). Building 45 was assessed as a large/elaborate building that was partially burned at

TABLE 8.4. Results of analysis of technological style represented in midden in Spaces 279 and 226

Space	Paddle-Marked Sherds (N = 407, 12%)	Regular Spacing (N = 76, 19%)	Irregular Spacing (N = 11, 2.7%)	Clusters, Overlapping (N = 32, 8%)
279 (N = 2,189)	331 (15%)	57 (17%)	11 (3.3%)	23 (7%)
226 (N = 1,328)	76 (6%)	19 (25%)	0	9 (12%)

abandonment. Building 45 contained an installed pot in the floor. With the exception of Buildings 60 and 45, which preceded the accumulation of these middens, the other buildings in the area were not identified as distinguished by size, elaboration, number of burials, or persistence over time as part of a history house.

A total of 3,517 sherds from these two middens, 407 of which were paddle marked (12%), were recorded. Paddle marking was noted on 331 of the 2,189 sherds from Space 279 (15%). Of 1,328 sherds from Space 226, only 76 (6%) were paddle marked (table 8.4). These are much lower proportions of paddle-marked sherds than identified in the previously recorded middens, where 24 percent of sherds, on average, showed paddle markings. The midden deposits in Spaces 226 and 279 overlap in time with those previously described, so this difference cannot be accounted for as a historical development. Further, there were no obvious differences in preservation that might account for this. While it is possible that the average sherd size in these middens was smaller, for the purposes of this analysis the difference is simply noted, and the lower proportion in this midden assemblage is used as the baseline to compare the middens.

There is a marked difference in the proportion of sherds with preserved paddle marking between the two middens. There was no systematic difference in preservation of sherd surfaces that would account for lower success in identifying signs of paddle marking in Space 226, and the number of sherds examined (1,328) is double the number in the largest midden deposit previously recorded. The same range of wares was present as in other deposits, in similar proportions. Yet only 6 percent of the recorded sherds in Space 226 showed preserved paddle marks. This suggests that the people responsible for the deposits in Space 226 were participating in use of pottery made with a significantly different dominant technological style. The low proportion of paddle-marked sherds here is paralleled in one midden in the previously described series, in Space 131. There, because the absolute number of sherds present was low, I was hesitant to suggest that the low frequency of paddle-marked sherds was significant. Like the midden deposit in Space 226, the midden deposit in Space 131 has few sherds with preserved paddle marks. None of the sherds in Space 131 had evidence of the technological style resulting in either irregularly spaced impacts or the clustering of impacts. In general, this is consistent with the dominance of regularly spaced paddle marks on the majority of sherds that preserve any paddle marking in Space 226, although there are a few sherds with clusters of paddle marks.

Discussion

Viewed as a series of deposits, the middens recorded showed four main patterns of variation in paddle marking. Two had very low proportions of paddle-marked sherds, with most of those present found regularly spaced in rows (Space 226, Space 131).

In most other midden deposits, paddle-marked sherds make up a similar proportion in each midden in a local area and show similar frequencies of different secondary finishing choices. The majority of paddle-marked sherds are regularly spaced impacts, about 4 percent are irregularly spaced impacts in rows, and a slightly higher proportion (9%–15%) have impacts in clusters, often overlapping.

One set of middens (in Spaces 260 and 261) had a much higher proportion of paddle-marked sherds than other middens in the same area. A very high percentage of these have paddle marks in clusters (21%), while the proportion with paddle marks in irregular rows was half the overall average for middens in this area.

Two other midden deposits, while having an average number of paddle-marked sherds, had higher proportions of sherds marked in irregularly spaced or clustering fashions. The most striking of these is the deposit in Space 259. Almost half (42%) of the paddle-marked sherds in this deposit were marked in clusters, and another 8 percent (twice the overall average) of marks occur in irregularly spaced rows. While at first the middens in Spaces 329 and 333 seem typical (with 24% of the sherds paddle marked and 3% of those in irregular rows), the proportion marked with impacts in clusters (18%) is higher than all but the middens in Spaces 260 and 261 and in Space 259.

From this overview, it becomes apparent that the relatively rare technique of overlapping impacts in clusters is present in varying amounts in different areas and is contributing the most to distinctions among midden deposits.

If we take adjacency to specific houses as a proxy for the possible residence of the people who contributed to specific middens, three of the five middens with distinctive patterns of technologies of paddle marking were associated with a history house sequence (Spaces 329 and 333, 260 and 261, and 131). This is proportional to the number of such middens in the 10 deposits recorded (6 of 10). Not all middens adjacent to the history house sequence showed distinctive patterns, there was no sustained pattern of distinctive technologies over time in the history house sequence, and the most distinctive midden in terms of the presence of the most unusual techniques was in Space 259—not associated with a history house.

Does this mean that technological knowledge was not a form of house property, a potential part of the estate of a history house? Unfortunately, it is not so simple. While a succession of rebuilt buildings with large numbers of burials exhibits the success of memory making, the formation of an identifiable history house is a product of specific sequences of events. The occupants of other buildings might have been interested in creating the same kind of persistence but experienced less success. This might be the case with individual houses with multiple burials (one form of history making) and those with various kinds of elaborations that also advanced the formation of memory, what have been described as elaborate buildings.

The midden deposits in Space 226 were located in an area where Building 45, a large/elaborate building that was partially burned at abandonment, had been located. This building was not rebuilt and thus did not become part of a history house sequence. Yet it exhibited features resulting from forms of history making and could have developed into a history house had events proceeded differently.

Building 45 also contained an installed pot in the floor. If we treat pottery as potentially a product of differentially held knowledge, installing pots in the floor of buildings can be seen as a possible form of history making not unlike burial of human residents or the placement of bucrania. Recording secondary forming techniques used on emplaced pots demonstrates a higher than expected use of distinctive and uncommon techniques, suggesting that these pots were selected from distinctive sources.

EVIDENCE OF POT INSTALLATIONS

The best argument for seeing the knowledge of how to make pottery as distinctive of residents of different buildings was already made in the most recent overview of the site's pottery by project ceramic analysts. Yalman, Tarkan

Özbudak, and Gültekin (2013, 182) conclude that "pottery might not have been used very much on a daily basis ... pots seem to have been used in certain social events, not necessarily feastings, but household gatherings ... the production and usage of pottery at Çatalhöyük was not a regular occurrence and may not have been available to everyone."

As part of the present analysis, data were recorded for sixteen whole or partially complete vessels (table 8.5).[4] This included four of seven Dark Line hole mouth jars set into the floors of buildings and eight of nine vessels placed in architectural fill or room infill. Seven of these were in a history house sequence, in a multi-burial building, or in an elaborate/large building. Two of these twelve vessels were bowls, the rest hole mouth jars.

With the exception of one bowl, all of the whole pots preserved evidence of the use of paddles, despite the majority being completely or almost completely burnished. On the eleven vessels with paddle marks, a disproportionate number showed irregular spacing or clustering of paddle marks (five of nine with evidence of mark spacing, 55% of the recorded vessels). Only the midden in Space 259 approaches this proportion of the generally less common forms of paddle marking (with 50% of sherds having clusters or irregular combinations of paddle marks).

Even more striking is the fact that many of the whole vessels were marked using larger paddles (six of eleven, 54%). These are also uncommon in the sherd samples (making up less than 4% of the sample). The whole vessels also showed the use of unusual burnishing techniques or tools compared to the entire recorded sample.

The records of these whole vessels were made at the first step of analysis, so there was no way the outcome of later analyses could have influenced recording. The divergence from the norms in the sherd collection as a whole strongly suggests that these pots were selected or obtained from different makers than the majority of the apparently similar bowls and jars that were used, broken, and discarded around them.

IMPLICATIONS

Does the analysis presented here prove that knowledge of pottery production techniques was immaterial property that could have formed part of and differentiated house estates at Çatalhöyük? That would be overstating the case. What I hope to have done is demonstrate that differential knowledge was involved in the making of pottery used at Çatalhöyük that we might be tempted to view as of uniform value because of the homogeneity of ware and form.

TABLE 8.5. Whole or partially complete pots with recorded paddle marks

Location	Object Numbers	Installed in Floor	Paddle-Mark Spacing and Paddle Dimensions
B.42	5430/S1	Yes	Regularly spaced 4.5 cm² paddle
B.44	11235 X1	Yes	Irregularly spaced common oval paddle
	10664 X1	No	Regularly spaced common oval paddle
	11446/S2	Yes	Irregularly spaced 3 cm² paddle
B.45	10044/S2	Yes	One mark 5 cm² paddle
	10061/S9	No	Clusters of common oval paddle marks
B.53	5849 X1	No	Irregularly spaced 3 cm² paddle
B.54	11924/S7, S8	No	Regularly spaced 3 cm² paddle
B.63	13970 X1	No	Regularly spaced 3.5 cm² paddle
	13925/S1	No	Irregularly spaced 2.5 cm² paddle
B.88	14103 X2	No	Clusters of common oval paddle marks
Space 440	17869 X1	No	Clusters of common oval paddle marks
Space 279	13159/S7	No	Clusters of common oval paddle marks
	13159/S8	No	Regularly spaced 2.5 cm² paddle
	14186/S66	No	Clusters of common oval paddle marks

Pottery produced by different groups using distinctive approaches to secondary finishing was used unevenly by the people who created different midden deposits, even at the same time. Pots selected for installation as visible parts of buildings and others deliberately placed within architectural fills were more apt to be finished with distinctive, less common approaches than pottery used in general.

Of course, technological style actually encompasses much more than a single finishing technique. Ideally, we would also assess the evidence of differences in burnishing techniques, for which data were collected. Paddle marking, however, is particularly interesting for an examination of knowledge property because paddles themselves have been owned as possessions in ethnographic and archaeological cases.

One archaeological example is provided by the burials from the Ekven site, where nine burials yielded paddles for finishing pottery (Arutiunov and Sergeev 2006). These are interpreted as possessions of families and as mnemonic devices for the recitation of origin narratives (Arutiunov and Sergeev 2006, 145). At least some of these paddles were worked in ways that would transfer markings to pots. In this case, paddles and marking of pots with paddles were arguably part of history making.

A second archaeological example is offered by the Swift Creek Complex of the southeastern United States. Movement of paddles and of pots produced from specific paddles can each be documented, forming separate trajectories accounted for by the actions of small social groups (Wallis 2008, 255–59). One archaeologically identified paddle was used for different vessels when it was freshly carved and, later, when it was worn and cracked. Paddles that were curated and possibly became heirlooms would fulfill the requisites for evidence of more conscious history making outlined by Hodder (this volume).

While other technologies with restricted distribution have been noted at Çatalhöyük, my choice of exploring pottery was not arbitrary. It contributes to challenging easy divisions archaeologists in the past made between the everyday and the ritualized, which work at the site has shown are of questionable utility. Archaeologists traditionally understood pots as pragmatic necessities responding to provisioning needs. Even when pragmatic motivations for making pots are strong, however, there always exist a range of ways to make a pot and a range of pots that can be made. Once created, fired clay vessels can mark particular moments in time, particular activities, or particular groups of people as distinctive. Creating pottery itself can even be ritualized, understood as an act of creation.

At Çatalhöyük, some buildings gather together unusual numbers and striking kinds of things, including murals, installations that we recognize as special (such as bucrania), and burials, which suggest concerns with cosmogony, religious practice, and the spiritual reproduction of the world. As forms of knowledge property, accounts of the way humans and other beings relate in the world and at its beginnings are accepted to have constituted bases for human action, including attracting affiliation by people whose relocation could shape a settlement at its beginnings and continue to provide its structure thereafter.

As ethnographic examples show us, such narratives need not be generalized for the entire community; they may articulate generalized propositions with details that distinguish one group from another (Bernardini 2008, 485–87, 491–92, 499). Other technologies that are less obviously engaged in world making could have played important roles in its everyday recapitulation and periodic restoration, as obsidian bifaces did and do in World Renewal ceremonies in Northern California.

This chapter has not explored these implications in any depth. What it has shown is that within the muted material distinctions of Neolithic villages there may exist traces of immaterial forms of difference that could have been socially salient, productive of prestige even before they were productive of hierarchy. The heuristic model of house societies is helpful in drawing our attention to the importance of immaterial property, something that may be harder for archaeologists to see but not, I hope I have demonstrated, impossible.

NOTES

1. There is evidence suggestive of production of pottery at Çatalhöyük, the focus of a separate paper in preparation.

2. Recording of sherds took place over the course of three seasons, in 2012, 2014, and 2015. In 2015 I was assisted by Russell Sheptak, who worked in the lab at the same time after being trained to make observations in the same way I did. A comparison of the data we each recorded in 2015 shows no significant differences that suggest interobserver error affected this analysis.

3. Identification of building technique was not a required step in analysis. Primary forming technique was noted when it was evident during examination of sherds for secondary forming techniques. Thus, these proportions are a minimal account of the presence of slab building. Only one possible example of coiling was noted. While coiling has been described as the likely main primary forming technique of pottery from Neolithic Çatalhöyük, this analysis does not support that claim.

4. Six other complete vessels were unavailable for recording.

REFERENCES

Abel-Vidor, Suzanne, Dot Brovarney, and Susan Billy. 1996. *Remember Your Relations: The Elsie Allen Baskets, Family, and Friends*. Ukiah, CA: Grace Hudson Museum.

Arutiunov, Sergei A., and Dorian A. Sergeev. 2006. *Problems of Ethnic History in the Bering Sea: The Ekven Cemetery*. Ed. and trans. Richard L. Bland. Anchorage: Shared Beringian Heritage Program.

Bernardini, Wesley. 2008. "Identity as History: Hopi Clans and the Curation of Oral Tradition." *Journal of Anthropological Research* 64 (4): 483–509. https://doi.org/10.3998/jar.0521004.0064.403.

Carmichael, Patrick H. 1986. "Nasca Pottery Construction." *Ñawpa Pacha: Journal of Andean Archaeology* 24: 31–48.

De la Fuente, Guillermo A. 2011. "*Chaîne operatoire*, Technical Gestures, and Pottery Production at Southern Andes during the Late Period (c. AD 900–AD 1450) (Catamarca, Northwestern Argentina, Argentina)." In *Archaeological Ceramics: A Review of Current Research*, ed. Simona Scarcella, 89–102. Oxford: Archaeopress.

Dillian, Carolyn. 2002. "More than Toolstone: Differential Utilization of Glass Mountain Obsidian." PhD dissertation, University of California, Berkeley.

Doherty, Christopher, and Duygu Tarkan Özbudak. 2013. "Pottery Production at Çatalhöyük: A Petrographic Perspective." In *Substantive Technologies at Çatalhöyük: Reports from the 2000–2008 Seasons*, ed. Ian Hodder, 183–92. Los Angeles: Cotsen Institute of Archaeology Press.

Farid, Shahina. 2014. "Timelines: Phasing Neolithic Çatalhöyük." In *Çatalhöyük Excavations: The 2000–2008 Seasons*, ed. Ian Hodder, 91–129. Los Angeles: Cotsen Institute of Archaeology Press.

Gillespie, Susan D. 2000. "Lévi-Strauss: Maison and société à maisons." In *Beyond Kinship: Social and Material Reproduction in House Societies*, ed. Rosemary A. Joyce and Susan D. Gillespie, 22–52. Philadelphia: University of Pennsylvania Press.

Gillespie, Susan D. 2007. "When Is a House?" In *The Durable House: House Society Models in Archaeology*, ed. Robin A. Beck Jr., 25–50. Carbondale: Center for Archaeological Investigations, Southern Illinois University.

Gosselain, Olivier P. 2000. "Materializing Identities: An African Perspective." *Journal of Archaeological Method and Theory* 7 (3): 187–217. https://doi.org/10.1023/A:102 6558503986.

Helliwell, Christine. 1993. "Good Walls Make Bad Neighbors: The Dayak Longhouse as a Community of Voices." In *Inside Austronesian Houses*, ed. James J. Fox, 45–66. Canberra: Australian National University Press.

Hodder, Ian. 2013a. "Becoming Entangled in Things." In *Substantive Technologies at Çatalhöyük: Reports from the 2000–2008 Seasons*, ed. Ian Hodder, 1–25. Los Angeles: Cotsen Institute of Archaeology Press.

Hodder, Ian. 2013b. "Introduction: Dwelling at Çatalhöyük." In *Humans and Landscapes of Çatalhöyük: Reports from the 2000–2008 Seasons*, ed. Ian Hodder, 1–29. Los Angeles: Cotsen Institute of Archaeology Press.

Hodder, Ian, and Craig Cessford. 2004. "Daily Practice and Social Memory at Çatalhöyük." *American Antiquity* 69 (1): 17–40. https://doi.org/10.2307/4128346.

Hodder, Ian, and Shahina Farid. 2014. "Questions, History of Work, and Summary of Results." In *Çatalhöyük Excavations: The 2000–2008 Seasons*, ed. Ian Hodder, 1–34. Los Angeles: Cotsen Institute of Archaeology Press.

Hodder, Ian, and Peter Pels. 2010. "History Houses: A New Interpretation of Architectural Elaboration at Çatalhöyük." In *Religion in the Emergence of Civilization: Çatalhöyük as a Case Study*, ed. Ian Hodder, 163–86. Cambridge: Cambridge University Press. https://doi.org/10.1017/CBO9780511761416.007.

Hugh-Jones, Stephen. 1993. "Clear Descent or Ambiguous Houses? A Re-examination of Tukanoan Social Organization." *L'Homme* 33 (126): 95–120. https://doi.org/10.3406/hom.1993.369631.

Joyce, Rosemary A. 2000. "Heirlooms and Houses: Materiality and Social Memory." In *Beyond Kinship: Social and Material Reproduction in House Societies*, ed. Rosemary A. Joyce and Susan D. Gillespie, 189–212. Philadelphia: University of Pennsylvania Press.

Kroeber, Alfred L. 1905. "Notes on 'The Obsidian Blades of California' by Horatio N. Rust." *American Anthropologist* 7 (4): 690–95.

Lechtman, Heather. 1977. "Style in Technology: Some Early Thoughts." In *Material Culture: Styles, Organization, and Dynamics of Technology*, ed. Heather Lechtman and Robert S. Merrill, 3–20. St. Paul, MN: American Ethnological Society.

Lévi-Strauss, Claude. 1982. *The Way of the Masks*. Trans. Sylvia Modelski. Seattle: University of Washington Press.

Nakamura, Carolyn, and Lynn Meskell. 2013 "Figurine Worlds at Çatalhöyük." In *Substantive Technologies at Çatalhöyük: Reports from the 2000–2008 Seasons*, ed. Ian Hodder, 201–34. Los Angeles: Cotsen Institute of Archaeology Press.

Regan, Roddy. 2014. "The Sequence of Buildings 75, 65, 56, 44, and 10 and External Spaces 119, 129, 130, 144, 299, 314, 319, 329, 333, 339, 367, 371, and 427." In *Çatalhöyük Excavations: The 2000–2008 Seasons*, ed. Ian Hodder, 131–89. Los Angeles: Cotsen Institute of Archaeology Press.

Sadarangani, Freya. 2014a. "External Space 226." In *Çatalhöyük Excavations: The 2000–2008 Seasons*, ed. Ian Hodder, 541–45. Los Angeles: Cotsen Institute of Archaeology Press.

Sadarangani, Freya. 2014b. "The Sequence of Buildings 53 and 42 and External Spaces 259, 260, and 261." In *Çatalhöyük Excavations: The 2000–2008 Seasons*, ed. Ian Hodder, 191–219. Los Angeles: Cotsen Institute of Archaeology Press.

Sather, Clifford. 1993. "Posts, Hearths, and Thresholds: The Iban Longhouse as a Ritual Structure." In *Inside Austronesian Houses*, ed. James J. Fox, 67–120. Canberra: Australian National University Press.

Vitelli, Karen D. 1998. "'Looking Up' at Early Ceramics in Greece." In *Pottery and People*, ed. James Skibo and Gary Feinman, 184–98. Salt Lake City: University of Utah Press.

Wallis, Neill J. 2008. "Networks of History and Memory: Creating a Nexus of Social Identities in Woodland Period Mounds on the Lower St. Johns River, Florida." *Journal of Social Archaeology* 8 (2): 236–71. https://doi.org/10.1177/1469605308089972.

Woodburn, James. 1982. "Egalitarian Societies." *Man* n.s. 17 (3): 431–51.

Yalman, Nurcan, Duygu Tarkan Özbudak, and Hilal Gültekin. 2013. "The Neolithic Pottery of Çatalhöyük: Recent Studies." In *Substantive Technologies at Çatalhöyük: Reports from the 2000–2008 Seasons*, ed. Ian Hodder, 147–82. Los Angeles: Cotsen Institute of Archaeology Press.

Yeomans, Lisa. 2014. "External Space 279 and Buildings 70 and 71." In *Çatalhöyük Excavations: The 2000–2008 Seasons*, ed. Ian Hodder, 527–39. Los Angeles: Cotsen Institute of Archaeology Press.

9

The Ritualization of Daily Practice

Exploring the Staging of Ritual Acts at Neolithic Çatalhöyük, Turkey

Christina Tsoraki

DOI: 10.5876/9781607327370.c009

INTRODUCTION: THE ARTICULATION OF RITUAL THROUGH DAILY PRACTICES

Theories about the development of the Neolithic in Anatolia and the Middle East gave prominence to the role of certain ritual practices for ensuring and sustaining long-term social cohesion and stability in Neolithic communities. Rituals such as the placement of burials in house platforms, the circulation of plastered human heads and the overall treatment of dead bodies, the installation of bucrania and other animal parts in structural elements such as platforms and walls, and the caching of human and animal remains within buildings have all been interpreted as symbolic practices that were actively used to construct social memory (Hodder 2012; Kuijt 2008; Rollefson 2000). In this case, house structures became the places where ritual practices intersected with productive/economic affairs (Hodder 2014a).

Over recent years, archaeologists have critically engaged with the use of the term *ritual*. Yet while we have come to acknowledge that the notion of "ritual—or ritualization (Bell 1992)—encompasses a range of activities from mundane acts of daily life to episodic or unique spectacles" (Hull 2014, 164), the widely held dichotomy between "ritual" and "domestic," although heavily critiqued (Bradley 2005), is still ingrained in the way materials and actions are discussed in the archaeological discourse. In this line of reasoning, certain practices have ritual affordances, and simplistic distinctions between quotidian and overtly symbolic materials are often drawn.

In this chapter, following a practice-based approach, I would like to reconsider the role and significance of daily activities and, more specifically, the role of grinding activities in ritual acts. There are various ways to approach "ritual," but the one adopted in the present study focuses on the notion of ritual as a "form of action," as a performance (Bradley 2005, 33; Hull 2014; Mills 2015). As Pels (2010, 258) concisely notes, "The profane part of daily activity will also to a large extent be used as a blueprint for anything extraordinary or transcendentally marked." Starting from the premise that ritual acts are "social projects" (Gell 1998) and as such they assemble material processes, objects, and people, I wish to consider how material practices embedded in daily life inform ritual processes, create social memory, and inevitably construct history. By focusing on the notion of performance, I explore "ritual" and "domestic" as a continuum of practice. Such a perspective shifts attention to the material processes entailed in the enactment of ritual acts and to the "ritualization of practice" (Bradley 2005, 33). To quote Bradley (2005, 34), ritualization is "a process by which certain parts of life are selected and provided with an added emphasis." Recently, Waterson (2013, 387) noted that "ethnography helps us to grasp the many ways in which everyday objects, and the house itself, can be put to use in ritual practice; the problem for the archaeologist of course is that the acts which create these layers of meaning often leave no discernible trace." While perhaps not all traces will be discernible, archaeologists have methodologies at their disposal that indeed permit exploration of meaning in ritual practices. Mills (2015, 252–53) has argued that in our quest to understand the materiality of ritual, "the location of deposits and what they might tell us about the size and structure of the social networks that participated in their production, use, consumption, and ultimate deposition are what we archaeologically have to work with."

Excavations at Neolithic Çatalhöyük offer ample evidence to explore this issue adopting a more contextualized perspective. Particular attention is paid to a recently excavated structure, Building 77, that contains rich in situ faunal, botanical, and lithic assemblages (House 2014; Taylor et al. 2015). The rich groundstone assemblage, the largest concentration of stone objects recovered from a single building since work on-site reconvened in the early 1990s, is the focus of this chapter. While the deposition of objects inside buildings, often in discrete concentrations (or "clusters," a term employed by the Çatalhöyük Research Project; Farid and Hodder 2014), during events associated with the creation or the abandonment of a building is a common practice (Hodder and Farid 2014; Nakamura 2010; Russell et al. 2014), no other burned or unburned structures at Çatalhöyük gave any comparable assemblages in terms of form, density, and structuring principles of depositional practices. Initial accounts of the groundstone assemblage from Building 77 interpreted the large volume of the material in terms of specialized production, unequal access, and social inequality (Wright 2013, 2014). The analysis presented here is based on data collected during the 2014 field season during which the complete groundstone inventory from Building 77 was revisited and studied in full (Tsoraki 2014).

IDENTIFYING RITUAL PRACTICES AT NEOLITHIC ÇATALHÖYÜK

Çatalhöyük, located in central Anatolia in Turkey, consists of two mounds: East and West. The East Mound, the focus of this chapter, was inhabited from the late Aceramic through to the Ceramic Neolithic for about 1,100 years (7100–6000 BCE). The site was first discovered in the late 1950s and was excavated by James Mellaart between 1961 and 1965 (Mellaart 1967). Since 1993, a new international team led by Ian Hodder has continued work on the site. The densely populated site is particularly well-known for its large size (13.5 ha), agglomerated pattern of settlement, and the spectacular wall paintings, sculpture, and other complex art found inside the houses (Hodder 2011; Hodder and Farid 2014).

One of the distinctive features of Neolithic Çatalhöyük is the presence of well-preserved and closely packed houses with rooftop entry and, in contrast to other Near Eastern Neolithic sites (Hole 2000), the lack of public architecture or ceremonial centers. While some of these structures were interpreted as shrines by Mellaart (1967), the current project interprets all excavated buildings as domestic houses, but with varying degrees of ritual elaboration (Hodder 2011, 937). The standard layout of the mudbrick houses comprises one main room, which accommodated an oven, a burial platform, and other platforms where different activities took place. The main room was flanked by one (commonly to the west) to three side rooms, which served as storerooms and food-processing areas. Prescribed norms in the use of internal spaces include the arrangement of various fixtures and fittings within the houses. Hearths and ovens are generally placed on the south side of the main rooms and are associated with dirty floors, that is, floor surfaces that contain micro-residues of different types of activities such as animal bone processing, charcoal rake-outs from cleaning the hearths, and micro-debris from the production and maintenance of stone tools (Hodder and Farid 2014). The northern platforms, which as a rule are whiter and cleaner, tend to be associated with different forms of symbolic expressions such as plaster reliefs, paintings, and other installations (Hodder 2011, 938–39; Hodder and Farid 2014, 21). Burials were located in the north and east areas of the rooms, although it is not uncommon for neonate and child burials to be found in the south part of the rooms near hearths and ovens or in the side rooms (Boz and Hager 2013; Pearson et al. 2015). A central sunken floor area, often accompanied by a hearth, is present in most houses and seems to represent a central activity area (Hodder and Farid 2014, 22). Multiple lines of evidence suggest that the buildings seem to have housed diverse productive activities that range from food processing and storage to bone and obsidian working. These structures also became the focus for symbolic elaboration and acted as the focal point for ritual activities.

Continuity encountered in building practices and depositional activities has led Hodder to suggest that two forms of history making were evident within the context of Neolithic Çatalhöyük (Hodder and Cessford 2004; Hodder, this volume). The first type refers to the repetition of activities within buildings as the result of habituated practices. In this case, continuity in practice is not based on conscious decisions but on embodied and routinized actions. The second type—commemorative history making—entails the curation, circulation, and deposition of objects through which concrete references between present and specific events in the past are established. The concentration of human and animal remains within buildings is among the practices that were actively used to construct social memory. Ritual was therefore closely linked to daily activities, and both sets of practices were firmly embedded in the house structures (Hodder 2005; Hodder and Cessford 2004).

BUILDING 77

One of the buildings excavated by the Çatalhöyük Research Project that stands out is Building 77, located in the North Area of the East Mound. The structure has been fully excavated, which allows for the exploration of depositional practices at different stages in the life cycle of the building (House 2014; Tung 2013, 2014). It is a large structure that was destroyed by intense fire (House 2014). The current consensus is that all fires at Çatalhöyük were intentionally set as part of the ritual practices involved in house closure (Hodder and Farid 2014, 18). The layout consists of a main room (Space 336), which has a sequence of six platforms and a bench and measures 4.4 m × 4.4 m, and a side room (Space 337), measuring 4.1 m × 2.0 m with evidence for storage facilities (House 2014). The building exhibits a high degree of elaboration with ritual and symbolic features concentrated mainly in the northern part of the building: a plastered calf head on the northern wall; two sets of bull horns in pedestals that surround the northeast platform, which also contains a long burial sequence; a frieze with multiple red handprints; other paintings on the walls around the platform, and a decorated niche on the north wall (Hodder and Farid 2014, 27). In many other respects, however, the building retains features usually associated with houses at Çatalhöyük, such as ovens and hearths, storage bins, basins, and working platforms. One more element that makes Building 77 stand out is the large quantity of material left on the floor and platforms of the structure prior to its abandonment. The cultural debris found in the building included large quantities of animal bones, cattle horn cores and antler, bone and stone tools, and concentrations of botanical remains. A three-phase stratigraphic sequence has been proposed for Building 77: the occupation phase (Phase C), the abandonment phase with primary deposits and material from primary structural collapse as a result of the burning event (Phase B), and the upper infill sequence (Phase A) (House 2014).

MATERIAL DEPOSITION AS A MEANINGFUL STATEMENT

With more than 2,100 worked and unworked stone artifacts[1] weighing over 360 kg in total, the groundstone assemblage from Building 77 is exceptionally large (table 9.1 and figure 9.1). While the vast majority of this assemblage (n = 1,653), as is frequently the case at Çatalhöyük, consists mostly of natural small-sized water-worn pebbles that derive from flotation samples mainly from room fill, the number of worked artifacts and products related to the groundstone technology (n = 413) is still significantly high when compared to other contemporary building assemblages (cf. Wright 2014, 21, tables 5 and 6).

Depositional patterns of groundstone material vary throughout the life cycle of the building. There is a statistically significant correlation between building phase and object category (χ^2 = 0.000, df = 14), with the vast majority of clearly used and intentionally modified objects occurring in Phase B (figure 9.2). The high peak in the frequencies of material in this phase suggests that the material was deposited on the floor and platforms of Building 77 at or prior to the burning event that destroyed it. Spatiotemporal visualization analysis by James Taylor and colleagues also confirms this "sudden burst" of activity just before the event of the fire, a pattern that is replicated in the deposition of faunal material and archaeobotanical remains (Taylor et al. 2015).

This contrasts greatly with the depositional patterns associated with the occupation levels (Phase C), which have very limited artifacts. This observation is in agreement with preliminary results that point toward a "low-level"

TABLE 9.1. Frequency of stone materials in the three phases of Building 77 (excluding indeterminate cases, n = 67).

Object Category		Building Phase			
		Phase C Occupation	Phase B Burning and Primary Collapse	Phase A Closure/ Infilling (demolition)	Total
Natural	Count	1,046	250	357	1,653
	% within phase	95.9%	44.2%	87.1%	80.0%
Edge tools	Count	0	12	0	12
	% within phase	0.0%	2.1%	0.0%	0.6%
Percussive tools	Count	1	7	6	14
	% within phase	0.1%	1.2%	1.5%	0.7%
Grinding/ abrasive tools	Count	7	224	18	249
	% within phase	0.6%	39.6%	4.4%	12.1%
Miscellaneous	Count	0	2	0	2
	% within phase	0.0%	0.4%	0.0%	0.1%
Multiple-use tools	Count	1	18	1	20
	% within phase	0.1%	3.2%	0.2%	1.0%
Ornaments	Count	0	1	0	1
	% within phase	0.0%	0.2%	0.0%	0.0%
Debitage/cores/ nodules	Count	36	51	28	115
	% within phase	3.3%	9.0%	6.8%	5.6%
Total	Count	1,091	565	410	2,066
	% within phase	100.0%	100.0%	100.0%	100.0%

distribution of artifacts and ecofacts for the majority of the occupation of the structure (Taylor et al. 2015, 146). Phase C has the highest frequency of natural small-sized pebbles, which, because of their extremely small size, could not have played any role in technological practices associated with groundstone technology and production. Among the very few tools attributed to this phase of the structure is a quern that was set into the central part of a large L-shaped platform (F.6058) located in the southwest corner of the main room near the entrance to the side room (figure 9.3). Microwear analysis shows that this quern was used as a tool for cereal processing. This in tandem with a botanical deposit of glume wheat grains found nearby suggests that the platform was used for food-processing activities. Although this botanical

FIGURE 9.1. *Stone artifact clusters in the main room (Space 336) of Building 77. Source: Çatalhöyük Research Project, photograph by Jason Quinlan.*

deposit is not in situ, its location next to the side room, which provides strong evidence for de-husking of the same type of glume wheat material, seems to confirm the use of these two areas for plant-processing activities (Bogaard et al. 2013, 123).

Material depositions were not equally distributed between the main room and the side room, with 71.9 percent of the worked stone deriving from the former space. The main room (Space 336) was littered with stone objects that were clustered in discrete concentrations mainly around the edges of the room (figure 9.4). With more than 100 stone artifacts weighing over 100 kg, the largest collection of groundstone was found on the surface of the northern platform (F.6062) (figure 9.5), which covers two thirds of the northern area in the main room (House 2014, 488). Cluster 17509 contains mainly querns and grinders, four quern rough-outs of elaborate form, newly manufactured grinding tools, a nicely shaped grooved polisher, a few hammerstones, as well as a large number of fragments from previously used grinding tools. Another interesting deposition of artifacts was located in the southeast corner of the main room (Cluster 17501) next to the ladder scar (House 2014) (figure 9.4). This artifact group contained a varied assemblage of materials: a polished

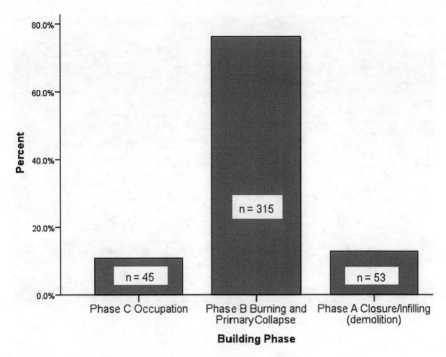

FIGURE 9.2. *Frequency of worked stone in Building 77, by phase.*

stone axe, a polisher used for smoothing and burnishing plastered surfaces, a quern rough-out of elaborate form, fragments of at least three newly manufactured or lightly used grinding tools, along with distinct types of faunal remains such as a dog skull, a cattle jaw, and caprine horn cores and scapula. The faunal team described these as "clearly special and unusual remains, both symbolically and food-wise" (House 2014, 495). During refitting analysis, one of the quern rough-outs from Cluster 17509 in the northern platform was refitted with the rough-out from Cluster 17501 to form the complete object (see also Wright 2013). A similar refitting pattern was encountered in the side room, where two conjoining fragments of another grinding tool were found once again on opposite sides of the room—one fragment comes from the north bin (F.3092) and the other from the south basin (F. 3091)—indicating that both examples broke prior to their deposition and that the conflagration did not facilitate the breakage of the objects.

The groundstone assemblage from Building 77 stands out not only in terms of quantity but also of quality. The largest concentration (n = 17, 68% of all

FIGURE 9.3. *Quern found embedded in a platform in Building 77 (scale measures 10 cm). Source: Çatalhöyük Research Project, photograph by Jason Quinlan.*

known examples) of carefully manufactured querns and quern rough-outs with regularly shaped margins and often body surfaces exhibiting a higher level of investment in their manufacture derives from this building. With only twenty-five identified examples to date, this more elaborate type of quern readily stands out when compared to the less carefully finished examples that occur widely at Çatalhöyük. Based on data collection from material excavated between 1993 and 2015, a limited spatial and temporal distribution is evident for this artifact type, which, in addition to Building 77, occurs only in Buildings 102, 49, 65, 68, 80, and 89 and the IST and TP excavation trenches dated to the middle and later part of the sequence. These artifacts were the products of a more formalized production sequence that placed emphasis on regularity and symmetry in form. Following the initial shaping of their margins by coarse flaking, carefully executed and regularly spaced pecking and grinding—both of which are time-consuming techniques—were employed to achieve the final form of the querns (see also Wright 2013). Consistency in the production sequence is also replicated in raw material selection, with a clear tendency for pink-colored fine-grained andesite with porphyritic texture (which contrasts with the gray-colored variant used more commonly at Çatalhöyük). The querns from

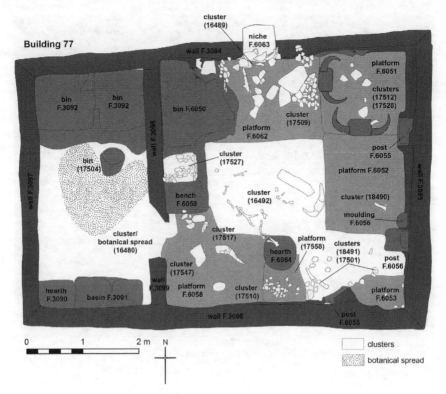

FIGURE 9.4. *Floor plan of Building 77 showing the location of stone clusters and other finds of Phase B. Source: Çatalhöyük Research Project, plan by Camilla Mazzucato, Cordelia Hall, and David Mackie.*

Building 77 also tend to be among the largest examples found on-site to date, with the heaviest tool weighing more than 50 kg. In addition, twelve polished stone axes come from this building, the highest such concentration in a single building, while other interesting objects—with a relatively limited presence within the temporal sequence at Çatalhöyük—include a nicely manufactured grooved polisher, two stone pestles, and the fragment of a marble bracelet. It is interesting that all occurrences of these distinct types are found exclusively in the deposits associated with the abandonment of the structure (Phase B).

In all, there is no great correlation between the stone assemblages of the main occupation levels and those associated with the abandonment of the structure in terms of diversity, form, or density. For instance, although elaborate querns are relatively common in Phase B, the quern found embedded in

FIGURE 9.5. *Cluster 17509 located on the northern platform of the main room in Building 77. Source: Çatalhöyük Research Project, photograph by Jason Quinlan.*

the platform of the main room and therefore associated directly with the use of the building, despite exhibiting intentionally modified margins mainly by coarse flaking and limited pecking, does not belong to the more elaborate type. Such variations in the depositional practices may well reflect changes in the practices associated with the different stages in the life cycle of the building.

STAGING RITUAL

But how are we to interpret these discrete concentrations of groundstone materials on the floor and platforms of Building 77?

A suggestion is that these deposits provide a snapshot of normal household activities and at least some of the clusters represent production areas (Wright 2014, 14; cf. Twiss et al. 2008); the presence of rough-outs and grinding tools at an early stage of their production cycle may indeed suggest that Cluster 17509 in the northern area represents an in situ stone-working area. The extremely low densities of debitage—which come from the primary stage of

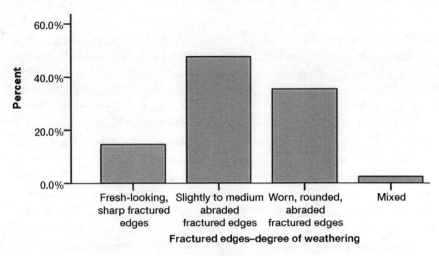

FIGURE 9.6. *Degree of weathering on fractured edges of grinding tool fragments found in Cluster 17509 (n = 82).*

the modification of unworked boulders—and the lack of micro-debris, however, do not support this argument. This is also corroborated by the fact that most of the materials found here are fragments of previously used grinding tools. During refitting analysis, very few conjoining fragments were identified, and no refits between these fragments and the larger querns were established. The lack of conjoining fragments, therefore, and, more important, variation in the degree of weathering of the fractured edges that ranges from fresh-looking, sharp edges to heavily worn ones (figure 9.6) suggests that these do not represent tools that broke in situ. Instead, these fragments have varied depositional histories and derive from different contexts prior to their abandonment in Building 77. The overall size of the fragments, which normally is fist-sized, does not correspond to breakage patterns that would be expected during the use-life of the tools. Moreover, the surviving thickness of the fragments (ca. 10 cm) clearly shows that they do not come from tools used until they reached exhaustion. Although deliberate breakage of tools could have resulted from a necessity to recycle larger-sized tools into smaller forms (i.e., tools intended to be used as handheld grinding tools), there are no indications that such recycling practices were a salient component of stone-working traditions of the Çatalhöyük community. Hence, we can plausibly suggest that the damage inflicted on large-sized grinding tools was not motivated by factors such as raw material economy or curation of seemingly valuable materials with an interest in prolonging their use-life. Instead, querns seem to have been deliberately decommissioned before they reached the end of their use-life.

The suggestion that these clusters do not relate to in situ production activities is further reinforced by a concentration of materials found on the central sunken floor area of the main room (Space 336). It was originally interpreted as a "craft production or working area" (House 2014, 493). The clustered assemblage included a grinding slab with a shallow basin and V-shaped grooves at one end, four stone axes, a figurine, two clay balls, red deer antler, horn cores, and bone (House 2014, 493–94, fig. 23.13). Microwear analysis suggests that the grinding slab was used for a combination of plant-processing activities concentrated on the shallow basin and the working of thin-edged implements at the opposed end of the tool's use-face. The wear traces evident in the interior of the grooves do not correspond, however, to the polishing of hard mineral materials such as stone axes. All axes found in association with this grinding slab have heavily used blunt edges, and they were not under modification at the time of deposition. Moreover, the structure of the deposition in this case, with the grinding slab placed in a protective position with the use-face facing toward the floor and the careful arrangement of four polished stone axes next to the grinding slab together with an antler, horn, core, and bone materials, suggests a deliberately placed deposit and not an in situ working area.

Bearing this in mind, the choice for the location of Cluster 17509 on the northern platform and in the northern part of the main room where intense symbolism was displayed warrants further consideration. At Çatalhöyük, lived space was carefully choreographed, and distinct sets of activities seem to have been associated with different areas of the main room. Although variations from the dominant pattern do exist, the southern areas of the house structures tend to be associated with productive activities (craft and food production, ovens, and hearths) and the northern areas with increased forms of elaboration and installations more explicitly symbolic in nature, whiter plastered floors and burials (Hodder and Farid 2014, 21, 23). The positioning of Cluster 17509 on the northern platform recalls the placement of bucrania and other animal installations in prominent locations within the house, in direct sight from the entry point to the house, therefore increasing the visibility of such deposits (Twiss 2012). Moreover, if, as Twiss and colleagues suggested, there is a distinction between the main living space and side rooms in their perceived character in terms of privacy/concealment (Bogaard et al. 2009; Twiss 2012; Twiss et al. 2008, 54), the choice of placing an assemblage of objects of distinct form and size in an area of the building with increased visibility suggests an interest in focusing attention on the act of deposition. Hence, the placement of this large

FIGURE 9.7. *Large quern found in the burned fill of Building 77 (Phase A: infilling/demolition) (scale measures 20 cm). Source: Çatalhöyük Research Project, photograph by Jason Quinlan.*

concentration of stone material in various states of preservation in an overtly symbolically significant location within the inscribed inhabited space further reflects intentionality in the depositional practices encountered in Building 77.

A certain degree of deliberateness and formality is expressed both in the choice of materials left on the floors and platforms prior to the final burning event and in their distinctive arrangement. These distinctive assemblages show signs of ritualized behavior, with elements of excess, wastefulness, and performance on display (cf. Russell et al. 2014). In this light, the subsequent deposition of two complete and still functioning elaborate querns (a roughout and a used quern [figure 9.7], which weigh more than 43 kg each)—the largest querns found on-site so far—together with large cattle bones in the otherwise clean and sterile burned fill of the building (House 2014; Russell et al. 2014), suggests intentional placement. As such, it might constitute a commemoration of the original events associated with the burning of the structure, but at a later stage of its life cycle (cf. Russell et al. 2014). Depositions such as this further reflect choreographed practices evoking the social memory of specific events and were actively employed in memory-constructing processes (commemorative memory, Hodder 2012; McAnany and Hodder 2009).

MATERIAL CONFIGURATIONS

The deposition of vast quantities of grinding tools and other stone objects in Building 77 was an act imbued with symbolic significance; these objects were firmly embedded in the ritual acts performed within Building 77 during the final events leading to its abandonment. While the structural principles of deposition are influenced by particular historical and social circumstances and thus their internal logic may differ, during the Neolithic the act of placing grinding tools in different structural elements seems to have consistently formed an integral component of ritualized practices associated with house construction and abandonment throughout Europe and the Near East (see, for example, Bradley 2005; Graefe et al. 2009). Yet the choices reflected in this act of deposition raise an important question: What are the *intrinsic qualities* of these objects that render them appropriate to be incorporated in ritual acts?

A close reading of material depositions in Building 77 suggests that a bundling of materials and practices is referenced in these clusters. Microwear analysis of the groundstone assemblage highlights the multitude of activities performed by these implements. Stone axes were used for shaping the wooden structural elements used in building construction. Grinding tools were incorporated not only in food-processing activities but also in the production of bone and stone tools and ornaments, in the processing of pigments used for wall paintings and clays for house construction activities, and in the smoothing and burnishing of plastered surfaces. More broadly, grinding tools played an important role in the workings of the subsistence economy, but they were also instrumental for the social reproduction of Neolithic communities. A close look at the *chaîne opératoire* of different technological practices clearly suggests that the act of grinding, an inherently transformative act, featured prominently in the production of a wide range of objects—among which were Spondylus bracelets, stone ornaments, and stone axes—that were incorporated in varied social practices (Tsoraki 2011).

The central role of grinding activities in the contexts of everyday life is also stressed by the strong correlation of grinding tools with buildings, often found embedded in the floors and platforms as exemplified by querns fixed in place in Building 77 in the North Area and Buildings 80 and 89 in the South Area (House 2014; Taylor 2014; Tsoraki 2015). Current research on the Çatalhöyük groundstone assemblage, and in agreement with Wright's previous interpretations (Wright 2013), suggests that grinding tools were part of a suite of stone implements that constituted the regular household toolkit. Grinding activities had strong associations with houses and their occupants and were among the activities that structured the perception and use of physical and social space.

These objects, therefore, played a vital role in those very acts that defined significant events in the life cycle of the house structures; they were embedded in the histories of houses and of those tied to them. Events with distinct temporalities—such as the shaping of wooden structural posts; the regular plastering of the walls, platforms, and other surfaces; the creation of wall paintings; food provisioning; wood working; and tool and ornament making—would have temporally punctuated the life of both buildings and users. The clusters, therefore, acted as material manifestations of the routines of everyday life.

Multifaceted research at Çatalhöyük has amply demonstrated that houses were the primary focus of a wide range of activities (Hodder 2013b, 2013c, and relevant chapters therein). People at Çatalhöyük spent considerable time inside houses, actively engaging in a network of activities that acted as mechanisms of socialization (Hodder 2011, 940; Hodder and Cessford 2004). Embedded in the social and economic life of Neolithic communities, grinding activities were among the daily practices that contributed to the socialization of the Çatalhöyük community. They provided a context for the transmission of knowledge about techniques and properties of materials and for perpetuating socially accepted ways of doing and making while allowing at the same time the reproduction of age and gender categories. More broadly, the symbolic potency of the act of grinding is suggested by the inclusion of grinding activities in rituals that signify a concern with the social transformation of younger (and as a rule female) members of the society to fully productive adult members (see, for instance, Kirk-Greene 1957). These daily activities may have a seemingly mundane character, but they were also a powerful mechanism of communication, an arena of social interaction that enabled the social reproduction of Neolithic communities on a daily basis. Thus, the identities of the house and the social entities tied to it were actively shaped through participation in these daily activities as much as through involvement in larger-scale ritualized events.

Hodder (2014a) suggested that at Çatalhöyük, houses were part of social networks that played out at different scales beyond the level of the individual house (neighborhoods, sectors, and sub-mounds), and the presence of different social groupings was manifested through shared material practices such as artistic representations and burial practices but also possibly through the appropriation of often rare but distinct items of material culture such as the flint daggers and items of bodily adornment (Hodder 2014b; Nazaroff, Tsoraki, and Vasic 2016). To this end, it is interesting to consider whether the more elaborate type of quern, the product of distinct technological investment, was one more material practice that articulated in material form associations of particular social networks that crosscut the community at Çatalhöyük. This observation is relevant to Joyce's suggestion that distinct technological knowledge may have been a form of immaterial property that was owned and appropriated by members of specific social houses during the performance of ceremonies (Joyce, this volume).

The density of materials found in the clusters in Building 77 may also contribute to this discussion and provide an indication of the size of the social group that participated in the events leading to the closure and final abandonment of the structure. The large number of querns and quern fragments found here far exceeds the number of grinding implements individual households tend to have in their possession. Ethnographic research suggests that in small-scale societies, households, independent of their economic status, have in their possession on average one to three querns, with ownership of one quern most frequent (e.g., Hamon and Le Gall 2013; Searcy 2011). It is plausible, therefore, that the deposition of material prior to the fire represents deposition by more than one household. This may also explain the presence of visibly used grinding tool fragments with different levels of weathering, indicating different depositional histories, post-breakage treatment, and, potentially, forms of curation. In this context, therefore, complete tools and fragments became material symbols and were employed to commemorate the contribution of different households to a symbolically charged event (Chapman 2000) that would have impressed itself vividly in the memory of both those performing and those observing. This is in agreement with the suggestion that feasts associated with the house-closure rituals at Çatalhöyük are "multi-household in scale" (Russell et al. 2014, 117).

Moreover, the deposition of approximately 300 kg of stone in one particular building with individual tools that weigh up to 50 kg brings to the fore the network of relationships and alliances involved in the procurement and subsequent transportation of material over significant distances. As the local landscape is dominated by lake limestones and marls (i.e., materials that could not have been used in groundstone production), the Çatalhöyük community sought to acquire appropriate materials from sources further afield. While the exact andesitic sources used for the production of the Çatalhöyük querns have yet to be identified, there are indications that appropriate materials in terms of physical properties and form (i.e., large boulders) would have been procured either from volcanic sources 35–40 km southeast of Çatalhöyük (Karadağ) or around 60 km west of the site from the Erenler Dağ-Alcadağ volcanic areas or potentially even from greater distances (e.g., volcanic outcrops to the east and northeast from which obsidian was procured) (Carter and Milić 2013; Doherty and Tarkan 2013; Wright 2013).

The distinct deposition of material in Building 77, therefore, can be seen as one more "component of collective action" (Hodder and Farid 2014, 33), a further expression of the communal ties that bound the Çatalhöyük community together. The sheer quantity of material deposited here could be interpreted as a symbolic reinforcement of the importance of the events associated with the burning of the building. The abandonment of numerous functional and rather difficult to procure objects is wasteful in character, but it is particularly evocative. It is a powerful statement about the ability of the social group associated with the building to not only access but also amass and ultimately dispose of such large quantities of material, which included rare types of stone tools.

Building 77 was clearly a building that at least at some stage in its life cycle became a receptacle for communally focused practices invested with symbolism; in that respect, it was implicated in wider social narratives. Fluidity in the materiality and history of the house, evident in part in the continual structural changes buildings underwent at Çatalhöyük (Hodder and Farid 2014), may have also been reflected in the sociality of the house and the social entities tied to it during different events in its life cycle. As Hofmann and Smyth (2013, 8) note, "The way in which social units framed by the house . . . could link with other potential groupings at other scales is crucial, and has a direct impact on the specific character of dwelling in a particular place at a particular time." Thus, our interpretations of activities associated with households need to allow for greater flexibility when considering the composition of the social entities appropriating household space at different times. This is also supported by the results of biodistance studies based on dental morphology of human remains (Pilloud and Larsen 2011), which show that biological kin was not the decisive factor in determining the location of internment in a particular house. Complex social ties, therefore, and not mere biological affiliation seem to have regulated house membership at Neolithic Çatalhöyük (Hodder 2013a). In the absence of public buildings and ceremonial centers, community-wide ritual events that were centered at the house structures, such as those evidenced at Building 77, would have ensured the reproduction and perpetuation of social bonds.

CONCLUSION

At Neolithic Çatalhöyük, houses materialized different worldviews of great significance for the social and economic vitality of the local community. Different types of social performances were enmeshed in the house structures; these houses became the focus of a meshwork of activities and performances that cannot be readily separated into "functional" and "ritual" spheres of life. In Building 77, spatial correlates between overtly ritual materials, such as the horned platform and wall paintings, and seemingly mundane ones further emphasize the lack of necessity to distinguish between these two spheres of social life. Interestingly, however, the opposite is suggested by the distribution of material in the contemporary and also burned Building 52 (Twiss et al. 2008, 53–54; Bogaard et al. 2009, 654). The artifact clusters in Building 77 represent an ensemble (cf. Knappett 2012) of materials, objects, people, and practices. As I have argued throughout, tools used in daily activities were *deliberately* chosen to be included in ritualized acts because they were capable of narrating household history. Embedded in habituated practices, objects of everyday life embodied social relationships, and through their curation and circulation they articulated these relationships in physical form. The placing of stone materials within Building 77 was carefully chosen to increase visibility of the act of deposition. Broken objects were assembled in discrete concentrations along the edges of the main room and served to create new material and social relations. In these deposits, fragments of objects came to stand for fragments of time and history; they acted as potent repositories of memory. The deliberate placement of groups of artifacts with varied histories at liminal stages in the life of houses and their occupants at Çatalhöyük should be seen as one component of memory-constructing processes at work at Neolithic Çatalhöyük. Events, such as the highly structured abandonment of Building 77 and associated materials, seem to blur the lines between the two forms of history making put forward by Hodder (this volume), suggesting that instead of seeing them as distinct forms of practice (and perhaps in opposition), they informed and facilitated each other.

Fluctuations and shifts in the structural histories of the house were visible not only in changes in the spatial arrangement of features but also in the social structure that came to be tied to particular buildings. As Souvatzi (2013, 47) succinctly notes, "Houses therefore exist within a number of different temporalities and spatialities, which ought to be defined analytically rather than being presupposed." To this we need to add socialities. The quantity of material found within Building 77 and the conditions of its final deposition indicate that membership in a particular house and the activities performed within its boundaries do not rely solely on residential criteria but also on affiliation based on social alliances. This form of deposition encompassing community-wide efforts seems to correspond with the "sodality-based form" of history making (chapter 1, this volume). If we accept that social groups are defined through a set of practices and associations, the commemoration of these practices in highly performative events, such as the intentional destruction of objects and the ritual closing down of structures, may have served to inscribe them in memory.

ACKNOWLEDGMENTS

I would like to thank Ian Hodder for inviting me to present this research at the symposium Religion, History, and Place in the Origins of Settled Life in the Middle East at the 80th Annual Meeting of the Society for American Archaeology (2015) and for his subsequent invitation to contribute to the current volume. I thank Marketa Stovickova, Mirjam van Saane, and Charlotte Spiering for assistance in data recording. I am indebted to Lisa Guerre and Jason Quinlan for providing invaluable on-site assistance during the analysis of the groundstone assemblage and to Ben Chan and the two reviewers for insightful comments on an earlier draft. Funding was provided by the European Union's Seventh Framework Programme (FP7/2007-13) under grant agreement no. PIEF-GA-2012-328862–Project CRAFTS and the Çatalhöyük Research Project.

NOTE

1. This does not take into account material from burials found inside the building and the large number of stone beads that are mainly associated with the burial assemblages.

REFERENCES

Bell, Catherine. 1992. *Ritual Theory, Ritual Practice*. Oxford: Oxford University Press.

Bogaard, Amy, Michael Charles, Alex Livarda, Müge Ergun, Dragana Filipovic, and Glynis Jones. 2013. "The Archaeobotany of Mid-Later Neolithic Occupation Levels at Çatalhöyük." In *Humans and Landscapes of Çatalhöyük: Reports from the 2000–2008 Seasons*, ed. Ian Hodder, 93–128. Los Angeles: Cotsen Institute of Archaeology, University of California.

Bogaard, Amy, Michael Charles, Katheryn C. Twiss, Andrew Fairbairn, Nurcan Yalman, Dragana Filipović, G. Arzu Demirergi, Füsun Ertuğ, Nerissa Russell, and Jennifer Henecke. 2009. "Private Pantries and Celebrated Surplus: Storing and Sharing Food at Neolithic Çatalhöyük, Central Anatolia." *Antiquity* 83 (321): 649–68. https://doi.org/10.1017/S0003598X00098896.

Boz, Başak, and Lori D. Hager. 2013. "Living above the Dead: Intramural Burial Practices at Çatalhöyük." In *Humans and Landscapes of Çatalhöyük: Reports from the 2000–2008 Seasons*, ed. Ian Hodder, 413–40. Los Angeles: Cotsen Institute of Archaeology, University of California.

Bradley, Richard. 2005. *Ritual and Domestic Life in Prehistoric Europe*. London: Routledge.

Carter, Tristan, and Marina Milić. 2013. "The Chipped Stone." In *Substantive Technologies at Çatalhöyük: Reports from the 2000–2008 Seasons*, ed. Ian Hodder, 417–78. Los Angeles: Cotsen Institute of Archaeology, University of California.

Chapman, John C. 2000. *Fragmentation in Archaeology: People, Places, and Broken Objects in the Prehistory of South-eastern Europe*. London: Routledge.

Doherty, Christopher, and Duygu Tarkan. 2013. "Pottery Production at Çatalhöyük: A Petrographic Perspective." In *Substantive Technologies at Çatalhöyük: Reports from the 2000–2008 Seasons*, ed. Ian Hodder, 183–92. Los Angeles: Cotsen Institute of Archaeology, University of California.

Farid, Shahina, and Ian Hodder. 2014. "Excavations, Recording, and Sampling Methodologies." In *Çatalhöyük Excavations: The 2000–2008 Seasons*, ed. Ian Hodder, 35–51. Los Angeles: Cotsen Institute of Archaeology, University of California.

Gell, Alfred. 1998. *Art and Agency: An Anthropological Theory*. New York: Clarendon.

Graefe, Jan, Caroline Hamon, Cecilia Lidström-Holmberg, Christina Tsoraki, and Susan Watts. 2009. "Subsistence, Social, and Ritual Practices: Quern Deposits in the Neolithic Societies of Europe." In *Du Matériel au Spirituel: Réalités Archéologiques et Historiques des "Dépôts" de la Préhistoire à nos Jours*, ed. Sandrine Bonnardin, Caroline Hamon, Michel Lauwers, and Bénédicte Quilliec, 87–96. Antibes, France: Éditions APDCA.

Hamon, Caroline, and Valerie Le Gall. 2013. "Millet and Sauce: The Uses and Functions of Querns among the Minyanka (Mali)." *Journal of Anthropological Archaeology* 32 (1): 109–21. https://doi.org/10.1016/j.jaa.2012.12.002.

Hodder, Ian. 2005. "Socialization and Feasting at Çatalhöyük: A Response to Adams." *American Antiquity* 70 (1): 189–91. https://doi.org/10.2307/40035278.

Hodder, Ian. 2011. "Çatalhöyük: A Prehistoric Settlement on the Konya Plain." In *The Oxford Handbook of Ancient Anatolia*, ed. Sharon R. Steadman and Gregory McMahon, 934–49. Oxford: Oxford University Press. https://doi.org/10.1093/oxfordhb/9780195376142.013.0043.

Hodder, Ian. 2012. "History-Making in Prehistory: Examples from Çatalhöyük and the Middle East." In *Image, Memory, and Monumentality: Archaeological Engagements with the Material Worlds: A Celebration of the Academic Achievements of Professor Richard Bradley*, ed. Andrew M. Jones, Joshua Pollard, Michael J. Allen, and Julie Gardiner, 184–93. Oxford: Oxbow.

Hodder, Ian. 2013a. "From Diffusion to Structural Transformation: The Changing Roles of the Neolithic House in the Middle East, Turkey, and Europe." In *Tracking the Neolithic House in Europe: Sedentism, Architecture, and Practice*, ed. Daniela Hofmann and Jessica Smyth, 349–62. New York: Springer. https://doi.org/10.1007/978-1-4614-5289-8_15.

Hodder, Ian, ed. 2013b. *Humans and Landscapes of Çatalhöyük: Reports from the 2000–2008 Seasons.* Los Angeles: Cotsen Institute of Archaeology, University of California.

Hodder, Ian, ed. 2013c. *Substantive Technologies at Çatalhöyük: Reports from the 2000–2008 Seasons.* Los Angeles: Cotsen Institute of Archaeology, University of California.

Hodder, Ian. 2014a. "Mosaics and Networks: The Social Geography of Çatalhöyük." In *Integrating Çatalhöyük: Themes from the 2000–2008 Seasons*, ed. Ian Hodder, 149–67. Los Angeles: Cotsen Institute of Archaeology, University of California.

Hodder, Ian. 2014b. "Çatalhöyük: The Leopard Changes Its Spots: A Summary of Recent Work." *Anatolian Studies* 64: 1–22. https://doi.org/10.1017/S0066154614 000027.

Hodder, Ian, and Craig Cessford. 2004. "Daily Practice and Social Memory at Çatalhöyük." *American Antiquity* 69 (1): 17–40. https://doi.org/10.2307/4128346.

Hodder, Ian, and Shahina Farid. 2014. "Questions, History of Work, and Summary of Results." In *Çatalhöyük Excavations: The 2000–2008 Seasons*, ed. Ian Hodder, 1–34. Los Angeles: Cotsen Institute of Archaeology, University of California.

Hofmann, Daniela, and Jessica Smyth. 2013. "Introduction: Dwelling, Materials, Cosmology—Transforming Houses in the Neolithic." In *Tracking the Neolithic House in Europe: Sedentism, Architecture, and Practice*, ed. Daniela Hofmann and Jessica Smyth, 1–17. New York: Springer. https://doi.org/10.1007/978-1-4614 -5289-8_1.

Hole, Frank. 2000. "Is Size Important? Function and Hierarchy in Neolithic Settlements." In *Life in Neolithic Farming Communities: Social Organization, Identity, and Differentiation*, ed. Ian Kuijt, 191–209. New York: Kluwer Academic/ Plenum.

House, Michael. 2014. "Building 77." In *Çatalhöyük Excavations: The 2000–2008 Seasons*, ed. Ian Hodder, 485–503. Los Angeles: Cotsen Institute of Archaeology, University of California.

Hull, Kathleen L. 2014. "Ritual as Performance in Small-Scale Societies." *World Archaeology* 46 (2): 164–77. https://doi.org/10.1080/00438243.2013.879044.

Kirk-Greene, Anthony H.M. 1957. "A Lala Initiation Ceremony." *Man* 57: 9–11. https://doi.org/10.2307/2795048.

Knappett, Carl. 2012. "Materiality." In *Archaeological Theory Today*, ed. Ian Hodder, 188–207. Cambridge: Polity.

Kuijt, Ian. 2008. "The Regeneration of Life: Neolithic Structures of Symbolic Remembering and Forgetting." *Current Anthropology* 49 (2): 171–97. https://doi.org /10.1086/526097.

McAnany, Patricia A., and Ian Hodder. 2009. "Thinking about Stratigraphic Sequence in Social Terms." *Archaeological Dialogues* 16 (1): 1–22. https://doi.org/10.1017/S1380203809002748.

Mellaart, James. 1967. *Çatal Hüyük: A Neolithic Town in Anatolia*. London: Thames and Hudson.

Mills, Barbara J. 2015. "Unpacking the House: Ritual Practice and Social Networks at Chaco." In *Chaco Revisited: New Research on the Prehistory of Chaco Canyon, New Mexico*, ed. Carrie C. Heitman and Stephen Plog, 249–71. Tucson: University of Arizona Press.

Nakamura, Carolyn. 2010. "Magical Deposits at Çatalhöyük: A Matter of Time and Place?" In *Religion in the Emergence of Civilization: Çatalhöyük as a Case Study*, ed. Ian Hodder, 300–331. Cambridge: Cambridge University Press. https://doi.org/10.1017/CBO9780511761416.011.

Nazaroff, Adam J., Christina Tsoraki, and Milena Vasic. 2016. "Aesthetic, Social, and Material Networks: A Perspective from the Flint Daggers at Çatalhöyük, Turkey." *Cambridge Archaeological Journal* 26 (1): 65–92. https://doi.org/10.1017/S0959774315000347.

Pearson, Jessica A., Scott D. Haddow, Simon W. Hillson, Christopher J. Knüsel, Clark S. Larsen, and Joshua W. Sadvari. 2015. "Stable Carbon and Nitrogen Isotope Analysis and Dietary Reconstruction through the Life Course at Neolithic Çatalhöyük, Turkey." *Journal of Social Archaeology* 15 (2): 210–32. https://doi.org/10.1177/1469605315582983.

Pels, Peter. 2010. "Temporalities of 'Religion' at Çatalhöyük." In *Religion in the Emergence of Civilization: Çatalhöyük as a Case Study*, ed. Ian Hodder, 220–67. Cambridge: Cambridge University Press. https://doi.org/10.1017/CBO9780511761416.009.

Pilloud, Marin A., and Clark Spencer Larsen. 2011. "'Official' and 'Practical' Kin: Inferring Social and Community Structure from Dental Phenotype at Neolithic Çatalhöyük, Turkey." *American Journal of Physical Anthropology* 145 (4): 519–30. https://doi.org/10.1002/ajpa.21520.

Rollefson, Gary. 2000. "Ritual and Social Structure at Neolithic 'Ain Ghazal." In *Life in Neolithic Farming Communities: Social Organization, Identity, and Differentiation*, ed. Ian Kuijt, 165–90. New York: Kluwer Academic/Plenum.

Russell, Nerissa, Katherine I. Wright, Tristan Carter, Sheena Ketchum, Philippa Ryan, Nurkan Yalman, Roddy Regan, Mirjana Stevanović, and Marina Milić. 2014. "Bringing down the House: House Closing Deposits at Çatalhöyük." In *Integrating Çatalhöyük: Themes from the 2000–2008 Seasons*, ed. Ian Hodder, 109–21. Los Angeles: Cotsen Institute of Archaeology, University of California.

Searcy, Michael T. 2011. *The Life-Giving Stone: Ethnoarchaeology of Maya Metates*. Tucson: University of Arizona Press.

Souvatzi, Stella. 2013. "Diversity, Uniformity, and the Transformative Properties of the House in Neolithic Greece." In *Tracking the Neolithic House in Europe: Sedentism, Architecture, and Practice*, ed. Daniela Hofmann and Jessica Smyth, 45–64. New York: Springer. https://doi.org/10.1007/978-1-4614-5289-8_3.

Taylor, James S. 2014. "Excavations in the South Area." *Çatalhöyük 2014 Archive Report*. Accessed January 11, 2018. http://www.catalhoyuk.com/archive_reports/2014.

Taylor, James, Amy Bogaard, Tristan Carter, Michael Charles, Scott Haddow, Christopher J. Knüsel, Camilla Mazzucato, Jacqui Mulville, Christina Tsoraki, Burcu Tung et al. 2015. "'Up in Flames': A Visual Exploration of a Burnt Building at Çatalhöyük in GIS." In *Assembling Çatalhöyük*, ed. Ian Hodder and Arkadiusz Marciniak, 127–49. Leeds: Maney.

Tsoraki, Christina. 2011. "Disentangling Neolithic Networks: Ground Stone Technology, Material Engagements, and Networks of Action." In *Tracing Prehistoric Social Networks through Technology: A Diachronic Perspective on the Aegean*, ed. Ann Brysbaert, 12–29. New York: Routledge.

Tsoraki, Christina. 2014. "Ground Stone." *Çatalhöyük 2014 Archive Report*. Accessed January 11, 2018. http://www.catalhoyuk.com/archive_reports/2014.

Tsoraki, Christina. 2015. "The Ground Stone Assemblage." *Çatalhöyük 2015 Archive Report*. Accessed January 11, 2018. http://www.catalhoyuk.com/archive_reports/2015.

Tung, Burcu. 2013. "Excavations in the North Area, 2013." *Çatalhöyük 2013 Archive Report*. Accessed January 11, 2018. http://www.catalhoyuk.com/archive_reports/2013.

Tung, Burcu. 2014. "Excavations in the North Area." *Çatalhöyük 2014 Archive Report*. Accessed January 11, 2018. http://www.catalhoyuk.com/archive_reports/2014.

Twiss, Katheryn. 2012. "The Complexities of Home Cooking: Public Feasts and Private Meals Inside the Çatalhöyük House." *eTopoi: Journal for Ancient Studies* 2 (Special Volume): 53–73.

Twiss, Katheryn, Amy Bogaard, Doru Bogdan, Tristan Carter, Michael P. Charles, Shahina Farid, Nerissa Russell, Mirjana Stevanović, E. Nurcan Yalman, and Lisa Yeomans. 2008. "Arson or Accident? The Burning of a Neolithic House at Çatalhöyük, Turkey." *Journal of Field Archaeology* 33 (1): 41–57. https://doi.org/10.1179/009346908791071358.

Waterson, Roxana. 2013. "Transformations in the Art of Dwelling: Some Anthropological Reflections on Neolithic Houses." In *Tracking the Neolithic House in Europe: Sedentism, Architecture, and Practice*, ed. Daniela Hofmann and Jessica Smyth, 373–96. New York: Springer. https://doi.org/10.1007/978-1-4614-5289-8_17.

Wright, Katherine I. 2013. "The Ground Stone Technologies of Çatalhöyük." In *Substantive Technologies at Çatalhöyük: Reports from the 2000–2008 Seasons*, ed. Ian Hodder, 365–416. Los Angeles: Cotsen Institute of Archaeology, University of California.

Wright, Katherine I. 2014. "Domestication and Inequality? Households, Corporate Groups, and Food Processing Tools at Neolithic Çatalhöyük." *Journal of Anthropological Archaeology* 33: 1–33. https://doi.org/10.1016/j.jaa.2013.09.007.

10

Virtually Rebuilding Çatalhöyük History Houses

NICOLA LERCARI

In 2012 Çatalhöyük became inscribed on the UNESCO World Heritage list because of its universal value and exceptionality. As a consequence, the site has gained additional visibility with both the general public through social media and with Turkish and international visitors who increasingly travel to Çatalhöyük in the late spring and summer seasons. Thus, providing visitors and internet users with interpretations of the site's archaeological heritage and explanations of the complex religious and social organization at Çatalhöyük has become even more crucial. The work of the Çatalhöyük Visualization Project on the Visitor Center and the interpretation of the archaeological record for visitors ("Çatalhöyük Visual Assemblage" 2017), as well as the on-site tour guide activities led by licensed professionals and tour guide students, the outreach initiatives on Çatalhöyük such as the Temper Project and the Çatalhöyük Summer Workshop led by Gülay Sert (Bartu-Candan 2007), and the web-based and social media communication directly managed by the Çatalhöyük Research Project team, are all initiatives that play or have played a central role in the wide dissemination of knowledge on Çatalhöyük and related public engagement.

Current 3D visualization techniques and digital archaeological methods make it feasible to further explore alternative means of meaning making in archaeology based on nonlinear narratives, three-dimensional perspective, and virtual simulation. Hence, this chapter

DOI: 10.5876/9781607327370.c010

discusses the use of digital technologies as a new, nonlinear way of conveying information on Çatalhöyük's art, religious ritual, and social organization to the general public.

In 2015 the Virtually Rebuilding Çatalhöyük Project was initiated as a joint effort among the University of California Merced, Stanford University, and the e-learning firm Corinth with the goal to simulate a sequence of Neolithic buildings in a data-driven, accurate, and engaging way.

More specifically, this chapter illustrates how 3D visualization and interactive data curation performed on tablet-based applications, namely Corinth Classroom ("Corinth Classroom" 2016) and Lifeliqe ("Lifeliqe for Windows 10" 2016; "Lifeliqe for iPad" 2016), can contribute to expand the debate on history making in early agricultural societies, involving an audience of nonspecialists, community members, and young students. In this regard, our digital archaeology initiative strives to represent in 3D the continuity of building practices in the stratigraphy of a number of Çatalhöyük houses located in the South Area on the East Mound that were rebuilt multiple times in the same place (figure 10.1).

Excavation on the Çatalhöyük East Mound documented the repetitive practices of rebuilding domestic features or entire houses in the same fashion over time as a manifestation of physical constraints on-site as well as social memory (Hodder and Cessford 2004).

The 3D visualizations and interactive data explorations discussed in this chapter also aim to provide both archaeologists and the general public with digital tools that enable a visual-interactive interpretation of the data collected in the excavation on the East Mound by both James Mellaart in the 1960s and the current Çatalhöyük Research Project. Our approach strives to foster an open and inclusive debate on the archaeological evidence that documents the conscious repetition of buildings and artworks at Çatalhöyük, as well as the intentional destruction of features in overlaying buildings, as an example of memory construction or history-making practices (Hodder and Pels 2010; Hodder 2016).

The virtual simulation of Çatalhöyük history houses—or more elaborated buildings that were rebuilt multiple times in the same place (Hodder and Pels 2010, 163–64)—aims to define a three-dimensional approach to archaeology that integrates a plurality of data in a visual-analytical environment where advanced interactive visualization techniques simulate the cosmology, building practices, material culture, and history-making aspects of Çatalhöyük.

The three-dimensional approach discussed in this chapter is based on the assumption that a data-driven, 3D reconstruction of an archaeological site or

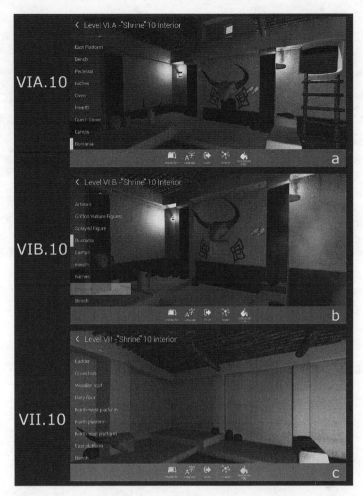

FIGURE 10.1. *Overlaying view of 3D reconstructions of (a) "Shrine" VIA.10, (b) "Shrine" VIB.10, and (c) "Shrine" VII.10 in Lifeliqe app for Windows 10. Source for a–c: author.*

building is a powerful tool for the spatial visualization and interactive exploration of the archaeological evidence. Thus, a 3D reconstruction of history houses has the potential to shed new light on the temporal depth of history making by presenting a new way to visualize, discuss, and interpret stratigraphic and spatial information related to the special type of buildings. For instance, the photorealistic approach to the 3D reconstruction of history houses used by

FIGURE 10.2. *View of a highly evocative 3D reconstruction of Çatalhöyük history house F.V.I, or the "Shrine" of the Hunters. Source: Artas Media, reconstruction by Grant Cox.*

Grant Cox produced highly evocative virtual simulations of Building F.V.I, or the "Shrine" of the Hunters. Cox's work proved that a data-driven 3D reconstruction can be successfully merged with subjective interpretations to produce new knowledge as well as aesthetically pleasant visualizations of the past (Cox 2011; ArtasMedia 2016a, 2016b) (figure 10.2).

Current technology facilitates 3D data exploration performed through inexpensive virtual reality (VR) displays. For instance, the head-mounted display Oculus Rift and the VR controller Oculus Touch ("Oculus Rift" 2016) enable new three-dimensional ways to visualize and interact with our data (Lercari et al. 2013, 2014, 2017). Thus, virtual simulation provides archaeologists, heritage practitioners, and historians with new tools that generate additional and redundant information on their case studies using a three-dimensional and spatial approach that goes beyond the textual dimension of a database or the bi-dimensional visualization of a traditional Geographic Information System (GIS) (Forte et al. 2012, 2015; Lercari et al. 2011; Lercari 2016a) (figure 10.3).

For the above reasons, this chapter strives to demonstrate that the interactive approach used in the Virtually Rebuilding Çatalhöyük Project makes a 3D reconstruction a valuable nonlinear tool for the interpretation of the past that aims to become instrumental to the study of history making at Çatalhöyük.

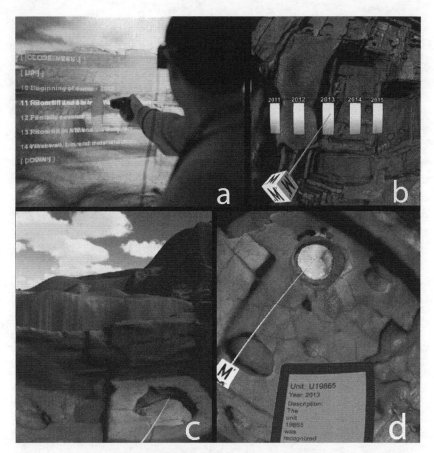

FIGURE 10.3. *View of (a) 3D models and immersive interaction of Building 89 in the Duke Immersive Visualization Environment (DiVE), (b) time line to filter 3D data by year of excavation in Dig@IT, (c) immersive interaction with B.89's burials in Dig@IT, and (d) interactive metadata browsing on a virtual tablet linked to the Çatalhöyük database server in Dig@IT. Source for a–d: author.*

INTERACTIVE VISUALIZATION AND 3D RECONSTRUCTION OF ÇATALHÖYÜK HISTORY HOUSES

The outcome of reconstructing Çatalhöyük history houses in 3D was first discussed at the symposium Religion, History, and Place in the Origins of Settled Life in the Middle East chaired by Ian Hodder at the SAA 80th Annual Meeting in San Francisco (Lercari 2015). The preliminary virtual simulations and 3D data presented at the event were created by the author of this chapter together with Ondrej Homola, Iveta Kalisova, David Motalik, and a team of 3D artists and engineers from the e-learning firm Corinth as part of the University of California (UC) Merced–led collaborative research project Virtually Rebuilding Çatalhöyük (Lercari 2017). As anticipated at the start of this chapter, the project's case study is the visualization of building variation at Çatalhöyük East Mound, with a specific focus on history houses and the practice of history making.

Rebuilding Çatalhöyük history houses in 3D is particularly relevant to this volume as our three-dimensional perspective on the archaeological record aims to start a debate on whether the stratigraphies of a site can tell us more than just chronological information related to sequences of buildings. In addition, our data-driven perspective on the 3D reconstruction of Çatalhöyük history houses strives to generate a more inclusive approach to the study of history making in early agricultural societies with the aim to engage the general public of museums, visitor centers, and the internet, as well as young students, on this topic.

This section of the chapter presents the preliminary visualizations and 3D reconstructions produced by the Virtually Rebuilding Çatalhöyük Project in the period 2015–16 (Lercari 2016b, 2016c). This initial phase of the project focused on the 3D reconstruction of three history houses (Buildings VIA.10, VIB.10, and VII.10) that belong to Mellaart's "Shrine" 10 sequence. This case study was selected because Mellaart's excavation in the 1960s and then Hodder's work in the last two decades testify that "Shrine" 10 is one of the longest and most repetitiously reconstructed buildings ever documented at Çatalhöyük (figure 10.4a). In particular, this preliminary work addresses the 3D reconstruction of "Shrine" 10 in Mellaart's Levels VIA, VIB, and VII, displaying three highly decorated overlaying buildings excavated by James Mellaart in 1962 and 1963 (Mellaart 1963, 1964, 1967).

Work on Çatalhöyük history houses identifies building variation as a key element to understand Çatalhöyük religious rituals by highlighting the fact that Mellaart used the parameter of architectural remaking among overlaying houses to identify a predominant religious function in "special" buildings defined as "shrines" (Hodder and Pels 2010, 163–64; Hodder 2016). More than twenty years of excavations by the current project on the East Mound have produced comprehensive interpretations of the repetition of architectural elements in buildings, providing evidence that most of the dwellings at Çatalhöyük had both domestic and ritual functions (Hodder 2000, 2005a, 2005b, 2007, 2014; Hodder and Ritchey 1996). Drawing on seminal work regarding building variation that provides thorough understanding of social organization, property, power, and religion in early settled life (Hodder 2006, 2010), this chapter

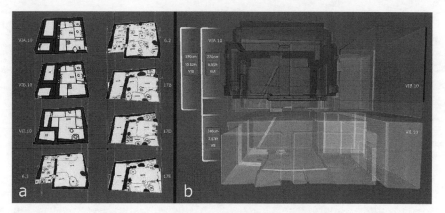

FIGURE 10.4. *(a) Drawings of the complete "Shrine" 10 sequence. Source (a): Çatalhöyük Research Project, drawing by Tim King. (b) View of Mellaart's "Shrine" 10 in Levels VIA, VIB, and VII rendered as transparent overlaying layers, including estimated elevation, in Corinth Classroom (author copyright).*

seeks to disseminate information and interpretations related to the concept of history making to the general public and young students.

To achieve this goal, the Virtually Rebuilding Çatalhöyük Project leverages the capabilities of Corinth Classroom, a user-friendly digital learning software that allows users to browse collections of 3D content and metadata, make annotations on the 3D models, take quizzes, capture snapshots and drawings, and participate in interactive discussions on the simulated material ("Corinth Classroom" 2016). In 2015 Corinth Classroom was provided to the UC Merced team by the e-learning firm Corinth under a memorandum of understanding and collaborative research initiative facilitated by the Çatalhöyük Project. Corinth Classroom is capable of displaying textual and visual information, such as 2D images and graphics, and 3D interactive content with incredible realism and advanced shading and lighting effects.

In 2016 the Virtually Rebuilding Çatalhöyük Project wanted to expand the reach of the 3D reconstructions of Çatalhöyük history houses developed by UC Merced and Corinth. Thus, we published the interactive data explorations discussed in this chapter in the Apple Store as free content for the mobile application Lifeliqe. This app is a powerful visual learning tool for mobile devices capable of interactively visualizing 3D content in high quality and detail. Lifeliqe is designed by the media company LifeLiQe to engage students in the K–12 curriculum in visual learning experiences. Lifeliqe supports both 3D interactive explorations, such as the 3D reconstructions of the

Çatalhöyük's history houses we designed, as well as augmented reality content that may constitute a feature development of our project.

iPad users can download and install Lifeliqe on their tablets free of charge ("Lifeliqe for iPad" 2016). Accessing Lifeliqe as "guests," iPad users can freely explore the 3D reconstructions of the three history houses created by our team.

As of June 2016, Lifeliqe became available to download for free from the Windows Store ("Lifeliqe for Windows 10" 2016). Thus, this app and our 3D reconstructions can also be installed on tablet PCs running Windows 10. The most noteworthy features of Lifeliqe are (a) its ability to display high-quality 3D interactive real-time visualizations, (b) its ability to support visual learning with a bilingual view (English and Spanish), (c) its ability to display background information and metadata side by side with the 3D reconstruction, (d) the possibility to interactively annotate the 3D models with comments and custom descriptions, (e) its ability to share snapshots and views of the 3D reconstructions displayed in Lifeliqe directly in a PowerPoint, (f) augmented reality capabilities that merge the 3D reconstructions with the real world, for instance, overlaying a 3D model of "Shrine" 10 in Level VI with the space where the building was excavated by Mellaart, and (g) its ability to zoom on the 3D data from a wide-angle view to a very close-range view of the 3D data.

HISTORY MAKING IN 3D

The 3D reconstructions and interactive explorations of history houses discussed in this chapter belong to the preliminary phase of a larger digital archaeology initiative. The overall goal of our project is to virtually rebuild the entire sequence of "Shrine" 10 from Mellaart's Building VIA to Building 17, including the different phases of such building that were excavated by Ian Hodder in the late 1990s and then again in 2015 (see figure 10.4a).

At first, the Virtually Rebuilding Çatalhöyük Project focused on the information included in Mellaart's report on the excavation season of 1962 when a "Shrine" 10 was identified in Level VI (Mellaart 1963, 70–73). Such a building had burials in the Central East and North East platforms and was highly decorated, with an abundance of ritual artwork and features.

Using the CAD drawings of Mellaart's levels that were produced by the Çatalhöyük Research Project team (figure 10.5), we reconstructed in 3D a large "Shrine" 10 in Level VIA measuring approximately 5.75 meters × 4.35 meters, with well-preserved walls approximately 2.7 m high and a roof access near the south wall and a crawl hole in the southern part of the east wall (Mellaart 1963, 70).

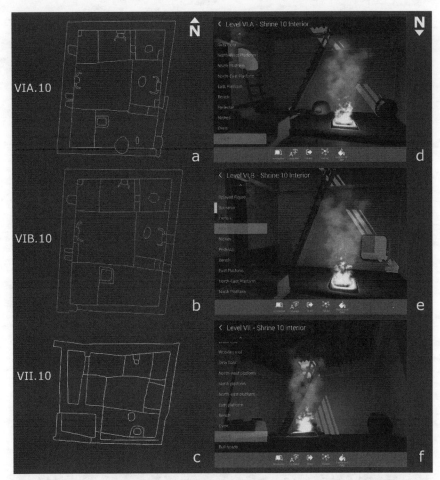

FIGURE 10.5. *Overlaying view of CAD drawings of (a) VIA.10; (b) VIB.10; and (c) VII.10 and overlaying view of house-based history making in the "Shrine" 10 sequence, displaying the repetition of the hearth in Level VIA. Source for a–c: Çatalhöyük Research Project. (d), VIB (e), and VII (f) rendered in Lifeliqe app for Windows 10. Source for d–f: author.*

The 3D reconstruction of this history house meticulously illustrates the northern platforms characterized by two lips dividing their surface into three parts, as well as by a bull pillar.

The north wall of VIA.10 was highly decorated with panels painted in red: a one-of-a-kind double-horn ram head in the middle and a plaster box for offerings. The 3D visualization of VIA.10's east wall also illustrates an

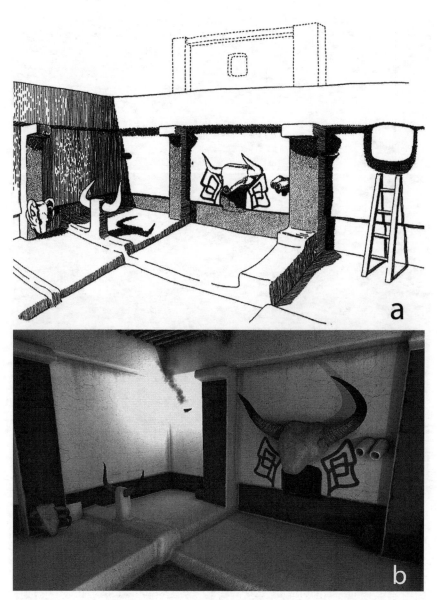

FIGURE 10.6. *Comparative view of (a) Mellaart's visual restoration of the eastern part of "Shrine" 10.VIB. Drawing by Grace Huxtable in Mellaart 1963, 72. (b) 3D reconstruction of the eastern part of "Shrine" 10.VIB in Corinth Classroom (author copyright).*

elaborated ritual composition made of two painted plastered posts, a painted bull head located between them, a wall painting surrounding the bucranium, and a painted niche underneath it (figure 10.6).

Most remarkably, the 3D reconstruction of this building includes "three superimposed bull heads" attached to the west wall and the approximately 1-meter-tall splayed figure that Mellaart interpreted as a "goddess giving birth to a ram." This artwork was discovered partially preserved at the bottom of the west wall. These splayed reliefs are interpreted by the current project as depicting a bear rather than a goddess. The size of the monumental splayed figure brought Mellaart to hypothesize the presence of a clerestory of about 3 meters in the central part of the building (Mellaart 1963, 70).

The 3D reconstructions discussed in this chapter also draw on the evidence and visual information provided in Mellaart's report on the excavation season during 1963. In this publication, Mellaart modified his initial interpretation of "Shrine" 10 in Level VI, identifying a later phase (VIA) and an earlier phase (VIB) (Mellaart 1964, 40–42).

Evidence in "Shrine" VIB.10 showed that burials in this earlier phase were located in the same Central East and North East platforms (see figure 10.5b). Unfortunately, Mellaart's reports do not provide detailed information on the burials documented in "Shrine" 10 during the 1962 and 1963 field seasons. The lack of information on voluntary retrieval of skeletal remains from burial pits in Mellaart's documentation does not allow our virtual simulations to include the dynamic aspects of active history making that occurred in Mellaart's Levels VI and VII.

Nonetheless, our 3D reconstructions of "Shrines" VIA.10 and VIB.10 stress the fact that these two phases are an outstanding example of the practice of history making based on the renewal of ritual artworks and features. The elaborated bucrania and wall painting that adorned the west, north, and east walls of VIB.10 were maintained in situ and renewed in VIA.10, even though the southern part of the house was modified. This evidence suggests a clear and intentional display of a common history among the people who occupied "Shrine" 10 in subsequent periods. Our 3D reconstructions of "Shrine" 10 in Levels VIA and VIB also include the splayed figure that decorated the west wall in both phases. Our virtual simulations also render two different reconstructions of the clerestory as proposed by Grace Huxtable, the illustrator working with Mellaart in the 1960s (figure 10.7).

This chapter's contribution to the interpretation of history making at Çatalhöyük is further represented by the ability of our virtual simulations to provide scholars, students, and the general public with a three-dimensional

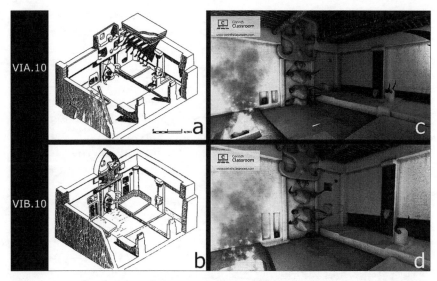

FIGURE 10.7. *Overlaying and comparative views of (a) Mellaart's visual restoration of "Shrine" 10.VIA. Drawing by N. Alckock and Grace Huxtable in Mellaart 1963, 71. (b) Mellaart's visual restoration of "Shrine" 10.VIB. Drawing by Grace Huxtable in Mellaart 1964, 51. (c) 3D reconstruction of "Shrine" 10.VIA in Lifeliqe for Windows 10, and (d) 3D reconstruction of "Shrine" 10.VIB in Lifeliqe for Windows 10.*

perspective on the stratigraphy of these buildings. This option makes it easier for our users to create mental connections among data, features, and areas of a building to find direct spatial relationships between the hearths documented in the history houses that we reconstructed in 3D. The overlaying view of the "dirty" areas of VIA.10 and VIB.10 (see figures 10.5d and 10.5e) visually depicts how the hearth occupies an almost identical location in VIA.10 even after the renewal of VIB.10.

Our 3D reconstructions of VIA.10 and VIB.10 also help users of Corinth Classroom or Lifeliqe to visually identify the discontinuity that characterizes the "dirty" areas of the two phases of "Shrine" 10. Major discontinuities between these two phases are substantiated by a reconfiguration of the access to storage rooms or other buildings through crawl holes located in the southern part of the east wall in "Shrine" 10.VIA. However, in the western part of the southern wall in 10.VIB, differences in the "dirty" areas are rendered mostly by the absence of an oven in 10.VIB and by the reduced size and number of the niches in the west wall of 10.VIA (see figures 10.7c and 10.7d).

It is important to highlight the renewal process of features and artwork that we meticulously created and rendered in our 3D reconstructions of "Shrine" 10. Hodder (this volume) argues that the compulsive repetition of new hearths in the same location for decades or even hundreds of years documented at sites such as Aşıklı Höyük and Çatalhöyük is evidence of embodied history-making practices in the Neolithic. To reinforce this assumption, we looked at the repetition of the hearth in the building below VIB.10 that was documented in Mellaart's 1963 excavation report. This publication discusses evidence of another "Shrine" 10 built in Level VII, just beneath VIB.10 (Mellaart 1964, 57). Such a building presents the same cosmology and usage of ritual space as the later phases of "Shrine" 10, but it is smaller and significantly less decorated, and it has one or two adjacent storage rooms or small buildings located west of the main environment. A crawl hole connects the storage room in the northwest part of the building to the main environment.

Most important for this discourse on history making in 3D, the hearth documented in VII.10 shares a very similar location with the hearths excavated in the later VIA.10 and VIB.10 (see figures 10.5d, 10.5e, and 10.5f). Again, we can see how our 3D visualizations of "Shrine" 10 help users of our virtual simulations picture a type of history making based on the repetition or renewal of buildings at Çatalhöyük.

The 3D reconstruction of "Shrine" VII.10 also shows that this building was less ornate compared to its later remakes in Levels VIA and VIB. Even if the plan of VII.10 strictly resembles the one of VIB.10, this earlier history house only had one plaster relief in the north wall and did not present evidence of wall painting. Mellaart's reconstruction of 10.VII shows another plaster artwork in the northeast corner of the building, specifically, "a stag on a rock" (Mellaart 1964, 57). This feature was consciously omitted from our virtual simulation of VII.10 because its interpretation was not adequately supported by the photographic documentation provided by Mellaart in a later publication (Mellaart 1967) (figure 10.8).

To conclude this section on history making in 3D, our work strived to render the continuity in building and ritual practices among houses that belong to the same stratigraphy with the aim to help interpret the practices of memory construction and history making that linked the inhabitants of "Shrine" 10.

At the time this chapter was written, only three of eight phases or reconstructions of "Shrine" 10 were rebuilt in 3D (see figure 10.4). Hence, future developments of the Virtually Rebuilding Çatalhöyük Project will need to continue the study of the history house excavated by Hodder in the late 1990s and again in 2015. The aim of future developments for this project would be to

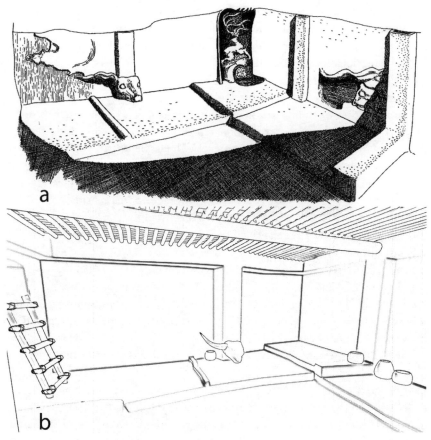

FIGURE 10.8. *Comparative view of (a) Mellaart's visual restoration of "Shrine" 10.VII. Drawing by Grace Huxtable in Mellaart 1964, 60. (b) isometric drawing of "Shrine" 10.VII automatically generated by Corinth Classroom (author copyright).*

complete the 3D reconstruction of the remaining phases of Buildings 6 and 17 that were documented underneath Mellaart's "Shrine" VII.10.

In addition, our goal is to conduct new research at UC Merced to develop a custom 3D visualization and interactive data exploration platform that will build off the work on house-based history making presented in Corinth Classroom and Lifeliqe. Our vision is to develop a custom 3D viewer using the 3D game engine Unity 3D ("Unity 3D" 2016) or similar technology, enriched by custom features specifically designed for the virtual simulation of different types of history making. For instance, the new 3D data curation platform will address in greater detail the visualization of active history making at Çatalhöyük with the goal to provide its users with a better understanding of the ritual practice of removing skeletal remains, stone tools, and ritual objects from a house and then replacing or reburying them in another building or in the foundation of a new building. The new virtual simulation platform will also feature interactive tools that allow users to simulate and display the conscious destruction of features and intentional burning of entire buildings that were documented in "special" history houses such as Building 77. To accomplish this goal, the UC Merced team will leverage virtual reality technologies, as well as custom animations, 3D diagrams, and hyperlinks between building components. The new platform will also display to users the connection between the 3D reconstructions of Çatalhöyük history houses and the related sources, metadata, or images that were used to create the 3D visualizations. In this regard, our team has already made significant progress on developing custom programs that link a 3D model and its components to the Çatalhöyük Database ("Çatalhöyük Database" 2016) or the Çatalhöyük Image Collection Database ("Çatalhöyük Image Collection Database" 2016). Further work still needs to be done to integrate our 3D reconstructions onto the Çatalhöyük Living Archive's website (Grossner et al. 2012; "Çatalhöyük Living Archive" 2016).

CONCLUSION

This chapter demonstrates that a contemporary discourse on the virtual simulation of the past and the 3D reconstruction of its relics must adopt a reflexive perspective and go beyond the discussion of technological improvements and methods that often characterizes digital archaeology scholarship. History making in 3D, virtual place making, the role of spatiality and temporality in a historical virtual environment, and the representation of multiple viewpoints on history are almost unchartered territories in a virtual simulation of the past that future research in this field should address.

The significance of virtually rebuilding Çatalhöyük history houses derives from the fact that a 3D reconstruction "attracts" people inside the archaeological context and involves them in a synesthetic process of meaning making in which both tangible and intangible elements of the past can be discussed, shared, and understood (Lercari 2010, 130). The three-dimensional approach to the study of "history making" presented in this chapter seeks to expand the potential of archaeological interpretation by producing new knowledge on the archaeological record both during the design and implementation phase of the virtual simulation and during the interactive exploration of its data performed by users.

The author of this chapter strongly believes that in the next few years, 3D GIS and other digital-visual analytical tools will push the boundaries of archaeology, creating novel methods for the interpretation of the past. Innovative 3D visualization technologies such as 3D game engines like Unity 3D, with analytical capability and real-world coordinates, have already started to contribute to the development of new paradigms for the digital visualization of the past. Such technologies present great potential because they are able to reduce the gap that still exists between the processes of data collection and interpretation, on one hand, and the dissemination of archaeological data, on the other. For instance, the users of a virtual simulation created in Unity 3D can take part in simulated religious rituals or social activities, embodying themselves as digital avatars that represent the people who were living in a specific place at a given time.

Phenomenology assigns a fundamental role to our body, arguing that the cognition and interpretation of the world in which we live occur through our sensorimotor system (Merleau-Ponty 1945). Drawing upon these theories, cognitive science emphasizes the importance of embodied mind, situated cognition, and enaction in the interpretation of complex data (Varela, Rosch, and Thompson 1992).

The virtual simulations of "Shrine" 10 presented in this chapter can thus be enhanced by an embodied interactive exploration of the 3D reconstructions, where users can experience the simulated scenario through the virtual bodies of their avatars while collectively reenacting the past and socializing with each other.

The power of virtual simulation in archaeology is consolidated by new, peculiar typologies of spatiality and temporality, typical of a 3D reconstruction. According to our perspective, such formal structures allow users to identify, analyze, discuss, and interpret the spatial and temporal dimensions of Çatalhöyük history houses with greater ease when compared to a traditional form of data curation such as a textbook or a photo collection.

Building off the theories on virtual place making discussed by Champion and Dave (Champion and Dave 2007, 333, 340–43), this chapter highlights the significance of visualizing and simulating the past by providing evidence of the epistemic value embedded in these new interactive and collaborative ways of interpreting the past. The proposed approach also strives to frame digital visualization and virtual simulation within the new revised reflexive methods used at Çatalhöyük (Berggren et al. 2015) and emphasizes the significance of promoting multiple viewpoints on history in the process of virtual recreation of Çatalhöyük history houses.

Beyond the pedagogical value of the proposed virtual reconstructions of "Shrine" 10 in Levels VIA, VIB, and VII, the significance of the preliminary results discussed in this chapter derives from the possibility of visualizing the three-dimensional relationship of features across multiple levels as rendered in Corinth Classroom (see figure 10.4b).

This chapter argues that a 3D virtual simulation allows users to better visualize and read the conscious or unconscious repetition of building patterns and the rebuilding or destruction of features in overlaying buildings. For instance, the users of our 3D reconstructions can verify the repetition of the hearth in three history houses belonging to the "Shrine" 10 sequence that were simulated in Lifeliqe (see figures 10.5d, 10.5e, and 10.5f). This type of comparative 3D visualization also informs its users of the estimated height of each building and correlates this information with a tri-dimensional perception of the stratigraphy of the visualized buildings.

As mentioned in the previous section, future work will finalize the simulation of the stratigraphy of "Shrine" 10, completing the 3D reconstruction of Building 6 in Level VIII and Building 17 in Level XI as they were excavated and documented by the current project (Hodder and Pels 2010, 170).

To conclude, our final aim is to conduct additional research on history making in 3D and to develop a custom 3D visualization and interactive data exploration platform that will build off the work on house-based history making presented in this chapter. Our ultimate goal is to design and develop a custom 3D data curation platform that leverages virtual reality and real-time computer graphics technologies—for instance, using the 3D game engine Unity 3D—and that is capable of enabling a more immersive and multi-vocal visualization of the past through the interactive exploration of different types of history making.

REFERENCES

ArtasMedia. 2016a. "Çatalhöyük—Decorated Structures." Accessed August 29, 2016. https://artasmedia.com/2015/02/23/catalhoyuk-part-4-decorated-structure-patterns/.

ArtasMedia. 2016b. "Çatalhöyük—the 'Shrine' of the Hunters (F.V.I)." Accessed August 29, 2016. https://artasmedia.com/2015/03/10/catalhoyuk-the-shrine-of-the-hunters-f-v-i/.

Bartu-Candan, Ayfer. 2007. "Developing Educational Programmes for Prehistoric Sites: The Çatalhöyük Case." In *Mediterranean Prehistoric Heritage: Training,*

Education, and Management, ed. Ian Hodder and Louise Doughty, 95–104. London: McDonald Institute for Archaeological Research.

Berggren, Asa, Nicolo Dell'Unto, Maurizio Forte, Scott Haddow, Ian Hodder, Justine Issavi, Nicola Lercari, Camilla Mazzucato, Allison Mickel, and James S. Taylor. 2015. "Revisiting Reflexive Archaeology at Çatalhöyük: Integrating Digital and 3D Technologies at the Trowel's Edge." *Antiquity* 89 (344): 433–48. https://doi.org/10.15184/aqy.2014.43.

"Çatalhöyük Database." 2016. Accessed August 29, 2016. http://www.catalhoyuk.com/research/database.

"Çatalhöyük Image Collection Database." 2016. Accessed August 29, 2016. http://server.catalhoyuk.com/netpub/server.np?base&site=catalhoyuk&catalog=catalog&template=search.np&showindex=true.

"Çatalhöyük Living Archive." 2016. Accessed August 29, 2016. http://catalhoyuk.stanford.edu/.

"Çatalhöyük Visual Assemblage." 2017. Accessed February 9, 2017. https://catalva.wordpress.com/.

Champion, Erik, and Bharat Dave. 2007. "Dialing up the Past." In *Theorizing Digital Cultural Heritage: A Critical Discourse*, ed. Fiona Cameron and Sarah Kenderdine, 333–48. Cambridge: MIT Press. https://doi.org/10.7551/mitpress/9780262033534.003.0017.

"Corinth Classroom." 2016. *Corinth*. Accessed August 29, 2016. https://www.ecorinth.com/.

Cox, Grant. 2011. "Photo-Realistic Reality: Focusing on Artistic Space at Çatalhöyük." MSc dissertation, University of Southampton, GB.

Forte, Maurizio, Nicolo Dell'Unto, Justine Issavi, Lionel Onsurez, and Nicola Lercari. 2012. "3D Archaeology at Çatalhöyük." *International Journal of Heritage in the Digital Era* 1 (3): 351–78. https://doi.org/10.1260/2047-4970.1.3.351.

Forte, Maurizio, Nicolo Dell'Unto, Kristina Jonsson, and Nicola Lercari. 2015. "Interpretation Process at Çatalhöyük Using 3D." In *Assembling Çatalhöyük*, ed. Ian Hodder and Arkadiusz Marciniak, 43–57. Leeds: Maney.

Grossner, Karl, Ian Hodder, Elijah Meeks, Claudia Engel, and Allison Mickel. 2012. "A Living Archive for Çatalhöyük." In *Proceedings of the 2012 Computer Applications and Quantitative Methods in Archaeology (CAA)*. Accessed August 29, 2016. http://catalhoyuk.stanford.edu/assets/CAA2014_A%20Living%20Archive%20for%20Catalhoyuk_Apr2014.pdf.

Hodder, Ian. 2000. *Towards Reflexive Method in Archaeology: The Example at Çatalhöyük*. London: McDonald Institute for Archaeological Research.

Hodder, Ian. 2005a. *Changing Materialities at Çatalhöyük: Reports from the 1995–99 Seasons*. London: McDonald Institute for Archaeological Research.

Hodder, Ian. 2006. *The Leopard's Tale: Revealing the Mysteries of Çatalhöyük*. New York: Thames and Hudson.

Hodder, Ian. 2010. *Religion in the Emergence of Civilization: Çatalhöyük as a Case Study*. Cambridge: Cambridge University Press. https://doi.org/10.1017/CBO9780511761416.

Hodder, Ian. 2016. "More on History Houses at Çatalhöyük: A Response to Carleton et al." *Journal of Archaeological Science* 67: 1–6. https://doi.org/10.1016/j.jas.2015.10.010.

Hodder, Ian, ed. 2005b. *Inhabiting Çatalhöyük: Reports from the 1995–99 Seasons*. London: McDonald Institute for Archaeological Research.

Hodder, Ian, ed. 2007. *Excavating Çatalhöyük: South, North, and KOPAL Area Reports from the 1995–99 Seasons*. Oakville, CT: McDonald Institute for Archaeological Research.

Hodder, Ian, ed. 2014. *Çatalhöyük Excavations: The 2000–2008 Seasons*. Los Angeles: Cotsen Institute of Archaeology.

Hodder, Ian, and Craig Cessford. 2004. "Daily Practice and Social Memory at Çatalhöyük." *American Antiquity* 69 (1): 17–40. https://doi.org/10.2307/4128346.

Hodder, Ian, and Peter Pels. 2010. "History Houses: A New Interpretation of Architectural Elaboration at Çatalhöyük." In *Religion in the Emergence of Civilization: Çatalhöyük as a Case Study*, ed. Ian Hodder, 163–86. Cambridge: Cambridge University Press. https://doi.org/10.1017/CBO9780511761416.007.

Hodder, Ian, with Tim Ritchey. 1996. "Re-opening Çatalhöyük." In *On the Surface: Çatalhöyük 1993–1995: Çatalhöyük Research Project*, vol. 1, ed. Ian Hodder, 1–18. London: McDonald Institute for Archaeological Research.

Lercari, Nicola. 2010. "An Open Source Approach to Cultural Heritage: Nu.M.E. Project and the Virtual Reconstruction of Bologna." *Cyber-Archaeology* 1: 125–33.

Lercari, Nicola. 2015. "Virtually Rebuilding Çatalhöyük History Houses." Paper presented at the annual meeting of the Society for American Archaeology, San Francisco, CA, April 15–19.

Lercari, Nicola. 2016a. "Simulating History in Virtual Worlds." In *Handbook on 3D3C Platforms*, ed. Yesha Sivan, 337–52. New York: Springer International. https://doi.org/10.1007/978-3-319-22041-3_13.

Lercari, Nicola. 2016b. "Virtually Rebuilding Çatalhöyük History Houses—'Shrine' 10 VIA." YouTube. Accessed August 30, 2016. https://www.youtube.com/watch?v=6ggPhwbNRAM.

Lercari, Nicola. 2016c. "Virtually Rebuilding Çatalhöyük History Houses—'Shrine' 10 VIB." YouTube. Accessed August 30, 2016. https://www.youtube.com/watch?v=mEKGHKwWonw.

Lercari, Nicola. 2017. "3D Visualization and Reflexive Archaeology: A Virtual Reconstruction of Çatalhöyük History Houses." *Digital Applications in Archaeology and Cultural Heritage* 6:10–17. https://doi.org/10.1016/j.daach.2017.03.001.

Lercari, Nicola, Maurizio Forte, David Zielinski, Regis Kopper, and Rebecca Lai. 2013. "Çatalhöyük at DiVE: Virtual Reconstruction and Immersive Visualization of a Neolithic Building." Paper presented at the annual meeting of the Digital Heritage International Congress, Marseille, France, October 28–November 1.

Lercari, Nicola, Stephanie Matthiesen, David Zielinski, and Regis Kopper. 2014. "Towards an Immersive Interpretation of Çatalhöyük at DiVE." Paper presented at the annual meeting of the American School of Oriental Research, San Diego, CA, April 19–22.

Lercari, Nicola, Emmanuel Shiferaw, Maurizio Forte, and Regis Kopper. 2017. "Immersive Visualization and Curation of Archaeological Heritage Data: Çatalhöyük and the Dig@IT App." *Journal of Archaeological Method and Theory*: 1–25. https://doi.org/10.1007/s10816-017-9340-4.

Lercari, Nicola, Elena Toffalori, Michaela Spigarolo, and Lionel Onsurez. 2011. "Virtual Heritage in the Cloud: New Perspectives for the Virtual Museum of Bologna." Paper presented at *VAST: International Symposium on Virtual Reality, Archaeology, and Intelligent Cultural Heritage*, Prato, Italy, October 18–21, 153–60. Goslar, Germany: Eurographics Association.

"Lifeliqe for iPad." 2016. *iTunes Store*. Accessed August 29, 2016. https://itunes.apple.com/us/app/lifeliqe/id1064873813?mt=8.

"Lifeliqe for Windows 10." 2016. *Microsoft Store*. Accessed August 29, 2016. https://www.microsoft.com/en-us/store/p/lifeliqe/9nblggh4rgt5.

Mellaart, James. 1963. "Excavations at Çatal Hüyük, 1962: Second Preliminary Report." *Anatolian Studies* 13: 43–103. https://doi.org/10.2307/3642490.

Mellaart, James. 1964. "Excavations at Çatal Hüyük, 1963: Third Preliminary Report." *Anatolian Studies* 14: 39–119. https://doi.org/10.2307/3642466.

Mellaart, James. 1967. *Çatal Hüyük: A Neolithic Town in Anatolia*. London: Thames and Hudson. http://journals.cambridge.org/abstract_S0003598X00033998.

Merleau-Ponty, Maurice. 1945. *Phénoménologie de la Perception*. Paris: Gallimard.

"Oculus Rift." 2016. *Oculus*. Accessed August 30, 2016. https://www.oculus.com/rift/.

"Unity 3D." 2016. *Unity Technologies*. Accessed August 30, 2016. https://unity3d.com/.

Varela, Francisco J., Eleanor Rosch, and Evan Thompson. 1992. *The Embodied Mind: Cognitive Science and Human Experience*. Cambridge, MA: MIT Press.

Contributors

KURT W. ALT
Center of Natural and Cultural History of Man, Danube Private University, Krems, Austria, and Department of Biomedical Engineering and Integrative Prehistory and Archaeological Science, Basel University, Switzerland

MARK R. ANSPACH
LIAS, Institut Marcel Mauss, École des Hautes Études en Sciences Sociales, Paris

ANNA BELFER-COHEN
Institute of Archaeology, Hebrew University of Jerusalem, Jerusalem

MARION BENZ
Department of Archaeological Sciences, University of Freiburg, Germany

LEE CLARE
German Archaeological Institute (DAI), Berlin/Istanbul

OLIVER DIETRICH
German Archaeological Institute (DAI), Berlin/Istanbul

GÜNEŞ DURU
Department of Political Science, Galatasaray University, Istanbul

YILMAZ S. ERDAL
Department of Anthropology, Hacettepe University, Ankara

NIGEL GORING-MORRIS
Institute of Archaeology, Hebrew University of Jerusalem, Jerusalem

IAN HODDER
Department of Anthropology, Stanford University, CA

ROSEMARY A. JOYCE
Department of Anthropology, University of California, Berkeley

NICOLA LERCARI
School of Social Sciences, Humanities, and Arts, University of California, Merced

WENDY MATTHEWS
Department of Archaeology, University of Reading, United Kingdom

JENS NOTROFF
German Archaeological Institute (DAI), Berlin/Istanbul

VECIHI ÖZKAYA
Archaeology Department, Dicle University, Diyarbakir

FERIDUN S. ŞAHIN
Archaeology Department, Dicle University, Diyarbakir

F. LERON SHULTS
Institute for Religion, Philosophy, and History, University of Agder, Kristiansand, Norway

DEVRIM SÖNMEZ
German Archaeological Institute (DAI), Berlin/Istanbul

CHRISTINA TSORAKI
Faculty of Archaeology, Leiden University, Holland

WESLEY J. WILDMAN
Boston University and the Center for Mind and Culture, Boston, MA.

Index

abandonment: ritual, 13–14, 20–21, 204; settlement, 15–16 'Abr 3, Tell, 171; stone objects from, 145, 146, *148*
Abu Hureyra, 11, 12, 21; hearths at, 186–87; house superpositioning, 17–18
accelerator mass spectrometer (AMS) dating, at Göbekli Tepe, 118
Acheulian era, 100
Aeneas, 190
agent-based models (ABMs), 35, 59
aggregation, 11, 23, 171; Epipaleolithic, 100–101, 107, 169
aggression, 151
agriculture, 3, 170, 180; in NSIM, 42, 45–46, 48, 54; and religion, 9–10; transition to, 36, 167; Zagros, 65, 83–84
agropastoralism, transition to, 36
Ahmarian period, 100
'Ain Ghazal, 16, 20
'Ain Mallaha: house superpositioning, 12–13; ritual abandonment at, 13–14. *See also* Eynan
Ainu, 189
Ali Kosh, 65, 82
almonds: in Zagros sites, 73
AMS dating. *See* accelerator mass spectrometer dating
Anatolia, 19, 24, 78, 132, 155(n5), 167, 188

ancestors, 5, 8, 9, 190; place making, 141–42, 152; veneration of, 20
animal pens, in Zagros sites, 73, *77*
animals, 144, 204, 238; depictions of, 131–32, 133, 150; domestication of, 3, 65, 170, 180; on Göbekli Tepe T-pillars, *124*, 129; ritual deposition of, 76, 81, 242, 245–47
archaeobotanical remains, 36; from Building 77, 242, 243–44
architecture: and burials, 105–6, 108; Çatalhöyük and Aşıklı Höyük, 194–95, 196–97; imitative domestic, 187–88; monumental and domestic, 119–20; mudbrick, 75, 240 artifacts, 38; transportable, 171–72. *See also* objects; *by type*
Asiab, 65, 70, 81, 82
Aşıklı Höyük, 8, 19, 108, 118, 163; at Çatalhöyük, 194–95; description of, 193–94; hearths at, 187, 196–203, 207(n7); public and private space at, 170, 172–73; repeated building at, 75, 192; ritual construction at, 23–24, 205
assimilation, and loss of identities, 176
Aswad, 142

Page numbers in italics indicate illustrations

Augustus Caesar, 190, 191
Aurenche, Olivier, 167
Aurignacian period, 100
aurochs, depictions of, 131–32
authority, ritual, 217
axes, in Building 77, 244–45, 247, 250, 252
Ayia Irini, 188
Azraq basin, 100

Balkans, house replacement in, 7
barley, 65, 73
Bar-Yosef, Ofer, 167
base camps, Natufian, 11
Batman Çayi, 139
beads, 144, 218
bear, as symbol, 38
Beidha, 12, 17, 19
Belfer-Cohen, Anna, 167
belief systems, 138, 167, 171–72
Bell, James, 35
benches, 15, 202
Bestansur, 66; repeated building at, 75–76, 83–85, 90
birthrate, and NSIM, 39, 49
Bismil, 142
Black Box, Neolithic Social Investment Model, 39, 40–50
boars, depictions of, 125; at 'Ain Mallaha, 13
Boncuklu, 23, 75, 78, 187, 194, 195
bone tools, in Building 77, 242
Boqer Tachtit, 100
Borneo, space ownership in, 217
botanical remains, in Building 77, 242, 243–44
Bouqras, 18, 20
Braidwood, Robert, 165
Britain, cremation in, 192
Bronze Age Crete, 187–88
bucrania, symbolic use of, 131
building(s), 20, 24, 71, 140, 163, 188, 193, 205, 216, 217; at Aşıklı Höyük, 172–75, 198–99, 205; at Çayönü, 18–19; communal, 80–81, 132–33; middens over, 227–28; object curation and circulation in, 7–8; repeated, 75–78, 119–20, 122–23; repetitive actions, 68–69, 88; 3D reconstructions of, 264–79; technologies, 6–7
Building F.V.I. ("Shrine" of the Hunters), 3D reconstruction of, 266
Building T (Aşıklı Höyük), 195, 196
Building 33 (Çatalhöyük), 204
Building 77 (Çatalhöyük), 241, 256; deposits in, 239–40, 254–55; groundstone in, 242–53
Building 80 (Çatalhöyük), 252
Building 89 (Çatalhöyük), 252, 267
bull baiting, 132
bulls, symbolism of, 38
burial(s), 5, 11, 14, 19, 20, 23, 80, 82, 173, 195; and architectural remains 105–6; and Aşıklı Höyük hearths, 200–202; at Çatalhöyük, 7, 9, 85–86, 87, 119, 188–89, 241; and hearths, 203, 205, 207(n5); at Körtik Tepe, 16, 140, 144, 145–50, 152, 153; of monumental structures, 120, 125–29, 130; Natufian, 12, 13, 103–4; placemaking through, 141–42; skulls cached in, 104–5; at Zagros sites, 76, 77, 81
Byrd, F. Brian, 179

caches, caching, 14, 20, 204, 238; skull, 104–5
Cafer Höyük, 21
Canhasan, 205
Cappadocia, 189, 194
caprines, in Building 77, 245. *See also* goats; sheep
Carmel, Mount, 106
carrying capacity, in NSIM, 40
castes, emulation of higher, 192
Çatalhöyük, 5, 6–7, 19, 21, 23, 26, 78, 99, 108, 118, 120, 131, 132, 188, 193, 218; Aşıklı Höyük and, 194–95; Building 77 at, 239–40, 241–42; burials at, 85–86, 87, 88 202; commemoration at, 8–9, 204–5; domestic architecture at, 119–20; groundstone deposits at, 242–51; hearths at, 187, 203–4; history houses at, 216, 220; house use at, 7, 188–89, 190, 192, 242–56; intensified occupation of, 73, 74; middens at, 225–29; NSIM, 45, 51–52, 55, 58, 59–60; pottery in house floors at, 224–25; pottery production at, 214–15, 219–21, 231–33; pottery styles at, 221–31; religion at, 10, 34, 37–38; repetitive building at, 22, 75, 79, 192; ritual practices at, 81, 240–41, 248–55; 3D visualization techniques, 263–79; and Turkish politics, 165–66
Çatalhöyük House Entanglement Model (CHEM), 59
Çatalhöyük Research Project, 241, 263, 264, 270
Çatalhöyük Visualization Project, 263

286 INDEX

cattle, in Building 77, 242, 245
Cauvin, Jacques, 167
Çayönü, 148, 165, 172, 195; building types in, 18–19; Skull Building at, 19, 21, 163
cemeteries, Natufian, 104–5, 106
Center for Mind and Culture, Modeling Religion Project, 34
Central Zagros Archaeological Project (CZAP), 66, 67
ceramics. *See* pottery
ceremony, and reproduction of identity, 69–70
Chalcolithic period, 205
ChangeRate parameter, NSIM, 42, 44
charnel houses, 21
Chia Sabz, East, 72, 89
chief's houses, 189–90, 191
children, burials of, 13–14, 20, 144, 204, 241
chlorite, Körtik Tepe objects of, 145–50
Chogha Bonut, 82
Chogha Golan, 72, 89
circular enclosures, PPNA, 16
collectivity, 178, 179; and material deposits, 254–55
commemorative behavior, 7, 131, 153, 201; at Çatalhöyük, 8–9, 204–5; ceramic manufacturing, 213–14
commodities, Epipaleolithic, 101
communal architecture, at Göbekli Tepe, 116–17, 119, 121–22
communality, 138
communal space, storage, 169–70
communication networks, regional, 176–77
community, communities, 5; collective actions of, 254–55; handling, 132, 133; long-term memory in, 100–106; monument construction and, 121–22; shared ideologies, 151–52; social interactions among, 176–78; social networks in, 253–54
computer simulation, in archaeology, 34, 35–36
concretions, at Çatalhöyük, 218
Corinth, Corinth Classroom, 264, 269, 274, 275, 279
cosmology, 77; at Çatalhöyük, 85–86, 87
courtyards, 17
Cox, Grant, 266
cranes, depictions of, 125
cremation, in Britain, 192
Crete, Bronze Age, 187–88
cult buildings, 188, 193

cults, 191; monumental, 137–38
cultural exchange, Epipaleolithic and Neolithic, 4–5
curation: of knowledge, 216–17; of objects, 7–8, 145–50, 254–55; of T-pillars, 122, 123–25, 133
CZAP. *See* Central Zagros Archaeological Project

daily activities: ritual and, 239
Dark Line Ware, 215, 231
Dayak, 217
delay of gratification, in NSIM, 46
Delphi, altar of Hestia, 190
depositional sequences, 70
despotism, transition to, 36–37
diet, 218; at Körtik Tepe, 139, 142–44
digital archeology, at Çatalhöyük, 263–79
Dikaios, Porphyrios, 202
DiVE. *See* Duke Immersive Visualization Environment
Diyarbakir region, 142
Dja'de el Mughara, 18, 21
dogs, skulls, 245
domestic activities, 198
domestication, 54, 169; as entanglement, 37–38, 45; plants and animals, 3, 36, 65, 170
domus/agrios, 172, 179
Driessen, Jan, 187
Duke Immersive Visualization Environment (DiVE), 267
dung, 72, 73, 74, 76, 82

East Chia Sabz, 72, 89
ecological data, 66
economy, Neolithic, 45, 48, 49, 164–65
egalitarianism, 36; at Çatalhöyük, 188–89
Egypt, pyramids in, 191
Ein Gev I, 11, 102
Ekven site, potters' paddles at, 232–33
elites, 132; emulation of, 191–92
emperors, imitation of, 191–92
emulation, of elites, 191–92
Enclosure C (Göbekli Tepe), 122–23, 128
Enclosure D (Göbekli Tepe), *124*, *129*, 131
Enclosure H (Göbekli Tepe), 129; radiocarbon dates, *127*; rebuilding episodes at, 123–25
entanglement, 42, 48; NSIM, 59–60; religion as, 37–39

INDEX   287

environment, 101, 139; and transition to sedentism, 166–67
Epipaleolithic traditions, 5, 10, 26, 137, 162, 180; aggregation during, 107, 168, 169; history making, 24–25; long-term memory and community, 100–106; resource use, 3–4; social interaction, 176–78. *See also* Natufian
EPPNB, at Göbekli Tepe, 117, 118
Erbaba, 19
Erq el-Ahmar, 105
estates, house societies and, 216–17
ethnoarchaeology, 68
ethnography, ritual practice in, 239
Euphrates River, 18, 167
exchange networks, 26, 101, 167, 168, 171; regional, 150–51
Eynan (Ain Mallaha), 102, *103*, 104, 105, 106

families, 120; and early village life, 178–79; pottery paddles, 232–33
fashion, imitation and, 187–88
faunal remains. *See* animals
fear, and permanent communities, 151
feasting, 70, 101, 120, 122, 131
figurines, 21, 80, 218
Fine Ware, distribution of, 192, 206
fire, 105, 107(n5); sacred hearth, 190, 197
fish, at Körtik Tepe, 139, 143
Flannery, Kent, on early village life, 178–79
floor plastering, 16, 78; Abu Hureyra, 17–18
floors, 78, 200, 240; pottery installed in, 224–25, 228, 230–31
food: processing of, *244*, 250; storage of, 73
foraging, foragers, 143, 151
foundation processes, settlement, 15–16, 20

Galilee, 106
Ganj Dareh, 65, 70, *71*, 76; ritual deposition at, 75, 81, 83
gazelle horns, 104
gazelle hunting, 12
Gesher Benot Ya'aqov, 100
Glass Mountain (California), 219
Global Land Use and technological Evolution Simulator (GLUES), 36
goats, 65, 204; skulls from, 82, 83
Göbekli Tepe, 16, 21, 26, 150, 166, 167, 168, 171, 176, 202; artwork at, 131–32; description of, 116–18; history making at, 115–16, 129–31;
monumental buildings at, 24, 120, 127–29, 137; T-pillars at, 9, 122, 123–25, *126*
Golden Triangle, 167–68
grasslands, expansion of, 72, 73
grave goods: at Körtik Tepe, 144, 145–50; Natufian, 103–4
graves: Hayonim Cave, 12; Mallaha, 13
graveyards, Natufian, 106
Greeks, sacred fire, 190
grinding activities, as ritual, 239
groundstone, grindstones, 140, 195, 218; production of, 246–47; ritual deposition of, 21, 46, 242–55
Gusir Tepe, 150

Habermas, Jürgen, 178
Hacilar, 21
Haji Firuz Tepe, 82
Hallan Çemi, 11; PPNA, 14, 15, 131; transportable artifacts, 171–72
Halula, Tell, 22, 83, 85, 90
Harran plain, 116
Hasankeyf Höyük, 142; stone objects from, 146, 150
Hatula, Natufian, 11–12
Hayonim Cave, 100, 106; decorated burials in, 103–4; Natufian, 12, 13; skull caching at, 104–5
headdresses, Natufian, 104
heads, 21, 238. *See* skulls
hearths, 12, 70, *141*, 195, 206; Abu Hureyra, 17–18; in Aşıklı Höyük, 196–202, 207(n7); in Çatalhöyük houses, 190–91, 203–4, 205; as fixed spaces, 186–87; Middle Paleolithic, 10–11; sacred, 189–90
heirlooms. *See* curation
Herero, sacred hearth and fire, 189–90
Hestia, altar of, 190
high-investment (HI) lifestyle: in NSIM, 39, 40–50, 52–54; religious beliefs in, 56–57
Hilazon Tachtit Cave, "shaman's" burial at, 104, 106
history houses, 6; at Çatalhöyük, 34, *74*, 120, 216, 220; virtual simulation of, 264–79
history making, 8, 14, 81, 99, 204, 212, 215, 241; Epipaleolithic and Neolithic, 24–25; at Göbekli Tepe, 115–16, 129–33; hearths as, 190, 201; houses and, 6, 7; through object deposition or manufacture, 21–22, 255;

288 INDEX

Paleolithic, 10–11; settlement structure and, 25–26; 3D reconstruction, 270–77
Holocene, Körtik Tepe, 140–41
Hopi, knowledge curation, 216–17
horn cores, in Building 77, 242, 245
households, 142, 199
houses, 23, 179, 194, 202, 213, 218, 238; abandonment and foundation processes, 15–16; at Aşıklı Höyük, 198–99; building technologies, 6–7; in Çatalhöyük, 59–60, 190–91; features in, 195–96; imitative architecture in, 187–88; inscribed space in, 250–51; pottery production, 214–15; remodeling, 16–18; ritual abandonment of, 13–14; ritual associations of, 69–70, 242–55; sacredness of, 189–90; as shrines, 188–89; social integration and, 59–60; superpositioning, 12–13, 14–15, 18, 22 house societies, 216–17; Yurok as, 218–19
Höyücek, 21
human remains, 16, 204, 207(nn5, 6), 238; ancestral, 5, 8. *See also* burial(s); skulls, human
hunter-gatherers, 44, 70, 138, 168, 176, 178; and transition to sedentism, 166–67; Upper Paleolithic-Epipaleolithic transition, 162–63
huts, Epipaleolithic, 101, 102

identity, 70, 80; communal, 9, 138, 163, 168; community, 102–3; Natufian burials, 103–4; social, 176–77; symbolism and rituals, 179–80
Ilisu Dam, 139
imitation: of high status, 191–92; in pottery making, 192–93; and tradition building, 152–53, 187–88
India, caste emulation in, 192
individualization, 177; of space, 170, 174–75
Indonesia, Megalithic tombs in, 121
Indus Valley, 36
infants, burials of, 205, 241
insects, depictions of, 150
Iran, 177; Neolithic sites in, 65–66, 71; occupational intensity in, 70, 72–73
Iraq, 14, 65–66, 155(n5)
'Iraq ed-Dubb, 14
Islam, in Turkey, 166
isotopic analyses, at Körtik Tepe, 142–44
Israel, Kebaran in, 11

Jani, 66, 79; occupational intensity, 73, 76, 78
Jarmo, 65, 75, 82
Jerf el Ahmar, 14, 16, 20, 149, 167, 169, 171
Jericho, 12, 20, 23; house superposition/ sequences at, 15–17, 19, 21
Jilat 6, 101, 168
Jordan, 11, 14, 17
Jordan Valley, 11, 12, 102

Karim Shahir, 82
Karuk, 219
Kebara Cave, 10–11, 100
Kebaran period, 10–11, 23
Keos, 188
Kfar HaHoresh, 20
Kharaneh IV, 11, 101, 168
Khirokitia (Cyprus), 202
kinship, neighborhoods/social groups, 163
Knossos, hearths at, 186
Knossos effects, 188
knowledge: material forms of, 216–17; ownership of, 212–13, 215; in pottery making, 219–21; property as, 218–19, 233; ritual, 21–22
Konya plain, 23, 194, 205; in NSIM, 52–54, 55, 58
Körtik Tepe, 16, 153, 172, 187; burials at, 141–42, 152; diet at, 142–44; exchange and communication networks at, 150–51; monumental cult buildings at, 137, 138; occupation at, 139–40; symbols used at, 145–50
Köşk Höyük, 21
Kozlowski, Stefan Karol, 167
Ksar Akil, 10
Kult Bau (Nevali Çori), 163

labor, for monumental construction, 121–22
landscapes, 73, 171
Lares, 191
Last Glacial Maximum (LGM), 101
leaderships, hearths and houses and, 189–90
Lebanon, 10, 11
legumes, 73
leopard, as symbol, 38
Levant, 14, 100, 168, 188; Kebaran period in, 10–11; Natufian in, 102, 137; PPNB, 16–17, 170
Lévi-Strauss, Claude, on social house, 213, 218
LGM. *See* Last Glacial Maximum
Lifeliqe, 264, 269–70, 274, 276
lifestyle conversion, 39; computer simulation, 40–50

INDEX 289

lineage-specific behavior, pottery use as, 220
lime-burning, at Çatalhöyük, *74*
living rooms, hearths as, 199–200
longhouses, in Saribas Iban, 217
low-investment (LI) lifestyle, in NSIM, 39, 40–50, 52–54

Maison des Morts, at Dja'de el Mughara, 21
Masraqan period, 101
mating networks, 107
meals, and social integration, 59–60
Melas, Manolis, 188
Mellaart, James, 165, 240, 264; on houses as shrines, 188–89, 268, *269*, 270, *272*
memory, memory construction, 7, 13, 19, 23, 108, 151, 201, 213, 256; at Çatalhöyük, 99, 120; community and, 100–104; skull caching and, 20, 104–5; social, 7, 138
memory destruction, at Göbekli Tepe, 130
Mesopotamia, 3, 137, 197
micro-analyses, 89
micro-archaeological analyses, on Zagros sites, 64, 65, 67
micro-histories, of place, 68–69
micromorphological analysis, 70, *74*, 76, 88
micro-stratigraphy, of burials, 86, 88
middens: at Çatalhöyük, 218, 225–29; pottery distribution in, 229–30
Middle Euphrates, 167
Middle Paleolithic, Kebara Cave, 10–11
migration, of Sankeçili, 162
mimetic processes, 152–53, 154, 191–92, 193
Minoan Hall, imitation of, 187–88
modeling and simulation (M&S), 34, 35, 36
Modeling Religion Project, 34
Modoc, 219
Monte Carlo experiment, in NSIM, 55
monuments, monumental buildings, 25, 188, 195; communal building and maintenance of, 132–33; at Göbekli Tepe, 116–17, 119; at Körtik Tepe, 137–38; life cycle of, 120, 121–29; ritual burial of, 125–29, 130; Upper Tigris and Euphrates, 153–54
morality, moral action: in lifestyle transitions, 46–47; ritual and, 57–58
mortality rates, domestication and agriculture, 180
mortars, 102, 140
mortuary practices, 80

Mousterian period, 100
mud-brick architecture, 75, 240
Mureybet, 11, 14–15
Musular, 195, 196, 205

Nahal Oren terrace, 105, 106
names, and physical things, 216
Natufian, 5, 23, 107, 137, 168, 171–72; community identity in, 102–3; decorated burials in, 103–4; memory construction in, 104–5; settlement patterns in, 11–13
Nebi Musa, 21
neighborhoods, 163, 195
neoliberalism, in Turkey, 166
Neolithic, 5, 7, 19, 26, 132, 192, 240; aceramic, 75; economy, 164–65; hearth use, 186–87; history making in, 24–25; resource use, 3–4; in Zagros, 65–66
Neolithic Social Investment Model (NSIM), 39–40; at Çatalhöyük, 51–52; limitations to, 58–60; religious variables in, 56–58; variables in, 41–50
Neolithic transition, 36–37, 39, 115
Neolithization, 115, 164–68, 169
neonates, burials of, 205, 241
networks, cultural and resource, 4–5
Nevalı Çori, 21, 163, 172, 195, 202; symbolic network, 150–51
niche construction theory, 164
nomadic groups, 162
Northwest Coast, house societies in, 216–17
NSIM. *See* Neolithic Social Investment Model

objects, 215, 252; circulation and curation of, 7–8, 23; symbolic, 138, 145–50, 153; transportable, 150–51, 171–72
obsidian, 168; at Çatalhöyük, 189, 204; Yurok production and ownership of, 218, 219
Oculus Touch, 266
Ohalo II, 11, *101*
ornaments, 173
ovens, 195, 205
ownership, of knowledge, 212–13, 215
Özdoğan, Mehmet and Aslı, 188

paddles, paddle marks: ownership of, 232–33; in pottery construction, 221–24, 228–29, 231
painted pottery revolution, 192–93, 206

Palatine Hill, 190, 191
Paleolithic, 100; repetitive practices, 10–11
pastoral intensity, NSIM, 45, 48
Penates, 191
performance, 69, 81
pestles, 145, 155(n5)
Phaistos, 186
phenomenology, 278
pigs, 65
Pillar 66 (Göbekli Tepe), 131
pistachios, 73
place, 18, 151, 153; burials and commitment to, 16, 152; micro-histories of, 68–69; repeated building and, 75–78; in Zagros, 70, 72
place making, at Körtik Tepe, 140–42, 153
Plantago lanceolata pollen, 73
plants, domestication and storage of, 3, 65, 73
plastering: at Neolithic sites, 78, 79; ritual importance of, 80–81, 200
platelets, stone, 147–50
platforms, in houses, 195
Pompeii, Cult of Vesta in, 191
pottery: access to, 219–20; at Çatalhöyük, 221–24; in house floors, 224–25, 228, 230–31; imitation in, 192–93; manufacturing techniques, 213–14; production and ownership of, 231–33, 234(n3); ritual knowledge of, 21–22, 214–15; spread of, 192–93
Pottery Neolithic, 18, 180
power, and monument building, 122
PPN Interaction Sphere, 167
Pre-Pottery Neolithic A (PPNA), 3, 11, 24, 25, 107, 121, 131, 132, 133, 163; belief systems, 171–72; communal buildings, 153–54; at Göbekli Tepe, 116–18, 119, 123; houses in, 14–15; at Jericho, 15–16; skull removal in, 105, 108
Pre-Pottery Neolithic B (PPNB), 3, 4, 24, 25, 107, 168, 172, 170, 177; burials, 141–42; communal building, 153–54; at Göbekli Tepe, 116, 117; Levant, 16–17; memory construction, 19, 23; residential continuity, 15, 17–18; skull removal in, 105, 108
prestige buildings, 193
private space, 170; activities in, 177–78; Aşıklı Höyük, 172–75
projectile points, 7–8, 189
property, 215; knowledge as, 218–19, 233
Proto-Neolithic, Jericho, 12

psycho-cultural changes, 167
public monuments/architecture. *See* monuments, monumental buildings
public space, 179; activities in, 177–78; Aşıklı Höyük, 172–75
pyramids, Egyptian, 191

Qadesh Barnea, 100
Qaramel, Tell, stone objects from, 145, 146, *148*, 172
Qermez Dere, 14
querns: breakage of, 248–50, *251*; in Building 77, 243, 244, 245, *246*, 247–48, 252, 254

radiocarbon dating, at Göbekli Tepe, 118, 123–24, *127*
Raqefet Cave, 106
reciprocity, and monumental architecture, 132–33
red pigment, ritual significance of, 80–81, 82
religion, 10, 132, 138, 187; and agricultural systems, 9–10; entanglement in, 37–39; in NSIM, 46–47, 56–58, 59; sedentism and, 34, 166
repetitive practices, 88; hearth building, 186–87, 196–97; in Neolithic, 68, 70, 72, 75–78; in Paleolithic, 10–11
residential continuity, Natufian-PPNA, 14–15
Rig-Veda, on sacred fire, 197
ritual, 3, 7, 23, 25, 47, 132, 138, 193, 205, 217; abandonment and foundation, 20–21; at Çatalhöyük, 38, 248–50; communal building and, 80–81; definition of, 238–39; group binding through, 57–58; houses in, 240–41; Natufian, 12, 13–14; and reproduction of identity, 69–70; and sub-identities, 179–80; in Zagros, 65, 78, 81–82, 86, 88
ritual buildings: at Göbekli Tepe, 120–29; public, 6, 19, 23–24
ritual complexes, at Aşıklı Höyük, 19, 23–24
ritual deposits, 21, 75; at Çatalhöyük, 245, 248–55; in Zagros sites, 82–83, 86–88
ritual knowledge, passing on of, 21–22
Rome, 190; imitation and status in, 191–92
Rosh Horesha, 168
round-oval structures/enclosures: at Göbekli Tepe, 116–17, 119; life cycle of, 121–25; purposeful burial of, 125–29

Sabi Abyad, Tell, 18, 20, 193
sacredness, of houses and hearths, 189–90
Saribas Iban, longhouses, 217
Sarıkeçili community, 162
seashells, exchange system, 168
seasonal congregation, Epipaleolithic, 100–101
seasonal cycles, 11
seasonal encampments, Natufian, 12
sedentism, 34, 169, 178, 179; Natufian 11–12, 107; transition to, 162–63; triggers of, 166–67
Servius, 191
Sesklo, 7
settlement(s), 6, 14, 102; abandonment and settlement processes, 15–16; PPNB, 170–71
settlement patterns: Natufian, 11–12, 102; PPNA, 132–33
settlement structures, 163–64; history making in, 25–26
shaft straighteners, decorated, 149
"shaman's" burial, Hilazon Tachtit Cave, 104, 106
sheep, 65, 81, 204, 218; skulls of, 71, 82, 83
sheep herding, 194
Sheikh-e Abad, 66, 89; building at, 76, 77; intensified occupation at, 70, 72–73; ritual and burial practices, 82–83
Shimshara, 66, 78, 79
Shkârat Msaied, 148
"Shrine" of the Hunters (Building F.V.I.) 266
shrines: hearths and fire in, 189–90; houses as, 69, 188–89, 240; 3D virtual reconstruction of, 265–79
"Shrine" 10; 3d virtual reconstruction of, 265–76
skeletal remains, 81, 207(nn5,6). *See also* burial(s); skulls, animal; skulls, human
Skull and Flagstone ceremonial buildings, 21
Skull Building (Çayönü), 19, 21, 163
skulls, animal: auroch, 131; goat, 81; sheep, 71, 81; at Sheikh-e Abad, 77, 82
skulls, human, 9, 82; circulation of, 14, 21; house histories and, 16, 23; ritual treatment of, 5, 13, 20, 104–5, 107–8
slab building: of pottery, 221–22
Social Brain Hypothesis, 164
social gatherings, Zagros, 70
social groups, neighborhoods, 163
social houses, 213; identities, 214–15; Yurok, 218–19

Social Identity theory, 177
social intensity variable, in NSIM, 41–42, 52–54, 57–58
social interaction/networks, 120; in Çatalhöyük, 253–54; Epipaleolithic, 176–78
social structure, 171; at Aşıklı Höyük, 173–75; Çatalhöyük houses in, 59–60; NSIM of, 47–48; village life, 178–79
socio-ritual contexts, at Göbekli Tepe, 118
sodalities, construction of, 25, 38
Soles, Jeffrey, 188
space, 163, 171, 217, 250; communal to individual transition, 169–70; hearths as fixed, 186–87; public and private, 172–74, 177–78; reorganization of, 174–75
Special Purpose Buildings Area (Aşıklı Höyük), 173–75
spirits, Neolithic engagement of, 37
stalactites, at Çatalhöyük, 218
standing stones, at Çayönü, 21
Stanford University, 264
status, 193, 206; imitation and, 191–92
stone objects, 155(n15), 218; decorated, 147–50, 153; symbols on, 50–51
stones, recirculation and reuse of, 21, 153
stone tools, in Building 77, 242–55. *See also* groundstone
stone vessels, in Körtik Tepe burials, 145–50, 153
storage, storage facilities, 71, 73; communal to private transition, 169–70
stratigraphy, vs. 3D reconstruction, 22–23
stress, in NSIM, 44, 55–56
sub-identity, 178, 179–80
Sulawesi, feasting in, 122
surveyance intolerance, in NSIM, 46, 50
Swift Creek Complex, 233
symbolism, symbols, 38, 131, 170, 187; communities, 168, 176; of groundstone depositions, 251–55; in Körtik Tepe burials, 145–50, 153; regional exchange systems, 150–51; and sub-identities, 179–80; on transportable artifacts, 171–72
Syria, 11, 14, 20–21, 101, 193
Syro-Mesopotamia, 188
systems-dynamics model, 9–10

target transition pathway, in NSIM, 52
Taurus, 11

technology: knowledge of, 219–21; and NSIM, 43–44, 54–55
temples, 191, 197
Templeton Foundation, John, 5
Tepe Guran, 75, 82
Terrazo Building (Çayönü), 165
Theoretic Culture, 164
3D reconstruction, 22–23; at Çatalhöyük, 263–79
Tigris River, Körtik Tepe on, 139
Todd, Ian, 193
tombs, 190, 191; West Sumba, 121, 122, 131, 132
Torajan highlands, 122
tortoise shell, in Körtik Tepe burials, 144
tourism, 166, 263
T-pillars, T-shaped pillars: construction and transport of, 121, 122; curation and reuse of, 123–25, 133; at Göbekli Tepe, 9, 16, 115, 116, 117, 118, *126*, 127–28, *129*, 131
tradition building, mimetic processes, 152–53
transcendental social, 138
Transjordan, 101
tradition, imitation and, 187–88
Tukanoan people, 213
Turkey, 11, 16, 19, 162, 188, 263; abandonment and foundation practices, 20–21; Neolithic transition in, 115–16; politics and archaeological sites, 165–66

UNESCO World Heritage list, 263
University of California Merced, 264, 268, 277

Upper Mesopotamia, 121, 137; Neolithization, 115–16
Upper Paleolithic, 35, 100, 162

Versailles effect, 188
vessels, stone, 145–50, 153
Vesta, Vestals, 190, 191
villages, early, 178–79
Virtually Rebuilding Çatalhöyük Project, 264, 266, 268, 269–70, 275–76

Wad, el-, burials at, 103, 104, 105
Wadi Hammeh 27, 12, 13, 14, 102, 105, 106
wall painting, at Çatalhöyük East, 131–32
warfare, 36, 46, 49
West Sumba, Megalithic tombs in, 121, 122, 131, 132
wheat, 73
White Deer Skin Dance, 218
whitewashes, ritual significance of, 80–81
Wiener, Malcolm, 188

Younger Dryas, 72, 137, 140, 143
Yurok, 215; as house society, 218–19

Zagros, 18, 167; low-density activities in, 70, 72; micro-archaeological analyses in, 64, 67; Neolithic sites in, 65–66, *71*; repeated building in, 75–78; ritual practices, 80, 81–82, 86, 88, 89–90
Zawi Chemi Shanidar, 11, 75, 81, 90
Zeribar, Lake, 66; pollen cores, 72, 73